T0299047

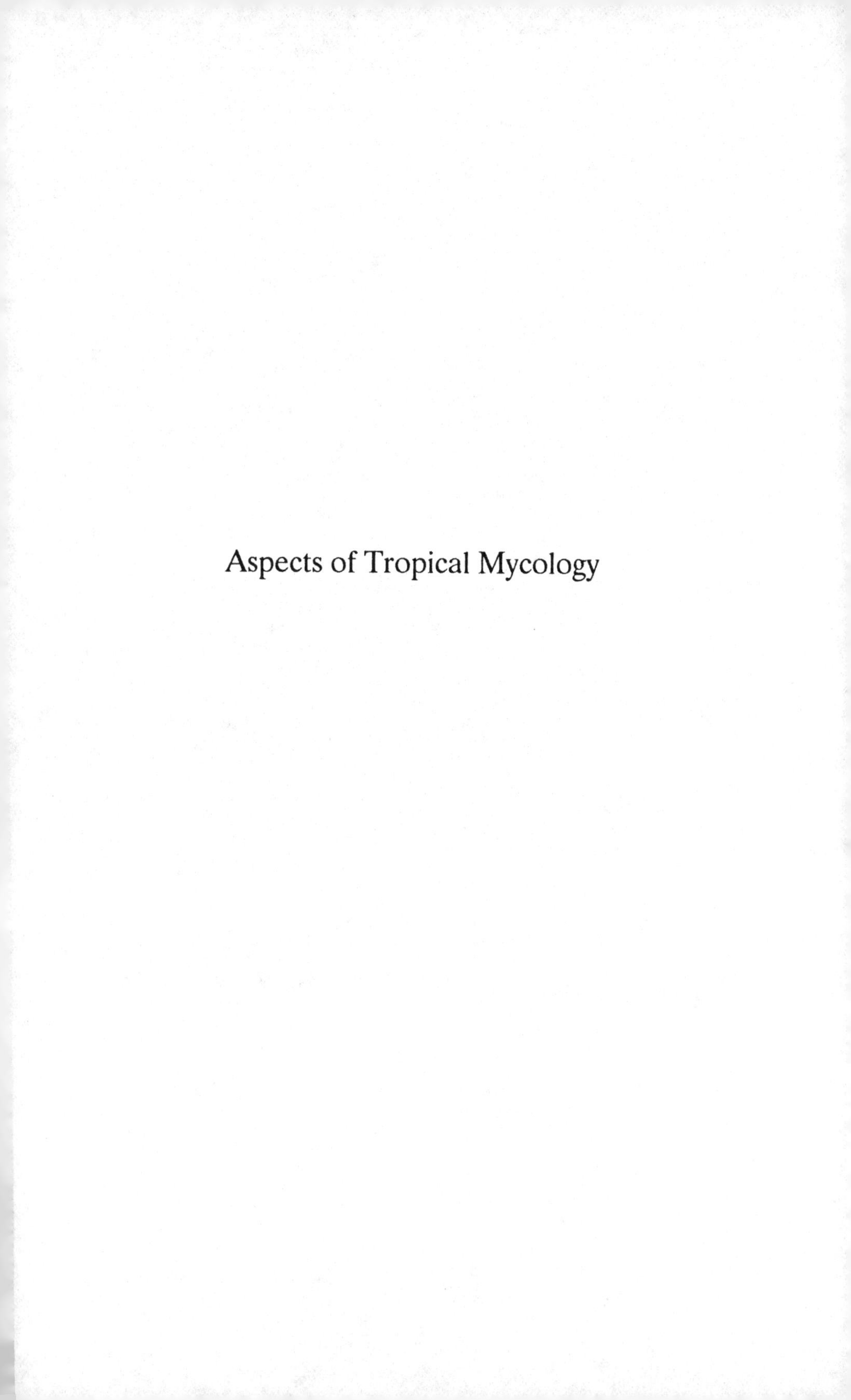

Aspects of Tropical Mycology

Aspects of Tropical Mycology

Symposium of the British Mycological Society held at the University of Liverpool, April 1992

Edited by
Susan Isaac, Juliet C. Frankland, Roy Watling & Anthony J. S. Whalley

Published for the British Mycological Society by

CAMBRIDGE UNIVERSITY PRESS
Cambridge, New York, Melbourne, Madrid, Cape Town, Singapore,
São Paulo, Delhi, Dubai, Tokyo

Cambridge University Press
The Edinburgh Building, Cambridge CB2 8RU, UK

Published in the United States of America by Cambridge University Press, New York

www.cambridge.org
Information on this title: www.cambridge.org/9780521450508

© British Mycological Society 1993

This publication is in copyright. Subject to statutory exception
and to the provisions of relevant collective licensing agreements,
no reproduction of any part may take place without the written
permission of Cambridge University Press.

First published 1993

A catalogue record for this publication is available from the British Library

ISBN 978-0-521-45050-8 Hardback

Transferred to digital printing 2009

Cambridge University Press has no responsibility for the persistence or
accuracy of URLs for external or third-party Internet websites referred to in
this publication, and does not guarantee that any content on such websites is,
or will remain, accurate or appropriate. Information regarding prices, travel
timetables and other factual information given in this work are correct at
the time of first printing but Cambridge University Press does not guarantee
the accuracy of such information thereafter.

Contents

Preface

The British Mycological Society first recognized the development of interest in tropical mycology at the 6th General Meeting of the Society several years ago, when a tropical workshop was organised. After the success of that session a tropical newsletter was launched, to keep interested parties in touch, with a promise to bring experts together for a future conference. This volume is a result of that promise being fulfilled as an International Symposium on Tropical Mycology held at the University of Liverpool in April 1992.

Tropical Mycology has been defined quite broadly in this volume in order to include the examination of the mycobiota of a wide range of ecosystems, e.g. deserts, sand dunes, rainforests, and mangrove swamps. The hottest regions of the world are usually considered to fall within latitudes 25 North and 25 South of the Equator, but many environmental factors interact to influence the ultimate geographical incidence of tropical organisms and the fungi are no exception. The palm lines, delimiting the natural occurrence of Palmae, may therefore be more useful general indicators of the distribution of tropical fungi (Kirk, Chapter 6).

In this selection of Chapters we have attempted to bring together all major groups of fungi, but we hope particularly to stimulate interest in the mycobiota of tropical sites, the attributes of these fungi and how they react in particular conditions. Our knowledge of tropical fungi is even more limited than that of many temperate regions, so the title of the volume reflects the fact that not all aspects could be covered. Although many of the Chapters may interlink, they have been arranged to cover the broader issues of ecology (Chapters 2-7), taxonomy (Chapters 8-11), experimental approaches (Chapters 12-14) and industrial interest (Chapter 15). To these have been added two further Chapters, one by a doyen of tropical mycologists, Professor E. J. H. Corner (Chapter 1) and the other covering the challenge to mycology in the future and diversity of fungi involved, by Professor D. L. Hawksworth (Chapter 16).

It is fully appreciated that nutrient cycling is an important phenomenon in the tropics, but there is a scarcity of data and information in the literature. Nutrient turnover in varied tropical environments, together with interactions between fungal species (Hedger *et al.*, Chapter 2 and

Lodge, Chapter 3) and with other organisms (Zak, Chapter 4) have been addressed from an ecological standpoint in this volume. Although biogeography is already a well documented discipline in plant and animal studies, the difficulties of identification and the paucity of collections have left mycogeography as a shadowy area. Investigations into the biogeographical distribution of marine fungi (Jones, Chapter 5), Zygomycetes (Kirk, Chapter 6) and Xylariaceae (Whalley, Chapter 7) have been included here, each with its own ecological bias. It is often assumed that tropical organisms may hold the key to understanding biological relationships. The important taxonomic criteria which have resulted from treatments of the tropical Heterobasidiomycetes (Oberwinkler, Chapter 8) and tropical Polypores (Ryvarden, Chapter 9) reflect that, for some fungi at least, this is true and some very exciting observations are reported. Fascinating information relating to the distribution of macromycetes (Watling, Chapter 10) and ectomycorrhizal fungi (Thoen, Chapter 11) arising from collecting in some specially selected tropical sites is given and demonstrates the enormity of the problems facing the tropical mycologist. Experimental aspects, putting some of this field work onto an analytical footing, are currently in progress and reported by Mohammed & Guillaumin (Chapter 12), Isaac *et al.* (Chapter 13) and Manners *et al.* (Chapter 14). Tropical fungi have long been used in natural culture as food and fermentation agents, but we are largely ignorant of the range of secondary metabolites which these fungi produce and it is pertinent, therefore, that commercial considerations and industrial interests have been raised by Fox (Chapter 15).

The Editors have included abstracts of the offered papers and the posters presented at the meeting in order that workers wishing to enter the field of Tropical Mycology may have the opportunity, in a single volume, to find contacts and identify current areas of work. Abstracts of papers presented by invited speakers who have not published a full contribution in this volume have been added for completeness even though some of the work covered will appear in expanded form in Journals.

The Chapter by Emeritus Professor E. J. H. Corner is a personal account of his experiences in many tropical areas especially whilst living and collecting in S. E. Asia and his subsequent visits. This contribution not only demonstrates his vast experience, particularly of the larger fungi, in the field but also the Chapter is a unique reflection of the development of tropical mycology, from the input almost totally from Europeans to the encouragement of young workers in tropical countries. This Chapter is a condensate of the British Mycological Society Benefactors' Lecture given by Professor Corner in order to share his experiences with those active in the field. This honour is awarded to distinguished Mycologists and Professor Corner is undoubtedly that in the field of Tropical Mycology.

He worked as Assistant Director, Garden Department Straits Settlement, at the Singapore Botanic Garden, pioneering studies on the fungi, from 1929-1945 then returned to England and took up a post in Cambridge, subsequently becoming Professor of Tropical Botany there.

As this volume goes to press, the British Mycological Society's first expedition to the tropics is about to set off. We hope this heralds a new era of mycological exploration.

Susan Isaac *Department of Genetics & Microbiology, University of Liverpool, P.O. Box 147, Liverpool, L69 3BX, U.K.*

Juliet C. Frankland *Institute of Terrestrial Ecology, Merlewood Research Station, Grange-over-Sands, Cumbria LA11 6JU, U.K.*

Roy Watling *Royal Botanic Garden, Inverleith Row, Edinburgh EH3 5LR, U.K.*

Anthony J. S. Whalley *Department of Microbiology, School of Biomolecular Sciences, Liverpool John Moores University, Byrom Street, Liverpool L3 3AF, U.K.*

Chapter 1

'I am a part of all that I have met' (Tennyson's Ulysses)

E. J. H. Corner

Emeritus Professor of Tropical Botany, University of Cambridge

Introduction

Ladies and gentlemen, to have been asked to deliver this Benefactors' Lecture is a great honour. It is a privilege which I have accepted in the humility of retirement and, yet, it has bewildered me. It is many years since I lectured and never before to a mycological audience except for some impromptu occasions in Japan. The trouble is that, as years mount, so do thoughts and, the older one gets, the more they jumble into incoherence. Certainly I planned in youth to be a mycologist but circumstances have so buffeted me that I have ended my professional career as a lecturer in angiosperm taxonomy. All along, however, mycology has been my hobby and recreation. Thus it is that I have chosen as the title for this lecture the words attributed to that classical wanderer Ulysses.

Last year Dr Roy Watling wrote to me 'Talk on what you will!'. This is a symposium on tropical mycology. I wish to show that it is but one aspect of tropical botany. It needs to be incorporated and the mycologist must have some understanding of the great range in stature and floristic variety of tropical vegetation for, without it, there would be no tropical mycology of equivalent magnitude. I think not only of advanced monographs but of introductory works to recruit mycologists who will be resident in the tropics. There is a pressing need for collecting and studying living fungi in the tropics from which the forest is fast receding.

Malesia is the vast region from the Malay Peninsula to the Solomon Islands. It holds the richest and most varied angiosperm flora and, in consequence, as I hope to show, the richest fungus biota, though it is the least known.

Chance took me into mycology. At the end of my first summer term at Rugby School I had two sixpences left from my weekly pocket-money. A year had been spent rather fruitlessly on the classical side and I was to move into natural science. I had a book on pond-life and a childish microscope with which a school friend and I had rejoiced at *Spirogyra*, *Paramecium* and *Vorticella* because they were our own discoveries. So, I went to Over's book-shop, which is still there in the Market Place, but uncertain if my coins could purchase anything of interest. I searched shelves without success until I came on a small pile of little, paper-backed booklets at 6d each. They had such titles as *Wild Flowers at Home* and the last two were *Toadstools at Home*. The horrid slimy things of childhood that I had seen in rotting trunks, and the fleshy stalks of things which I had kicked over in disgust swam into my mind. I wondered who could be interested in writing about them. As I turned the pages and looked at the photographs, curiosity grew. Intrigued, I bought them and that summer and early autumn I became a mycologist. I remember the shout of joy on Dartmoor 'I've found a *Boletus*!'. The author was the medical doctor and politician Somerville Hastings, an account of whose life has been given to us recently, as if in premonition of this lecture, by Dr Ainsworth in last year's issue of *The Mycologist*. My father, who was a surgeon, joined in my enthusiasm. We made the acquaintance of Somerville Hastings and graduated thence to the dusty world of the loquacious John Ramsbottom at the British Museum (Natural History). In 1922 we went to the autumn foray of the British Mycological Society at Keswick where we met the great authorities Carlton Rea, Elsie Wakefield and Reginald Buller. So my zest for mycology was encouraged into respectability. I like to recall a few incidents.

At school I had Berkeley's wooden *Outlines of British Fungology* and Massee's *Handbook of British Fungi* (in four volumes). Unable to identify a small violet toadstool, I sent it to Mr Ramsbottom, but no answer! I sent him a stamped postcard for reply, and it came with '*Omphalia fibula* var. *swartzii*'. I learnt to persevere and to be more careful because I had overlooked it. Unhappily, a prefect whose intellect was accommodated in the Army Class had a down on me. Finding a tin under my arm one day, he insisted on my opening it and discovered a large specimen of the Lawyer's Wig about to deliquesce; in consequence, he administered a painful slippering that night in the dormitory. I learnt to dodge. We met, years later, in Singapore when he was a dreary military captain and I a thriving botanist. One up I thought for mycology.

At the Keswick foray, Carlton Rea was explaining to me one evening how to recognise a small agaric. We were joined by two young men whom I knew as senior demonstrators at Cambridge. As they seemed to question his point, Rea told them to look at the stem with a pocket-lens, which neither had. 'What!' exclaimed Rea, the amateur, champing his dentures. 'Two botanists from Cambridge and no lens!' One became Sir Frank Engledow, Professor of Agriculture, and the other Sir Samuel Wadham, of the Agricultural Ministry in Australia. I knew both in later years but never reminded them. I wonder how many plant scientists at Cambridge, now, could pass the test.

One day at Keswick, a group of mycologists had gathered round a crouching figure scaping lightly at the humus in a wood. It was Reginald Buller of Winnipeg University, searching for truffles. When asked how he knew where to look, he continued scraping in silence until, finding an *Elaphomyces*, he looked up to say:

> There was a young girl who said 'Why
> Can't I look in my ear with my eye?
> If I put my mind to it,
> I'm sure I could do it.
> But you never can tell till you try.'

The concourse dispersed. The professor rose up and said to me 'Young man, there was a fragment on the ground partly eaten by a squirrel. That's the clue.' Those piercing blue eyes have been with me ever since and that little rhyme was his creed.

My copies of *Toadstools at Home* have disappeared. I never saw others until 1983 when I had given a talk to the Japanese Mycological Society in Tokyo. I had mentioned the little books and, next day, Rokuya Imazeki showed me his battered copies; they had started him on his mycological career. He and I had long been friends and, when he died last year, I thought again of Somerville Hastings and how such books as his were needed in the tropics.

As an undergraduate, I was bored by the botanical lectures at Cambridge. I had struggled with Thalassiophyta and The Somatic Organisation of the Phaeophyceae at school. It happened that two intending mycologists had come from Oxford to finish off at Cambridge under the tutelage of F. T. Brooks. They were Robert Leach and E. B. Martyn, whom some of you may recall. They took me to meet A. H. Church at Oxford. We lived, then, at Great Missenden in Buckinghamshire and I often drove over to Oxford to renew these meetings. Church taught me to draw accurately and, often, concluded with the exhortation 'Go to the tropics!' I still prefer the accurate line-drawing to the photomicrograph which is too frequently blurred with scraps of hyphae or their walls among other blobs, and even to SEM photographs which usually show only surfaces. I remember Church's astonishment when I showed him *Clavaria pistillaris* (L.) Fr., *Hydnum erinaceus* (Bull.) Fr. and *Cortinarius praestans* (Cordier) Sacc., for he had no idea that such splendid fungi grew anywhere near Oxford.

My first Long Vacation at Cambridge in 1924 was spent in learning German in Vienna. We had had a smattering at school. John Ramsbottom arranged that I should stay with Dr Karl von Keissler, mycologist and lichenologist, who was director of the Botanical Department of the Natural History Museum in Vienna. There I met Handel-Mazzetti, the authority on the flora of China, and Heinrich Lohwag whose long articles on gastromycetes I had endeavoured to read. They were published in the *Arkiv für Protistenkunde* which had seemed very odd until I learnt that he was a friend of the editor. Both these experts were tall imperious men who looked down on 'der kleine Englander'. Long after, Donk and I dissected Lohwag's theses and I came to correct most of Handel-Mazzetti's identifications of Chinese *Ficus*.

As a research student under F. T. Brooks, I was set a tedious problem with mildews, but I was delighted to discover how the conidia, germinated on water, sent a long germ-tube straight into the air without rolling over, and how the conidia, sown on fern leaves, caused an orange-yellow patch to develop in the epidermal wall around the point of attempted penetration. A German mycologist has, since, followed up the aerial growth of the germ-tubes and attributed it to the giving off of certain unspecified substances by the conidia (Domsch, 1954); the effect on the fern leaves seems to have excited no interest though a matter of cellulose composition. However, that year of research was relieved by the study of discomycetes which I had begun with W. D. Buckley who lived near us at Slough. Though I did not complete the Ph.D. course, which in its early years was designed for those unable to get jobs, I did write the papers on discomycetes. Doctor I have never been, though it is idle to remonstrate. John Dandy, of the British Museum (Natural History), used to say that any one who had been a botanist for ten years deserved to be a professor.

In 1929 I went to Singapore with the now archaic title of Assistant Director of the Gardens Department of the Straits Settlements. A new world opened. I became my own master. I followed in the wake of T. F. Chipp who had left to become Assistant Director of the Royal Botanic Gardens at Kew. The belief then was that tropical forest was distinguished in its evergreen growth by the absence of any spectacular display of fungi, such as the autumn run of basidiomycetes in temperate climates. I used to walk through the Gardens Jungle of merely some 7 ha, though with many original forest tress, two or three times a week in search of higher fungi, and I would visit Bukit Timah forest and the Reservoir Jungle at least once a week. In three years I discovered the mistake. Whenever heavy and prolonged rain followed a spell of fine weather sufficient to dry the humus and top-soil, there would follow an unbelievable sprouting of fruitbodies in such incredible variety that, in the course of fifteen years, I was never able to study all in the living state, even the commoner; they decomposed so rapidly. These runs happened twice a year in Singapore, March-May and August-October. Any one species would appear merely for a few days, to be replaced by others (Corner, 1935). If one was not on the spot for those brief occasions, that fungus would be missed. Afterwards I found that this regularity held generally throughout the lowlands of the Malay Peninsula and, also, in Sarawak. Meteorologists find no place for this in their equations, just as they ignore the precision of many tropical trees, but I continue to advise mycologists intending to collect in the tropics to study meteorological records and time their visit accordingly; few do and court failure. In June last year, I had a surprise telephone call from Kuala Lumpur. An American mycologist wanted to know where he might find certain boleti and, as I was supposed to be the expert, he asked me to tell him because so far he had failed. It was the wrong time of year and he had been looking for lowland fungi in mountain forest of wrong composition. I fear all his expenses were wasted.

During those early years in Singapore, I witnessed the felling of all but one of the few remaining tracts of original forest in the island. That one which I secured and persists, though much damaged during the Japanese Occupation, is the historic collecting ground of Bukit Timah. There were much more extensive fellings on the mainland. It shocked me to think that the primeval forest was being destroyed without botanical record. In the intervals between the fungus seasons I turned to forest botany in general. It led to my study of the freshwater swamp forests in danger of conversion into rice-fields (Corner, 1978a). It led to my employment of coconut monkeys to help, as botanical collectors, with the great trees so far out of reach, and I note that the only mycological companion that I had in Malaya in those years was one of my monkeys (Corner, 1992b). It led to the book on *Wayside Trees of Malaya* to enable the public to appreciate the countryside (Corner, 1940). It led to my study of *Ficus* to learn from that natural genus how woody plants had evolved into lofty trees, climbers and epiphytes, for *Ficus* itself could build a forest (Corner, 1965; 1978b). This study revived the interest in fig-insects which is now a vogue as part of what I call sycology. But the Japanese invasion put an end to my career in Singapore. Nevertheless, in those years of captivity I had the leisure to consolidate my thoughts (Corner, 1981). The Durian Theory came upon me as the means of encompassing the great variety of tropical vegetation (Corner, 1949), and I worked out a geometry of the basidium-unit that is too intricate to be understood (Corner, 1947; 1972a). Indeed, I doubt if any one can, as yet, conceive how fungi without sense organs are put together by filaments working in unison; leastways, textbooks are silent. I began, also, in those war years the systematic investigation of seed structure (Corner, 1976) and largely completed the monographs on clavarioid and cantharelloid fungi, as well as that on Malaysian *Boletus* (Corner, 1950; 1966b; 1972b).

After the war, I joined UNESCO and spent two years in Latin America, mostly in Brazil. My task was to initiate an institute for scientific research in Amazonia with its seat at Manaus. When that purpose was about to succeed, political faction in UNESCO destroyed it. I returned to Cambridge to lecture on the botany of flowering plants. Thus, I have never professed mycology but it has been a light even in the despair of captivity and the loneliness of Manaus. It has been the hobby in my wanderings, ever rewarding with discovery.

A conclusion that I have drawn from my travels is that the Malesian region holds not only the richest of angiosperm vegetation but, also, the richest fungus biota. It is the least known of fungus biotas and, both practically and theoretically, the most difficult to elucidate. For names one must search the fungus biotas of the world because *Amanita vaginata* (Bull.: Fr.) Vitt., and *Clavaria acuta* (Low: Fr.) can be found together with *Termitomyces* and *Lachnocladium*. To discover the fungi which occur in a neighbourhood, one must search almost daily because fruit-bodies commonly appear for so short a time and many but rarely. Besides an enormous number of undescribed species, there are many which over-ride the limits of established genera. This is so aggravating for authority that the tendency is to neglect Malesian fungi, although one could hardly neglect its angiosperms in systematic synthesis.

There are, at least, some 10,000 kinds of tree in Malesia and, probably, as many kinds of woody liane. Is any one kind of fungus peculiar to one kind of tree or liane? Who knows, because who can identify any of them? And what fungi grow upon all the many kinds of palm, pandan, bamboo, ginger and tree-fern in Malesia? Do you suppose that *Elatostema* (Urticaceae) with its multitude of Malesian species has the same kinds of fungus as *Urtica* of like habit, or that the multitude of Malesian Vitaceae have the same as the grape-vine? A distinction is commonly made between species of fungi which grow on coniferous remains and those on angiosperm, but there is little regard for the enormous variety of tropical woods, probably because they, too, cannot be identified in the forest where botanist and mycologist are surrounded with unknowns. Is poisonous anacardiaceous wood decayed by the same fungi as aromatic dipterocarpaceous? Many agarics grow only on wood in the last stages of decomposition and these agarics are so scattered in the forest as to suggest that they are connected with similarly scattered timbers. There are said to be 55 species of hymenomycete associated with the three species of Fagaceae in Great Britain. In Malesia there are some 200 species of Fagaceae, so does this imply some three or four thousand attendant species of hymenomycete? I collected on Mt Kinabalu in North Borneo roughly 2000 species of hymenomycete within a radius of half a mile from our base camp situated in rich fagaceous forest, and the further I explored more novelties turned up because, in my experience, Malesian forest is so complex that it may never repeat itself in entirety. As for the actual numbers of known species, Horak (1980) has described 55 species of *Entoloma* from the Malay Peninsula (about the size of England) and I have described about 140 species of *Boletus*; these are probably not half the real totals. I note, too, that one of the largest fruit-bodies known in *Entoloma* grows in the Solomon Islands. I have collected a hundred or so species of *Cortinarius* from western Malesia, and they have been with Dr Moser for 20 years with scarcely an identification.

As a baffling genus, I will mention the agaric *Trogia*. It raised hue and cry when I first exposed it in the work on cantharelloid fungi (Corner, 1966b), but I have a revision soon to be published (Corner, 1992a). *Trogia* has a very characteristic hyphal structure which can be seen most readily in the living stem. It is a pantropical genus with species misplaced in *Collybia*, *Mycena*, *Hemimycena*, *Hydropus*, *Gerronema*, *Omphalina*, *Marasmius* and others because mycologists do not study stem-structure in sufficient detail, and the dried material with which they work is generally unsuitable for the purpose. Therefore the consensus is that I must be wrong. If I cannot study the living, I preserve material in alcohol-formalin which is almost as good. Roger Heim was the only other mycologist that I knew to be aware of this necessity.

Several, if not many, Malesian hymenomycetes are rare or appear to be so because they seldom fruit. In 1929, when I first walked through Bukit Timah forest, I found a *Russula* which I have not collected again, though there are many species in western Malesia. I found the massive polypore *Buglossoporus magnus* Corner once only on a large and slowly decaying, fallen trunk in that forest, and yet for several years I had inspected that trunk. Likewise, I found once only the large and conspicuous *Sarcodon conchyliatus* Maas G., described by Maas Geesteranus (1971), by a path near the summit of Bukit Timah where I had often looked for fungi. Then, some 25 years later, I met almost identical specimens, once only, in the Solomon Islands. In fact, I never visited Bukit Timah forest, where I collected methodically over 12 years, without meeting a fungus new to me. It makes me realise that we may never know the full fungus biota of Malesia. During one week in November 1930, the floor of the forest near Tembeling in Pahang was covered with fruit bodies of some ten species of *Geaster*, as far as the eye could see, and so thickly that one could not tread without stepping on them; the air was pungent with their spores. This outburst may have sprung from a multitude of dead roots caused by an unprecedented flood some years before.

Geaster raises a more general problem. Its elaborate fruit bodies display in two or three hundred species all sorts of specific devices, as it seems, for lifting the spore-bag and opening it with different kinds of peristome. They are supposed, maybe tacitly, to be improvements for better spore dispersal, but can that really be so? Very few of the millions of spores survive. *Lycoperdon* and *Scleroderma* succeed just as well without such elaboration. Thus, I often think of natural perfection rather than natural selection, as the force which drives the hyphae to elaborate to the utmost of their combined effort. We simply do not know what it is that impels and controls them. Likewise, in view of the enormous wastage of spores, it does not seem to have been the survival of the fittest that changed decurrent gills to adnate and free, from crowded to distant, or to reticulate and tubular. As for the specific colours, spore size and shape, or those of cystidia, I can see no effective survival value through natural selection. *Russula* and *Clitocybe* may fruit side by side with *Boletus* and *Hydnum*. The fruit bodies of fungi seem to have evolved intrinsically into the array of specificity, genus by genus, rather in the way proposed by Motoo Kimura (1985) in his neutral theory of molecular evolution. It is often held that the possession of a stem has selective advantage in aiding spore dispersal. Yet the number of sessile and resupinate species must exceed greatly the number of stipitate and, in many genera, there are pleuropodal and sessile fruit bodies of equal success, to show that the stem is being lost, not evolved. The stem appears as part of the original endowment, or equipment, of the fruit body, to disappear finally in the resupinate and subterranean.

I have met in the Malesian forests immense clusters of fruit bodies in such diverse genera as *Hydnum*, *Tricholoma*, *Bondarzewia*, *Inonotus* and *Scleroderma*. The largest that has been found seems to have been that of *Tricholoma crassum*, discovered recently in Japan, with a fresh weight of over 80 kg. From what have these enormous constructions arisen? How did the mycelium muster the reserves from humus or dead roots? One is forced to consider the bulk of dead matter in the soil. At the other extreme, tiny solitary fruit bodies develop from mere fragments such as a petiole, a twig, a fruit husk or a rootlet. The individual fruit bodies of the great caespitose *Hydnum*, which I found in Malaya, were almost identical with those of the solitary *Hydnum repandum* (L.: Fr.); and the other genera have their solitary counterparts. I am led to conclude that the caespitose habit is primitive in luxuriant forest and from it the small and solitary have been derived as mycelia specialised in restricted habitats. I note that the huge sclerotia of *Lentinus* (*Panus*) *tuber-regium* (Fr.: Fr.) and *Polyporus* (*Lignosus, Trametes*) *sacer* in Malesia and which are commonly found on the ground and supposed to have formed there, are developed in large rotten fallen trunks from which they eventually dislodge, often to lie in rows as evidence of their origin.

I come to the major problem of specific abundance and why tropical forest is so rich in species and genera. I give two examples from my own experience. On Mt Kinabalu in North Borneo some 70 species of *Ficus* grow in the forest at altitudes of about 1200-1800 m. On Bougainville Island, east of New Guinea, I came upon a small valley, partly cleared, where roughly every fifth tree was a species of *Ficus*; at an altitiude of about 500 m, I collected in that valley over 40 species of *Ficus*. Ashton (1988; 1989) in his account of the dipterocarp trees of Borneo discovered similar congregations of species. The ecologists have studied mainly the parts of the plants above ground and from such subaerial features they have proposed explanations for this richness. Roots are not considered, chiefly, I suppose, because it is impossible to identify and trace them in the dense tangle in the soil. Fungi come to our assistance. Some terricolous fungi certainly appear to be distributed at random but many are so local, often in circles, as to indicate an association with specific trees; indeed, the presence of one species of fungus may prove the presence of one kind of tree concealed from view in the dense canopy. The apparent absence of *Amanita*, *Russula* and *Boletus* from the forests of the Solomon Islands may be ascribed to the absence of Fagaceae while the abundance of those fungi in western Malesia may be ascribed to the abundant variety of Fagaceae. One must remember, too, that for any green plant on land, whether a new species or not, the first essential is to establish a root-system. I think that it is accomplished in the tangle of roots in the forest by differential activity.

I was taught that roots absorb water and mineral salts and that was enough but it hardly seemed to account for the tangle of different roots in tropical forest. Now it is known that roots can absorb and exude various organic compounds, and absorption may not be the simple process of diffusion as once supposed. The litter in tropical forest, derived from a multitude of species, must be the most complex mixture of chemicals in varying stages of disintegration. Hyphae are reagents with certain sets of enzymes according to their species, and it is unreasonable to suppose that one species can deal with all this complex litter. Rather, the hyphae attack parts, perhaps in relays, in the breakdown of the organic matter. Similarly, while roots have their common requirements in water and mineral salts, they may also have special requirements of organic matter, to be absorbed either by their own means or through mycorrhiza. What one sees is that the loose litter is beaten down by rain and falling objects to become felted with all sorts of mycelium. Into this felt grow a multitude of rootlets, many fixed by their root-hairs to decomposing bits of leaves, twigs, animal matter and so on. When, therefore, a new species of green plant has established itself, it will add new material to the litter for other, probably newly specific, hyphae and rootlets. Further, one must not forget the larger roots deeper in the soil. There are always many moribund and dead trees in primary forest and, accordingly, a multitude of moribund and dead roots in the soil. In the tropics termites may decompose most of these larger roots but there are many kinds of termite, all with their differential activity, and they add to the complexity of the humus and soil; a new species of termite will add new organic material for hyphae and rootlets. In effect, the more varied the litter and humus, the more kinds of mycelium and roots it will support and, in consequence, the more chance of evolving new species to augment the complexity.

The specific richness of tropical forest comes from its diverse litter. It should become richer and richer in species if it were not that species are always being lost. Of course, once this litter or humus is destroyed by felling, burning and rain-wash, the fertility of the soil, as the life of the forest, is lost; if it can recover, it will be a matter of centuries of slow rehabilitation. There was abundant evidence for this on the barren laterite hills of Singapore, deforested last century, denuded and eroded. Yet, there were lumps of resin in the loose earth which had come from the heartwood of dipterocarp trees in proof that, once, there had been high forest.

As an instance of the complexity of life in forest soil, I give that of *Cordyceps*. I asked Dr Yoshio Kobayasi, the authority on the genus, how it was that some species grew on *Elaphomyces*. They switched, so he said, from insect in the soil to fungus. Indeed, on Mt Kinabalu I found in a small part of the forest a perithecial *Cordyceps* on *Elaphomyces* and, adjacent, fluffy conidiophores on deeply buried cicada larvae, as if the *Cordyceps* had different organic requirements for the two stages in its life cycle. Intermixed with these stages of *Cordyceps* there grew the small angiospermous root parasite *Mitrastemon* (Rafflesiaceae) to add to the problems of soil-life.

Before I close this mycological ramble, I will explain why, almost alone now, I follow A. H. Church in his theory that the higher fungi evolved from multifilamentous algae (Church, 1919; 1920a; Mabberley, 1981). First there is the morphological fact that the clavarioid fruit-body, especially such as *Clavulina* with corticate stem, is remarkably like the multifilamentous algal soma, the evolution of which, as Church explained (1920b), had been through the environment of marine phytobenthos; that is submerged in the littoral and sublittoral zones. It is unreal to suppose that the same construction could have been evolved by diffuse hyphae in the quiet of saprotrophic existence on land; that it happens today is not more proof than to suppose the backbone was the outcome of animal life on land in as much as land animals develop it. Moreover, I am not aware that anyone has attempted to show in any convincing manner how else the clavarioid fruit body evolved. Second, it was not until green vegetation was established on land that the debris for saprotrophic life began to accumulate. Third, the multifilamentous soma is not strong enough to compete with the parenchymatous in the upward struggle of the green vegetation on land. As that vegetation gained height, so the multifilamentous soma was relegated to the shade with the increasing accumulation of debris; with internal and descending hypha-like filaments it was almost preadapted. Presumably, these hyphae began to take precedence in the life cycle as the means of absorbing nutrient material, and the multifilamentous soma was retained as the hereditary means of reproduction; in the course of the saprotrophic evolution the soma was gradually reduced until it has practically disappeared from some modern resupinates or entirely, as some believe, from the yeasts. Lastly, if this was not the course, what became of the multifilamentous algae of the landward transmigration?

Ideally, as many maintain, there can be no real evidence of the course of evolution without a fossil record. Alas, however astonishingly detailed the fossil record may be of some plants, as a whole it is notoriously imperfect. For the fungi and most lowly organisms one must rely on comparative morphology of the living, to which there may be added experimental genetics. Perhaps the most remarkable experiment, because it seems to have been a natural one, is the addition of photosynthetic cells to the fungus soma. The result is the algal forms to be seen in the larger lichens. The fact was emphasised by Church in his theory of the lichen transmigrant as the reconstituted 'skinned alga' (Church, 1921). I was once on the eastern slopes of the Bolivian upland and I found myself in a landscape of such lichens, as if the seashore had been uplifted - a fact which Darwin established long ago for the Andes.

I doubt if the evolution of any large genus of fungi has been thought out from the evidence of all its species. I doubt if it can be done with any satisfaction until more is known of the Malesian fungus biota. In 1966 I found in the estuarine mud of a river in the Solomon Islands the clavarioid fungus that I called *Paraphelaria* (Corner, 1966a). It had been described over 100 years previously as a species of *Thelephora* and it had been found several times in Malesia without record of its habitat. The much branched fruit bodies suggested a transmigrant alga already saprotrophic and supplied with auriculariaceous basidia. It could be regarded as a highly elaborate *Eocronartium* which is parasitic on mosses. What, then, has happened to all the intermediates needed to connect these two so different in build and habitat? Where are the fossils? And, sadly now, where is the primeval forest about an estuary that has not been drastically logged? May be it all happened long ago when the transmigrant vegetation of the seashore was beginning to evolve fungi and bryophytes, and we are left, luckily, with these two relics. So I close with the exhortation to go on exploring before it is finally too late, or, as Church would say, 'Go to the tropics!'

References

Ashton, P. S. (1988). Dipterocarp biology as a window to the understanding of tropical forest structure. *Annual Review of Ecology and Systematics* **19**, 347-370.

Ashton, P. S. (1989). Species richness in tropical forests. In *Tropical Forests*, (ed. L. B. Holm-Nielsen, I. C. Nielson & H. Balslev), pp. 239-251. Academic Press: London.

Church, A. H. (1919). Thalassiophyta and the subaerial transmigration. *Oxford Botanical Memoirs* no. 3. pp 95. Oxford University Press: Oxford, U.K.

Church, A. H. (1920a). Elementary notes on the morphology of fungi. *Oxford Botanical Memoirs* no. 7. 29pp. Oxford University Press: Oxford, U.K.

Church, A. H. (1920b). The somatic organisation of the Phaeophyceae. *Oxford Botanical Memoirs* no. 10. 11pp. Oxford University Press: Oxford, U.K.

Church, A. H. (1921). The lichen life-cycle. *Journal of Botany (London)* **59**, 164-221.

Corner, E. J. H. (1935). The seasonal fruiting of agarics in Malaya. *Gardens' Bulletin, Straits Settlements,* **9**, 79-88.

Corner, F. J. H. (1940). *Wayside Trees of Malaya*, volume 1 and 2. 3rd edition. Malayan Nature Society: Kuala Lumpur.

Corner, E. J. H. (1947). Variation in the size and shape of spores, basidia and cystidia in Basidiomycetes. *New Phytologist* **46**, 195-228.

Corner, E. J. H. (1949). The Durian Theory or the origin of the modern tree. *Annals of Botany,* NS, **13**, 367-414.

Corner, E. J. H. (1950). A monograph of Clavaria and allied genera. *Annals of Botany Memoir,* 1. 740pp.

Corner, E. J. H. (1965). Check-list of *Ficus* in Asia and Australasia, with keys to identification. *Gardens' Bulletin, Singapore*, **XXI**, part 1. 186pp.

Corner, E. J. H. (1966a). *Paraphelaria,* a new genus of Auriculariaceae (Basidiomycetes). *Persoonia* **4**, 345-350.

Corner, E. J. H. (1966b). A monograph of cantharelloid fungi. *Annals of Botany Memoir*, 2. 255pp.

Corner, E. J. H. (1972a). Studies in the basidium. Spore-spacing and the *Boletus* spore. *Gardens' Bulletin, Singapore*, **XXVI**, 159-194.

Corner, E. J. H. (1972b). *Boletus in Malaysia.* Singapore: Government Printing Office.

Corner, E. J. H. (1976). *The Seeds of Dicotyledons*, Volumes 1 & 2. Cambridge University Press: Cambridge, U.K..

Corner, E. J. H. (1978a). The freshwater swamp-forest of south Johore and Singapore. *Gardens' Bulletin, Singapore*, Supplement 1. 266 pp.

Corner, E. J. H. (1978b). *Ficus dammaropsis* and the multibracteate species of *Ficus* sect. *Sycocarpus. Philosophical Transactions of the Royal Society of London*, B. **281**, 373-406.

Corner, E. J. H. (1981). *The Marquis: a tale of Syonan-to*. Heinemann Educational Books (Asia) Ltd: Singapore.

Corner, E. J. H. (1992a). *Trogia* (Basidiomycetes). *Gardens' Bulletin, Singapore*, Supplement 2. 97 pp.

Corner, E. J. H. (1992b). *Botanical Monkeys*. The Pentland Press: Edinburgh, U.K. (in press).

Domsch, K. H. (1954) Keimungsphysiologische Untersuchungen mit Sporen von *Erysiphe graminis. Archiv für Mikrobiologie* **20**, 163-175.

Horak, E. (1980). *Entoloma* (Agaricales) in Indomalaya and Australasia. *Beiheft zur Nova Hedwigia*, **65**. 352pp.

Kimura, M. (1985). *The Neutral Theory of Molecular Evolution*. Cambridge University Press: Cambridge, U.K.

Maas Geesteranus, R. A. (1971). Hydnaceous fungi of the easter Old World. *Verhandelingender Kononklijke Nederlands Akademie van Wetenschappen, AFD. Natuurkunde*, 2nd series, 60, no. 3. 175pp.

Mabberley, D. J. (1981), *Evolutionary Botany. Thalassiophyta and other essays of A. H. Church*. Oxford Science Publications, Oxford University Press: Oxford, U.K.

Cornell, J. H. (1986). [illegible] species [illegible] *Philosophical Transactions of the Royal Society of London* [illegible] 3[illegible]–406.

Corner, E. J. H. (1964). *The Life of Plants*. [illegible]

[illegible] (1972a). [illegible] *Supplement* [illegible]

[illegible] (1972b). [illegible] Cambridge University Press. [illegible]

[illegible] in press.

[illegible] (1955). [illegible] *Proceedings of the Royal Society* [illegible]

[illegible] (1986). [illegible] Australian [illegible]

Kimura, M. (1983). *The Neutral Theory of Molecular Evolution*. Cambridge University Press, Cambridge.

[illegible] 2nd series, [illegible] 1–17 pp.

[illegible] (1981). [illegible]

Chapter 2

Litter-trapping by fungi in moist tropical forest

John Hedger, Peter Lewis & Habiba Gitay*

*School of Biological Sciences, University of Wales, Aberystwyth SY23 3DA, U.K. and *Ecosystem Dynamics Group, RSBS, Australian National University, PO Box 475, ACT 2601, Australia*

Introduction

Moist tropical forests often occupy ancient weathered oxisols and latosols where leaching may have removed contact with the underlying bedrock. Jordan (1985) estimated that only 6% of the soils in the Amazon basin have no nutrient limitation and Caatinga or black water forests are essentially oligotrophic, occurring on highly weathered soils. Nutrients in such forests are concentrated in the plant biomass, although inputs in dust and rain are of considerable importance (Jordan, 1985). Leaf litter produced by the plant subsystem contains nutrients, although as pointed out by Lodge (Chapter 3) key nutrients such as phosphorus may be withdrawn from leaf litter prior to abscission to a greater extent in tropical than in temperate forests. Even so, small litter contains a greater proportion of the nutrients than does woody debris–in studies in Panama 75% of the phosphorus flux and 41% of the potassium flux occurred in the leaf litter fall (Golley, 1983). The importance of leaf litter in tropical forest nutrient cycling is thus much greater than in temperate forests. Golley (1983) averaged data from studies on tropical and temperate forests and concluded that wood production in the two biomes was similar at 7 t ha^{-1} y^{-1} but that small litter production was some three times greater in tropical forests at around 9 t ha^{-1} y^{-1} as against 3 t ha^{-1} y^{-1} for temperate forests. In addition, small litter production in moist tropical forest is more or less continuous, although discontinuity increases with seasonality of rainfall.

The environment of the moist tropical forest is also different from that of the temperate forest. Temperature remains uniform, usually in the range 24–30°C (Myers, 1980), humidity is high and rainfall frequent, usually in excess of 150–200 cm y^{-1}. This can lead to rapid leaf litter decomposition at soil level (Swift, Heal & Anderson, 1979) although there is considerable resource- related variation, e.g. between sclerophylls and non-sclerophylls (Hedger, 1985).

Golley (1983) suggested that the water content of small litter may be sufficient for considerable decomposition to take place within the forest understorey and canopy. Our studies have shown that there is a widespread community of litter-trapping, decomposer fungi in the understorey of moist tropical forest. This chapter shows how this species assemblage traps the litter, and discusses adaptations to the understorey environment, with particular reference to field work carried out at two sites: Varirata National Park in Central Province, Papua New Guinea, in primary transitional lowland/lower montane forest, and at Reserva de Producción Faunistica Cuyabeno, Napo Province, Ecuador, in primary and disturbed lowland Amazonian Tierra Firme forest. A general account of resource relations of fungi at Cuyabeno can be found in Hedger & Gitay (1993).

Litter-trapping in moist tropical forest

A feature of all moist tropical forest is the presence of masses of litter, mostly leaves and small branches, trapped in the canopies of treelets and understorey trees. This litter is held in place as a result of either physical trapping by the plants, or retention by litter-trapping fungi. The former may include fortuitous temporary trapping but a number of forest epiphytes and treelets have evolved trapping systems that retain litter masses and exploit the nutrients released during decomposition by adventitious root production. In Malaysian forests, examples are: the epiphytic fern *Asplenium nidus* L. (Ng, 1980) and the recently described litter-trapping treelet *Barringtonia corneri* Kiew & Wong (Kiew & Wong, 1988).

Most of the litter retained in the understorey of the forest is held there by litter-trapping fungi. Trapped leaves and twigs are held together in masses in the canopy by mycelial connections, which attach litter to living canopy leaves and branches. Such connections can arise from mycelium within debris or from rhizomorph systems. Hedger (1985) and Lodge & Asbury (1988) have both drawn attention to mycelial connections formed in the litter on the forest floor, but did not define their structure. Ainsworth & Rayner (1990) described a similar system of mycelial transfer in the canopy of *Corylus avellana* L. by the fungus *Hymenochaete corrugata* (Fr.: Fr.) Lév. They termed the adhesive structure that bound branches together a 'pseudosclerotial plate' and considered that the surface structure of thickened pigmented hyphae protected the inner mycelium from desiccation. We have found that many of the larger connections betweeen branch debris in the canopy were similar to pseudosclerotial plates and had a hard sclerotised structure. However, there were many intermediates between pseudosclerotial plates and the non-sclerotised tufts of mycelium which formed connections between leaf litter masses and extended from rhizomorphs into litter. Jacques-Félix (1967) described the mycelial connections between rhizomorphs of *Marasmius* species and the substratum as 'byssi', drawing a useful parallel with the structures which anchor shells of the marine bivalve *Mytilus*.

We propose to use the general term 'contact zone' to define all aerial mycelial connections but anticipate that this terminology will need amendment once the structure is better understood. In using this term, we emphasise that these structures are differentiated following contact, in a thigmatropic-like response, between internally colonised litter and a new resource, or rhizomorph length and a resource. We believe that contact is responsible for initiation of new hyphal growth and subsequent differentiation of the mycelial connection.

For the purposes of the studies reported in this chapter, litter -trapping systems in the understorey canopy of the study sites were divided into two categories: contact zone systems and rhizomorph systems. In the former, mycelial connections cemented masses of litter to the canopy branches and leaves; in the latter, networks of fine (0·1 to 0·5 mm diameter) rhizomorphs held the litter together by byssi. We shall show that a major difference between these systems is the method by which litter was trapped and the ways in which the fungi progressed through the canopy.

Identity and structure of contact zone- and rhizomorph-forming fungi

Identity of contact zone-forming fungi

At the sites in Papau New Guinea and Ecuador, the larger pseudosclerotial plate-like contact zones on twig and branch debris were sometimes associated with fruit bodies of Aphyllophorales, especially the families Polyporaceae and Hymenochaetaceae. Smaller contact zones on leaf litter were sometimes associated with agaric fruit bodies, especially the genera *Marasmius* and *Crinipellis*. However, at both sites, the commonest leaf litter trapper was a fungus which was exceptional in forming a thin white mycelium over the litter surface (Fig. 1a). SEM studies (Hedger & Baxter, this volume p. 305) showed that the surface whiteness was caused by dicotomously branched, cystidium-like structures (0·1 to 0·2 mm in diameter), embedded in a layer of mucilage (Fig. 1b). These resembled the dicohyphidia found in the corticioid family

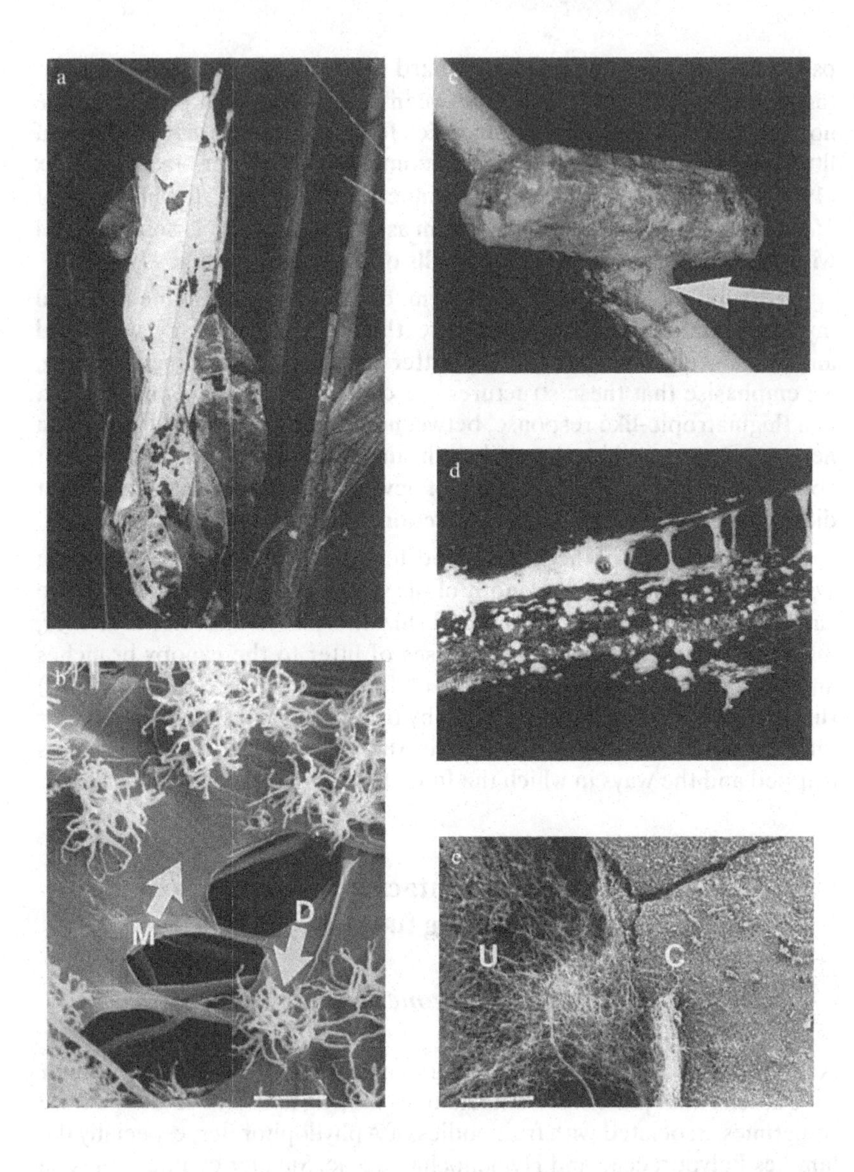

Fig. 1. Contact zone-forming fungi in the understorey of tropical moist forest, Ecuador. (a) Leaf litter trapped in a palm apex, probably by a *Vararia* species; ×0·25. (b) SEM of a sample taken from the litter shown in (a), arrows indicate dicohyphidia (D) and mucilage matrix (M); bar = 10 μm. (c) sclerotised contact zone (arrowed) between a dead branch (broken) and living canopy branch; life-size. (d) non-sclerotised mycelial connections forming contact zones on dead branch; ×0·25. (e) SEM of initial stages in the formation of a non-sclerotised contact zone between colonised (C) and uncolonised (U) leaf surface (see text for full explanation); bar = 10 μm.

Dichosteriaceae, (Pegler, Young & Henrici, 1992). Basidia, to confirm the identity, were not found with this mycelium, which appeared to belong to a pantropical litter trapper, but it was probably a species of *Vararia*. M. Nuñez (pers. comm.) found this genus to be a common coloniser of debris in the forest canopies of Puerto Rico and Cameroon.

Structure of contact zone-forming fungi

The size and colour of the contact zones studied in Papua New Guinea and Ecuador varied considerably. Tree branches up to 50 mm in diameter were linked by larger pseudosclerotial plate-like zones up to 5 mm in thickness, with a hard surface layer. Attempts to remove branches usually resulted in breakage of the branch rather than the pseudosclerotial plate as can be seen in Fig. 1c. Contact zones between small litter components were similarly pseudosclerotial plate-like, especially those of marasmioid mycelium, but delicate non-sclerotised connections were also common (Fig. 1d).

Studies on leaf litter samples placed in contact with debris in shrub canopies at Jatun Sacha, Ecuador (J. N. Hedger & F. Fox, unpublished data) showed that the non-sclerotised contact zones were differentiated within a week. SEM of sequential samples showed that within 2 days there was an outgrowth of a dense mass of hyphae from the canopy debris, which radiated out onto the leaf surface (Fig. 1e), and this was followed by secretion of mucilage over the mycelium within a week. Eventually the whole of the contact area was covered with a mucilage layer. Fracturing of the tissue revealed that subsequent hyphal growth occurred within the resource tissue. Formation of the larger contact zones was not studied.

Identity of rhizomorph-forming fungi

Canopies of understorey treelets and shrubs at Varirata and Cuyabeno were often occupied by systems of black and brown rhizomorphs (Fig. 2). At Varirata we found three common systems and at Cuyabeno we were able to identify up to eight distinct types, differing in diameter, colour and surface structure. In most cases the identity of the rhizomorph-producing taxa remained unclear, since basidiomes were not found. However, most appeared to be marasmioid. Of the taxa which produced basidiomes, *Marasmius crinisequi* F. Muell.: Kalbr. occurred at both sites. However, at Cuyabeno, the common rhizomorph-former was *Micromphale brevipes* (Berk. & Rav.) Sing. which formed black, pilose rhizomorphs.

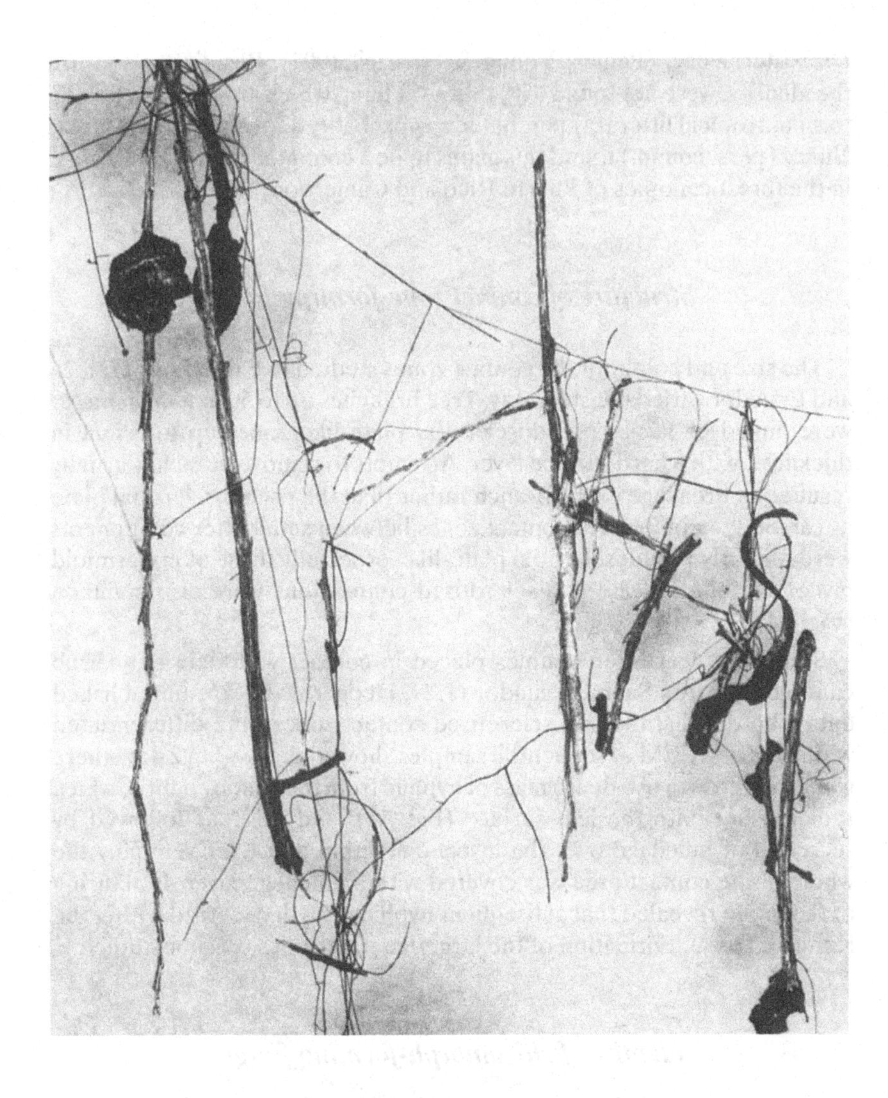

Fig. 2. Part of a rhizomorph system of *Micromphale brevipes* with trapped small litter, from Cuyabeno, Ecuador; × 0·5.

Rhizomorphs were usually found at 1-5 m height, but at Varirata there was a relatively unbranched marasmioid type arising from woody resources at 0·5 m height. At Cuyabeno, a similar system was found associated with fruit bodies of *Polyporus leprieurii* Mont. This taxon and other rhizomorph-forming *Polyporus* species occur in similar environments in Costa Rica (M. Nuñez pers. comm.).

Internal structure of rhizomorphs

All the marasmioid understorey rhizomorph systems we studied had a number of features in common. Transverse sections revealed many similarities with soil rhizomorphs, e.g. those of an *Armillaria* species (Townsend, 1954). We define here two types. Type 1 had a medulla of thin-walled cells surrounding a central lumen in sections both proximal to the apex and at some distance from it. An example of type 1 belonged to *Micromphale brevipes* (Fig. 3a.). Surrounding the medulla was a cortex of thick-walled cells and a conspicuous surface layer (1 to 2 cells thick) which was impregnated with a dark pigment, probably melanin (Fig. 3a). Type 2 had a similar cortex. A medulla layer was present only in some of these rhizomorphs, but in all type 2 specimens the central region was occupied by wide (6 μm), thin-walled hyphae (Fig. 3b). One identifiable type 2 rhizomorph system belonging to *Marasmius crinisequi*, has recently been investigated by Cairney (1991), who also concluded that the rhizomorph was solid, except for a small central lumen close to the apex.

Apical structure of rhizomorphs

In marasmioid systems, extension occurred at the apex, which possessed a globose to campanulate cap 15 to 400 μm in length (Fig. 3c). Below the apex (up to 10 to 20 mm), the colour of the rhizomorphs was lighter and they were less rigid in comparison with the rest of the structure. Longitudinal sections (Fig. 3e) showed that beneath the apical cap was a central area (arrowed in Fig. 3e) in which hyphal elongation was initiated in a group of tightly packed hyphae and continued in the non-pigmented, sub-apical region. From studies in Papua New Guinea it was shown that rhizomorphs from which the apical cap was removed began to re-form new cap structures within as short a time as one day. Initial development consisted of the growth of hyphae at the apex to form a conical mass of diffuse mycelium. It was not clear how the process proceeded after this stage, but complete new caps formed within 3 to 4 days resembling the originals in shape and size, and subsequent growth reverted to hyphal elongation in the sub-apical region. Occasionally, removal of the cap and apical 5 to 10 mm resulted in the production of two new apices rather than

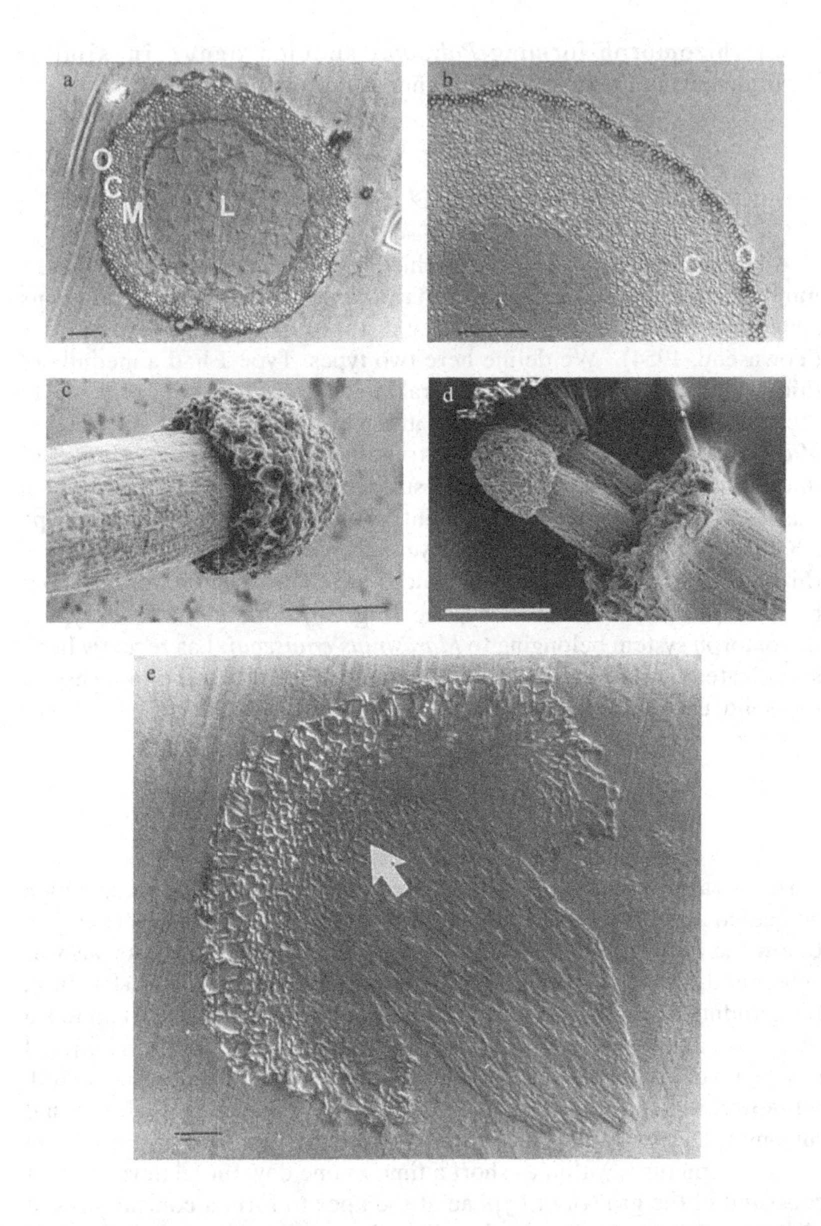

Fig. 3. Structure of rhizomorphs. (a) TS of type 1 rhizomorph, *Micromphale brevipes.* L = lumen, M = medulla, C = cortex, O = outer layer; bar = 100 μm. (b) TS of type 2 rhizomorph, taxon unknown; bar = 100 μm. (c) SEM of pileate apex, taxon unknown; bar = 0·5 mm. (d) SEM of regenerated apices, taxon unknown; bar = 0·5 mm. (e) LS of pileate apex, taxon as 3d, arrow indicates centre of hyphal elongation; bar = 100 μm. All light microscopy by interference-contrast.

one (Fig. 3d), with each growing point extending at the normal rate, 2 to 5 mm day^{-1} on average.

Examination of the rhizomorph systems in Papua New Guinea and Ecuador showed that many branches had lost their apices. Others had a jointed appearance suggesting that apical regeneration had occurred. Monitoring of the systems showed that apices disappeared in the night, presumably owing to feeding by canopy microvores. The capacity to regenerate apices appeared to be an adaptation to this grazing pressure, and we found that they regenerated in the absence of rainfall, provided the average r.h. remained at 80 to 95%. However, apices from which more than 10 mm had been removed failed to regenerate.

Jacques-Félix (1967) working with material of *Marasmius rotalis* Berk. & Br. and *M. rotula*(Scop.: Fr.)Fr.used the term 'stipes degrades' (modified stipes) or 'télépodes' to describe vegetative organs from which basidiomes arise. He suggested that such rhizomorphs represent stipes where elongation persists and in which basidiome development is stalled. Such a view is supported by Lewis & Hedger (this volume p. 301), who have shown that marasmioid rhizomorphs produced *in vitro* respond to gravity, light, carbon dioxide and nutrient levels in a similar way to stipes of basidiomycete fruiting bodies (Gooday, 1985).

Studies on litter-trapping fungi in the understorey canopy

The connections produced by contact zone fungi and rhizomorph formers enable them to effect aerial mycelial transfer into newly fallen litter, holding it in place in the understorey during subsequent colonisation. Although tropical forest is usually considered to be wet and humid, surface wetness after rain does not persist and, in dry periods, r.h. can fall to as low as 30% (Windsor, 1990). Aerial transfer, followed by internal colonisation and protection by mucilage secretion, appears to be an adaptation to the exposed aerial environment.

Litter trapping by contact zone fungi

The amounts of litter captured by contact zone systems is considerable and will be discussed later in this chapter.Their population structure and origin within the understorey canopy is intriguing. Surveys in both Papua New Guinea and Ecuador showed them to be common, but sporulation was rare. Much litter was retained by contact zone formation from existing

Table 1. Colonisation by contact zone-forming fungi of sterilised leaves and branch sections tied to living canopy or trapped litter in the forest understorey at Cuyabeno, Ecuador

Test resource	Forest type	Attachment site	Test resources (%) with contact zones after: 14 d	28 d	56 d
dead leaves	primary	trapped litter	44	48	56
		living foliage	4	18	9
	disturbed	trapped litter	48	64	35
		living foliage	8	4	2
branches	primary	trapped litter	12	20	35
		living branches	4	4	4
	disturbed	trapped litter	25	65	28
		living branches	0	0	0

trapped debris. In other instances litter was attached by contact zones to living stems or leaves. This litter could have been colonised prior to fall; alternatively the fungus could have been present as a cryptic endophytic phase in the living plant tissue.

A simple field experiment was carried out at Cuyabeno, Ecuador to investigate contact zone establishment in treelet canopies over 56 days. Twenty five labelled, heat sterilised dead leaves and the same number of 10 cm length branch sections were tied to living leaves and branches, and also to trapped canopy litter with obvious contact zones. Table 1 shows that, although there were differences between the two sites examined (primary and disturbed Tierra Firme forest), contact zone formation became well established on leaves and branch sections which were touching trapped debris within 14 days, and had increased further after 28 days. The surprising result was that contact zones were also formed between living leaves and branches and the sterilised material, although less frequently. This latter result confirms the possibility that these fungi may exist as an endophytic phase, which can grow into debris from living canopy leaves and branches.

An interesting feature of this experiment was that in some instances numbers of contact zones had declined after 56 days (Table 1). The agency of their removal was not discovered, but was likely to have been microvore feeders, perhaps the large millipedes which are frequently seen in the understorey canopy at night. These animals probably play a role in the dynamics of the litter-trapping fungi, by grazing the contact zones, and thus releasing litter from the canopy.

Litter-trapping by rhizomorph-forming fungi

We measured different rhizomorph systems in the field in Papua New Guinea and Ecuador and found branches grew at 3 to 6 mm d^{-1}, although extension slowed or stopped in dry periods. Rhizomorph branches usually grew upwards, away from the trapped debris. Simple experiments carried out at Varirata in which tips were tied down in a horizontal or downward position showed that they responded by negative gravitropism to begin upward growth again (Fig. 4a). This response appeared to be mediated at the sub-apical elongation zone rather than the tip.

Hedger (1991) observed a rhizomorph system in Ecuador and was able to show that these vertical branches were eventually displaced sideways and downwards under their own weight or by rainstorms. They usually started upward growth again forming a series of loops (Fig. 4b). Continued, negatively gravitropic growth of the branches, combined with branch formation and formation of byssi, gave rise to a network which was effective in trapping small resources such as leaves and twigs in the canopy of the plants (Fig. 4c). A feature of some rhizomorph systems was the production of new apices from contact zones giving rise to a stoloniferous growth habit (Fig. 4d), where again the loops increased the trapping capacity of the network. However, other systems branched by differentiation of new laterals.

The colour of these byssi varied, but most were oval to circular in shape, 0·5 to 2 mm diameter. They were produced by the rhizomorph after apical or, more usually, lateral contact with plant material, and differentiation was in two stages. The first was the outgrowth of fine (1 to 2 μm diameter) hyphae, which grew up to 2 mm away from the rhizomorph (Fig. 5a). The second stage was the production of extracellular material, probably mucilage, which may have had a role in adhesion and protection against desiccation (Fig. 5b). Sections through byssi revealed that the mucilage enclosed and protected a space containing interwoven hyphae and a hyphal layer on the leaf surface, forming a pseudosclerotial plate-like structure (Fig. 5c).

Microscopy of byssi sections on dead litter showed that hyphal penetration and saprotrophic exploitation of the resource had occurred; formation of byssi on living plant tissue did not appear to involve penetration. However, in some instances, especially with *Marasmius crinisequi,* local plant tissue death seemed to be associated with byssi, and

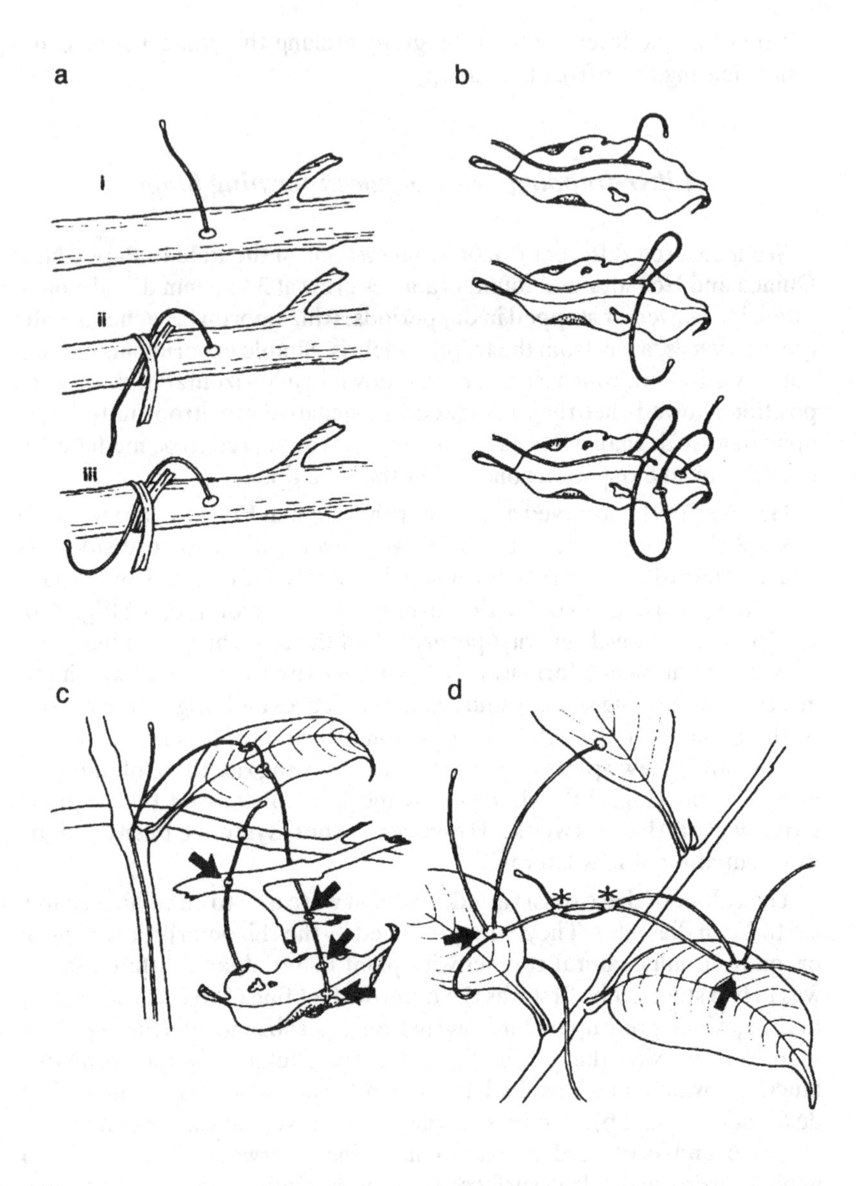

Fig. 4. Development of rhizomorph networks. (a) Response of a growing rhizomorph tip (i) to repositioning with a wire tag (ii) after 4 days growth (iii). (b) Formation of loops by continual apical growth over a period of 14 days. (c) Capture of litter by formation of byssi (arrowed). (d) Stoloniferous growth of rhizomorphs by production of new apices from byssi (arrowed). Note fusion of rhizomorphs (asterisks).

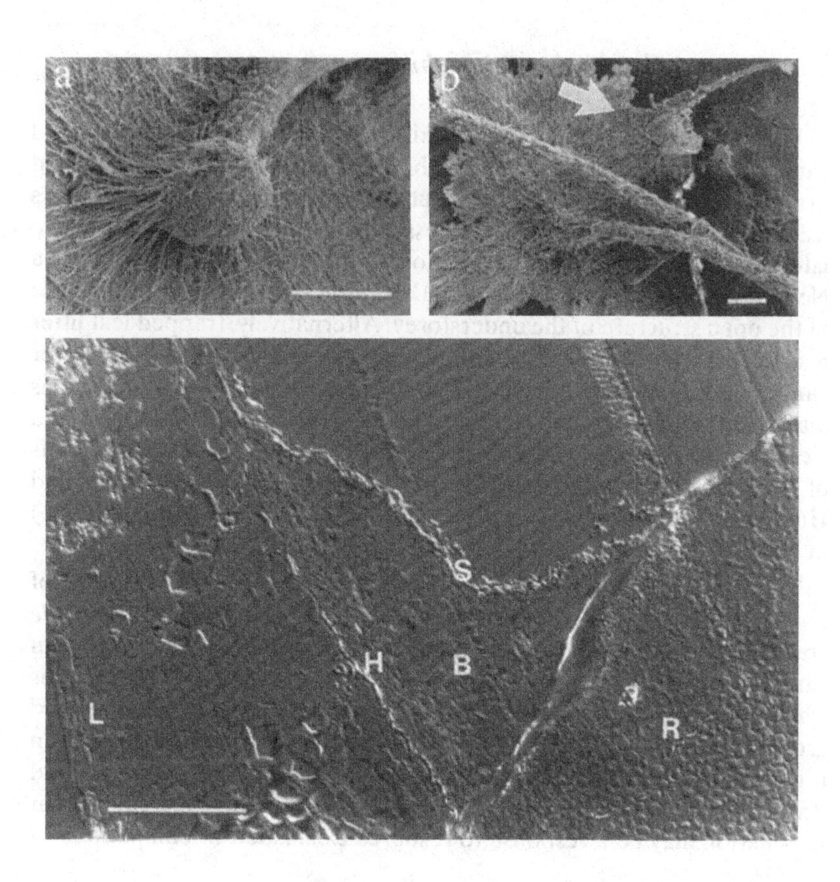

Fig. 5. Formation of byssi by rhizomorphs in contact with leaf litter. (a) SEM of initial outgrowth of hyphae from a rhizomorph; bar = 0·5 mm. (b) SEM of fully differentiated byssus, arrow indicates mucilage matrix; bar =0·5 mm. (c) section through a fully dlfferentiated byssus, R = rhizomorph, L = leaf tissue, S =sclerotised/mucilaginous byssus surface, H = hyphal layer on leaf surface, B = space within byssus containing loosely packed hyphae; bar = 100 μm. All light microscopy by interference-contrast.

---

this fungus sometimes acted as a foliar pathogen, particularly on suppressed treelets. *M. crinisequi* was first described from Sri Lanka by Petch (1915) and causes horse-hair thread blight in the canopies of tree crops such as tea and cocoa (Thorold, 1975).

Establishment of rhizomorph systems

Upward growth can maintain a rhizomorph network in an individual canopy, and existing systems were observed to keep pace with new leaf production in understorey palms, but the means by which new systems were established in canopies is less clear. Rhizomorphs may move laterally from canopy to canopy, but our work in primary forest in Papua New Guinea and Ecuador indicated that this was relatively rare, because of the open structure of the understorey. Alternatively, trapped leaf litter may fall from above with attached rhizomorphs and establish in the lower canopy, as was observed in Papua New Guinea. Dispersal of rhizomorphs may be aided by birds and we observed, in both Papua New Guinea and Ecuador, the widespread use of rhizomorphs in nesting material by Birds of Paradise in Varirata (Hopkins, pers. comm.) and Weaver birds and Humming birds in Cuyabeno, also reported by Wright & Ferraro (1986) in N. Argentina.

To test whether displaced rhizomorph systems were capable of regrowth, 10 samples of equal weight (5 g) of individual *Micromphale brevipes* systems, with or without attached litter, were transferred between canopies at 1·5 m height at Cuyabeno (Table 2). After 60 days, systems which had been moved together with trapped litter had recommenced growth and byssus formation, but those from which the litter had been removed, or which had been placed on the ground, failed to regrow and start litter capture. The latter did produce basidiomes, indicating that sporulation may be a response to resource exhaustion or competition in

Table 2. Transfer of living rhizomorph systems of *Micromphale brevipes* between canopies of understorey treelets at Cuyabeno, Ecuador.

Systems transferred*	No.† of systems with		
	byssi	growing apices	basidiomes
Rhizomorphs with attached litter	4	4	3
Rhizomorphs without attached litter	0	1	10
Rhizomorphs with attached litter (placed on ground)	0	0	5

* removed from host treelets (1 to 2 m above ground). Rhizomorph systems (5 g) unravelled and tied to 30 cm lengths of twig (1·.5 m above ground in canopy, or on ground).

† assessed 60 days after start of experiment (n = 10).

such fungi. The resulting basidiospores may establish new networks or even produce endophytic infections, as in the tropical canopy agaric *Crinipellis perniciosa* (Stahel) Sing. (Griffith, 1989).

Comparison of litter trapping by contact zone and rhizomorph systems

As plant litter is trapped and invaded by fungi in the understorey, positive feedback would be expected to stimulate further development until the weight of trapped litter is greater than the tensile strength (holding capacity) of the fungus. On the other hand continued lignocellulose degradation of the litter by these fungi would reduce the resource weight and render it so fragile that it would fall as a modified resource to the forest floor.

In order to estimate the quantity of litter that might be held by litter-trapping fungi in the lower understorey (treelets and palms to a height of 2 to 3 m), surveys of 14 randomly-situated sites were carried out in Tierra Firme forest at Cuyabeno, Ecuador. Litter trapped in the canopy and present at the soil level was collected, dried and weighed. Individual rhizomorph and contact zone systems were counted. Too few rhizomorph systems were encountered by this method to enable us to comment on the distribution and most litter was trapped by contact zone-forming fungi. The amount of litter trapped appeared to increase with increasing numbers of contact zone systems, and regression analysis showed a significant and positive relationship between the weight of trapped litter and numbers of contact zone systems. The average weight of ground litter was 178·8 g m^{-2} (1·78 t ha^{-1}). Trapped litter averaged 15·2 g m^{-2} (0·15 t ha^{-1}) representing 7·4% (range between sites 1 to 16%) of the total small litter.

In a second survey we attempted to map the distribution of individual rhizomorph and contact zone systems in the understorey canopy by intensive study of four 10 × 10 m plots, two in primary Tierra Firme forest, two in an area disturbed by a storm 6 years previously. Fungi were mapped as individuals in the understorey up to 2 m height at each 100 m^2 site, and for each fungal system the litter was collected and weighed. Rhizomorphs were separated and classified into taxa according to colour, width and surface hairiness. Contact zones were also separated and distinguished according to colour. As shown in Table 3, this more intensive sampling revealed that rhizomorph systems were present in the understorey at all four sites but, with the exception of one primary forest site, analysis of variance showed that they trapped significantly less litter than the contact

Table 3. Comparison of numbers of contact zone and rhizomorph systems in understorey canopies (up to 2 m height) at four sites (100 m^2) in Tierra Firme forest Cuyabeno, Ecuador

Litter-trapping system, forest type and site number	No. of individual systems per 100 m^2	No. of taxa per 100m^2	Average weight of litter per system (g)	Total weight of trapped litter (g $100m^{-2}$)
Contact Zone Systems				
primary 1	25	3	30·3	757·8
primary 2	33	4	16·4	539·6
disturbed 1	42	6	34·1	1433·6
disturbed 2	32	6	39·7	1269·8
Rhizomorph Systems				
primary 1	12	5	10·1	121·2
primary 2	30	2	15·6	468·8
disturbed 1	26	7	9·8	254·5
disturbed 2	40	5	10·9	437·6

Primary sites were in undisturbed Tierra Firme forest; at **disturbed** sites the forest had been damaged by storm 6 years previously.

---

zone systems both in total per site, and per system. The contact zone systems had also trapped significantly more litter in the disturbed forest than in the primary forest, again both in total per site and per system. There were also a higher density of taxa and larger numbers of systems of both contact zone- and rhizomorph-forming fungi in the disturbed forest. Using a χ^2 goodness of fit test, none of the sites showed non-random distribution of the litter-trapping systems, but using the variance: mean ratio and t test (Greig-Smith, 1983) both rhizomorph and contact zone systems showed a significantly clumped distribution in the disturbed forest although not in the primary forest.

This data set indicates that trapping of canopy litter can be more important in disturbed forest, possibly because production of small litter is greater in seral forest (Jordan, 1985), the increase in resources supporting more systems. Lateral transfer of systems may also be easier in the denser tangled understorey, as indicated by the clumped distribution of systems. However, the difference in amount of trapped litter between the two habitats was almost entirely due to contact zone systems (equivalent to 135·2 kg ha^{-1} in disturbed forest versus 64·9 kg ha^{-1} in primary forest) rather than rhizomorph systems (equivalent to 34·6 kg ha^{-1} in disturbed compared with 29·5 kg ha^{-1} in the primary forest).

These data sets were also used to compare the lengths of rhizomorphs in the lower understorey in the two forest types. In disturbed forest, the average was 32·0 km ha^{-1} compared with 19·6 km ha^{-1} in the primary forest. However, the greater length in the disturbed forest was largely due to the higher number of systems, since the average length of rhizomorph per system in the two habitats was not significantly different, being 9·7 m per system in primary forest and 9·4 m per system in secondary forest (t test $P = 0 \cdot 93$). It may be that the dynamics of the recruitment and loss of the population of canopy rhizomorph systems are related to the volume of the canopy treelets and palms and the amount of litter that can be retained.

Resource exploitation by litter-trapping fungi

Examination of trapped leaf and woody debris in Papua New Guinea and Ecuador showed that extensive decomposition had occurred. This was easiest to see in leaf litter, where zones of bleaching extended from contact zones or adhering rhizomorphs, presumably indicating internal invasion by the lignocellulose-degrading mycelium. Cultures were obtained from a number of the rhizomorph types in Papua New Guinea and Ecuador by surface sterilisation of sections followed by plating on a selective medium of Benlate (0·01%), malt extract (2%) and agar (4%). All cultures were able to degrade ordered cellulose although free cellulases were often difficult to detect in culture filtrates indicating that they may be partly cell wall bound in these fungi. Lignin degradation was investigated using dye analogues (Glenn & Gold, 1983) and all the cultures showed activity similar to a control white-rot fungus, *Coriolus versicolor* (L. ex Fr.) Quél. These rhizomorph-forming fungi can thus all be considered as lignocellulose utilisers and it is likely that the contact zone fungi are similar, given their aphyllophoralean affinities.

Two other aspects of the exploitation of the trapped resources are also important: nutrient and water availability. As noted in the introduction, the nitrogen and other nutrient content of plant litter is low and nutrient limitation is likely to be increased by leaching and canopy drip. Throughfall and stem flow studies have indicated that leaching from canopy litter and leaves may make a significant amount of minerals available to canopy organisms (Veneklaas, 1990). Soil rhizomorphs and mycelial cords have been shown to accumulate minerals from their surroundings whilst leaching from these structures is minimal (Stark, 1972; Jennings, 1990). In a similar way, canopy rhizomorphs may be able to utilise minerals present in solution in the aerial habitat.

Canopy litter is subject to wide fluctuations in water content. Our own observations in Papua New Guinea and Ecuador showed that trapped leaf litter had begun to dry in the daytime within 1-2 hours cessation of rain, and surface wetness had disappeared completely within 24 h without rain. In the absence of rain, the litter was still in equilibrium with an atmosphere of high relative humidity: we found r.h. values greater than 80% at 2 m height at Varirata and at Cuyabeno, even after 3 days without rain.

It is attractive to postulate that litter-trapping fungi show adaptation to the water-stressed conditions in the understorey canopy. One such adaptation may be the copious production of mucilage, noted earlier. A second might be the ability of mycelia to grow at low water potential and for rhizomorph extension to continue in dry air. The ability of isolates of rhizomorph-forming fungi from Ecuador and Papua New Guinea to grow at reduced water potential was investigated using malt extract agar (water potential adjusted using a range of solutes). The lower limit for mycelial growth of these fungi ranged from -4 to -8MPa, and rhizomorphs were produced from some cultures at -2MPa. If these results are compared with those of non-rhizomorphic, cocoa canopy fungi (Bravo, 1989), the effect of water potential on mycelial growth rates are seen to be similar. In the field however, it is more likely that r.h. and dew-point values will affect the growth of rhizomorphs. Laboratory experiments, growing rhizomorphs at different r.h., showed that rhizomorphs were unable to extend at values lower than 80%. However, rhizomorphs were able to tolerate r.h. as low as 33% for 3 days and to recommence growth after transfer to water-saturated air (Lewis & Hedger, 1992). It would therefore appear that rhizomorph systems may not grow under water stress but that these conditions can be tolerated.

Conclusions

The studies we have carried out in Papua New Guinea and Ecuador have shown that litter-trapping fungi exist in the understorey canopy of moist tropical forest, capturing resources by contact zones and aerial rhizomorphs. These mechanisms retain significant quantities of litter and prevent its fall to the decomposer community of the forest floor. The species assemblage has similarities to the fungi colonising the upper, drier layers of the forest floor litter where fungi forming mycelial connections and rhizomorph systems also dominate, being replaced by less disturbance-tolerant fungi in the lower leaf litter (Hedger, 1985). Lodge & Asbury (1988) have shown that this assemblage of litter-binding fungi retained forest litter on steep slopes (30%) in the El Verde reserve in Puerto Rico. We consider that the strategy of the understorey canopy species assemblage is the same, delaying export of litter to the forest floor, and their assemblage included at least one taxon, *Micromphale brevipes*, which we also found. It is also clear that although many fungi common in the upper litter in tropical forest bind litter, e.g. *Marasmius haematophalus* (Mont.) Fr. (Hedger, 1985), most do not colonise canopy debris, or, if they do, sporulation does not occur. The canopy species assemblage is physiologically and behaviourally adapted to its environment.

The canopy litter-trapping fungi offer a series of exciting research challenges. Within the nutrient dynamics of the forest we need to know the effect of this filtration of the litter fall on decomposition rates and nutrient recycling. Aerial cross-connection by contact zones and rhizomorphs also offer excellent model systems for the study of nutrient mobilisation by translocation between resources by fungi; litter-trapping systems could be used to study nutrient retention/export with better definition of input from canopy throughfall, and loss via leaching than soil litter which is in contact with the soil nutrient pool.

Individual litter-trapping systems can be readily identified in canopies by rhizomorph type or contact zone colour, which enables the size of individuals and the structure of populations to be studied. In Papua New Guinea, some rhizomorph communities were made up of several genets, whereas other systems represented clonal populations. Nuñez & Ryvarden (this volume p. 307) have recently shown that species assemblages of fungi in the upper canopy of forest in Cameroon included the genera *Hymenochaete* and *Vararia* that we found to be able to effect aerial mycelial transfer by contact zones. We do not know what the upper limits of rhizomorph systems are likely to be but we have observed them as high as 15 to 20 m in forest in Papua New Guinea.

Finally the mechanism of litter-trapping by these fungi remains extraordinarily interesting. The 'filtration' action of the network of rhizomorphs developed by the negatively gravitropic growth of the apices is unique, and it is a matter of debate whether these structures are analagous to soil rhizomorphs or indefinitely-extending stipes. We consider them to be functionally the former, but developmentally the latter. The suppressed pileus at the tips may protect the apex against desiccation, so affording a striking example of a fungal reproductive structure which has evolved a completely new role.

Litter-trapping fungi are probably present in all moist tropical forest. The use of rhizomorphs by birds to make nests seems to be ubiquitous, and there are reports of their decorative use by indigenous peoples in Ecuador (G. Paz y Mino, pers. comm.). In Papua New Guinea, we found an opplionid (harvestman) which lived within rhizomorph systems, apparently mimicking the rhizomorphs, even possessing irregular nodes on the legs resembling sites of apex regeneration – perhaps the only example of fungal mimicry by an arthropod.

Acknowledgements We are very grateful to Denise Nicholls for providing the illustrations, to Carol Parry for the light microscopy sections, Maria Nuñez for the identification of *Polyporus leprieurii*, and, in the field, to members of the Biological Sciences Department, University of Papua New Guinea and Depto. de Ciencias Biologicas Pontifica Universidad Catolica del Ecuador for their enthusiastic support. We are especially grateful to the Siona-Secoya tribes of Cuyabeno for permission to use their forest. P.L. acknowledges the financial support of SERC, and logistical and financial support of Xenova Ltd.

References

Ainsworth, A. M. & Rayner, A.D.M. (1990). Aerial mycelial transfer by *Hymenochaete corrugata* between stems of hazel and other trees. *Mycological Research* **94**, 263-266.

Bravo Velasquez, M. E. (1989). *Interactions between* Crinipellis perniciosa *and other Micro-organisms Isolated from Witches' Broom of Cacao*. PhD Thesis, University College Wales, Aberystwyth.

Cairney, J. W. G. (1991). Rhizomorph structure and development in *Marasmius crinisequi. Mycological Research* **95**, 1429-1432.

Glenn, M. H. & Gold, J. K. (1983). Decolourisation of several polymeric dyes by the lignin degrading basidiomycete *Phanerochaete chrysosporium. Applied and Environmental Microbiology* **45**, 1741-1747.

Golley, F. B. (1983). Decomposition. In *Tropical Rain Forest Ecosystems, Structure and Function*, (ed. F. B. Golley), pp. 157-166. Elsevier Scientific Publishing Company: Amsterdam, Oxford & New York.

Gooday, G. W. (1985). Elongation of the stipe in *Coprinus cinereus*. In *Developmental Biology of Higher Fungi*, British Mycological Society Symposium volume 10, (ed. D. Moore, L. A. Casselton, D. A. Wood & J. C. Frankland), pp. 311-331. Cambridge University Press: Cambridge, U.K.

Greig-Smith, P. (1983). *Quantitative Plant Ecology*. Third Edition. Blackwell Scientific Publications: Oxford, U.K.

Griffith, G. (1989). *Population Structure of the Cocoa Pathogen* Crinipellis perniciosa *(Stahel) Sing*. PhD Thesis, University College Wales, Aberystwyth.

Hedger, J. N. (1985). Tropical agarics, resource relations and fruiting periodicity. In *Developmental Biology of Higher Fungi*, British Mycological Society Symposium volume 10, (ed. D. Moore, L. A. Casselton, D. A. Wood & J. C. Frankland), pp. 41-86. Cambridge University Press: Cambridge, U.K.

Hedger, J. N. (1991). Fungi in the tropical forest canopy. *The Mycologist* **4**, 200-202.

Hedger, J. N. & Gitay, H. (1993). Fungal assemblages, their resource utilisation and response to climatic factors in tropical rainforest of Cuyabeno Reserve. In *Ecologia de la Amazon del Ecuador. El Noreste Amazonica y la Reserva Faunistica Cuyabeno*, (ed. E. Asanza & T. de Vries), in press. Pontifica Universidad Catolica del Ecuador.

Jacques-Félix, M. (1967). Recherches morphologiques, anatomiques, morphogénétiques et physiologiques sur des rhizomophes de champignons supérieurs et sur le déterminisme de leur formation. I. Observations sur les formations 'synnémiques' des champignons supérieurs dans le milieu naturel. *Bulletin Trimestrial de la Société Mycologique de France* **83**, 5-103.

Jennings, D. H. (1990). The ability of basidiomycete mycelium to move nutrients through the soil ecosystem. In *Nutrient Cycling in Terrestrial Ecosystems*, (ed. A. F. Harrison, P. Ineson & O. W. Heal), pp. 233-245. Elsevier Applied Science: London.

Jordan, C. F. (1985). *Nutrient Cycling in Tropical Forest Ecosystems, Principles and Their Application in Management and Conservation*. Wiley: Chichester.

Kiew, R. & Wong, K.M. (1988). *Barringtonia corneri* (Lecythidaceae), a remarkable new litter-trapping species from Johore, Malaysia. *Malayan Nature Journal* **41**, 457-459.

Lodge, D. J. & Asbury, C. E. (1988). Basidiomycetes reduce export of organic matter from forest slopes. *Mycologia* **80**, 888-890.

Myers, N. (1980). *Conversion of Tropical Moist Forests*, National Academy of Science: Washington, D.C.

Ng, F. S. P. (1980). Litter-trapping plants. *Nature Malaysiana* **5**, 26-32.

Pegler, D. M., Young, T. W. K. & Henrici, A. (1992) *Vararia ochroleuca* in Britain. *The Mycologist* **6**, 31-37.

Petch, T. (1915). Horse-hair blights. *Annals of the Royal Botanic Garden of Peradeniya* **6**, 43-68.

Stark, N. (1972). Nutrient cycling pathways and litter fungi. *Bioscience* **22**, 355-360.

Swift, M. J., Heal, O. W. & Anderson, M. J. (1979). *Decomposition in Terrestrial Ecosystems*, Studies in Ecology, *vol 5*. Oxford University Press: Oxford, U.K.

Thorold, C. A. (1975). *Diseases of Cocoa*. Oxford University Press: Oxford, U.K.

Townsend, B. B. (1954). Morphology and development of fungal rhizomorphs. *Transactions of the British Mycological Society* **37**, 222-233.

Veneklaas, E. (1990). *Rainfall Interception and Above Ground Nutrient Fluxes in Colombian Montane Tropical Rainforest*. WOTRO (Amsterdam, Netherlands Foundation for the Advancement of Tropical Research), subsidie W84-236.

Windsor, D. M. (1990). Climate and moisture variability in a tropical forest, longterm records from Barro Colorado Island, Panama. *Smithsonian Contributions to the Earth Sciences* number 29. Smithsonian Institution Press: Washington D.C.

Wright, J. E. & Ferraro, L. I. (1986). Hebras fungicas como principal componente de nidos de boyero en el ne Argentino. *Facena* **6**, 5-16.

Chapter 3

Nutrient cycling by fungi in wet tropical forests

D. Jean Lodge

U.S.D.A. Forest Service, Forest Products Laboratory, Center for Forest Mycology Research, PO Box B, Palmer, Puerto Rico 00721, U.S.A.

Introduction

Fungi are primarily responsible for the recycling of mineral nutrients through decomposition of organic matter (Swift, Heal & Anderson, 1979) and the uptake and transfer of these nutrients into plants via mycorrhizal fungi (Janos, 1983). In addition, fungi and other soil microorganisms serve alternately as sources and sinks of labile nutrients that are necessary for plant growth (Marumoto, Anderson & Domsch, 1982; Yang & Insam, 1991). Thus, fungal and microbial biomass can control significant fractions of the labile nutrient pools in some humid and wet tropical forests (Marumoto *et al.*, 1982; Lodge, 1985; Yang & Insam, 1991), and regulate the availability of nutrients that may limit plant growth (Jordan, 1985; Hilton, 1987; Singh *et al.*, 1989; Lee, Han & Jordan, 1990; Behera, Pati & Basu, 1991; Yang & Insam, 1991). The term biomass normally refers only to living organisms, but it is used more broadly in this chapter to refer to dead as well as living microorganisms as both contain nutrients.

Although humid tropical forests often have large stature and an abundance of vegetation, their growth and productivity is frequently limited by the availability of mineral nutrients. A diversity of soils occurs in the wet tropics, so it is difficult to generalise about nutrient limitation. However, the availability of phosphorus to higher plants is generally limited because phosphorus combines with aluminium and iron oxides in the highly weathered soils to form insoluble complexes (Sanchez, 1976). Elements such as nitrogen and potassium can be leached from ecosystems if the soil has little cation and anion exchange capacity (e.g. sands and highly weathered clay soils with low cation exchange capacity), and their availability may thus be quite low in a few tropical forests with high rainfall. Among wet tropical forests, limited availability of nitrogen appears to be most frequent in high elevation montane sites (Vitousek, 1984), but the causes are unknown. Other elements such as calcium and magnesium can be limiting in wet tropical forests (Cuevas & Medina,

1988), depending on the characteristics of the soil parent material and degree of weathering. Cuevas & Medina (1988) found greater root production in response to small scale applications of nitrogen in Tall Caatinga and Low Bana forests, to phosphorus in Bana forest, and to both phosphorus and calcium in Tierra Firme Amazonian forests of Venezuela. Vitousek (1984) hypothesised that phosphorus availability limited litterfall in some wet tropical forests based on the ratio of litterfall mass to litterfall phosphorus content. The few experimental plot fertilisation studies made in wet tropical forests have shown increased leaf litter mass and sometimes wood production with nutrient addition in Venezuela (Tanner, Kapos & Franco, 1992; N and P), Hawaii (P. M. Vitousek, pers. comm.), and Puerto Rico (Zimmerman *et al.*, 1992; complete nutrients), supporting the hypothesis that tree productivity is often limited by nutrient availability in these forests. The degree to which fungi and other microorganisms regulate the availability of limiting nutrients depends on the size of the labile nutrient pool, the quantity of fungal and microbial biomass, the fluctuations in fungal and other microbial biomass and its nutrient content through time (Hunt, Elliott & Walter, 1989).

In this Chapter, data on fungal and microbial biomass and nutrient contents, mostly from wet tropical forests, are compared. The tightness of nutrient cycling in wet and seasonally wet tropical forests is also discussed but with the emphasis on the role of saprotrophic rather than mycorrhizal fungi.

Fungal biomass and nutrient stores

Fungal biomass and total biomass of all microorganisms contain significant fractions of the labile nutrients in the forest floors of some tropical ecosystems (Marumoto *et al.*, 1982; Srivastava & Singh, 1988; Yang & Insam, 1991). The ability of decomposer fungi (Watkinson, 1971; 1984) and soil microbial biomass (Srivastava & Singh, 1988) to concentrate and store nutrients such as phosphorus when they are in short supply can accentuate the role of fungi and other microorganisms in regulating nutrient availability in nutrient depauperate sites. Furthermore, nutrients tend to be immobilised and thereby conserved by fungi and other microorganisms in their biomass during periods of high rainfall that cause leaching (Jordan, 1985; Hilton, 1987; Singh *et al.*, 1989; Behera *et al.*, 1991; Yang & Insam, 1991). Growth of higher plants in such systems may be dependent on nutrients released upon death and mineralisation of microbial biomass. Again, microorganisms can have relatively large effects relative to their biomass because microbial activities and products can maintain phosphorus in plant-available forms in highly weathered ultisols with phosphorus-fixing clays (Lee *et al.*, 1990).

Fungal and total microbial biomass

Relatively few publications contain estimates of fungal or total microbial biomass in wet tropical forests, and a variety of methods has been employed to make these estimations, none of them perfect under all conditions (see West, Ross & Cowling, 1986; Frankland, Dighton & Boddy, 1990). In addition, estimates made using different methods are often not directly comparable. For example, estimates obtained by fumigation followed by incubation (for measurement of carbon mineralised from the dead microorganisms; Jenkinson & Powlson, 1976) or extraction and measurement of microbial nitrogen (Brookes *et al.*, 1985) are thought to reflect both active and dormant microbial biomass, whereas estimates obtained through the substrate-induced respiration method (SIR, Anderson & Domsch, 1978) are thought to reflect only the active microbial biomass. Again, the selective inhibition method (Anderson & Domsch, 1973) is thought to measure biomass of only the active fungi or bacteria, whereas the most commonly employed direct observation methods detect active, dormant and dead fungi and bacteria. However, the various means of obtaining estimates of fungal and bacterial biomass are often within a factor of ten, and are useful in giving an indication of where the true value lies.

In this study, total (live plus dead) fungal volume was measured in litter and surface soil at two week intervals in subtropical wet forest located at El Verde in the Luquillo Experimental Forest of Puerto Rico (alt. 350 m) by direct observation using an agar film method (Lodge & Ingham, 1991; modified from Jones & Mollison, 1948). Soil samples were collected using a 12 cm diam. steel cylinder driven to 9 cm depth. Five samples were taken at preselected random coordinates on each date within a 1 ha plot. Initial samples were collected on 26 November 1984, and further samples were taken every two weeks from 11 February to 3 June 1985, with an additional sampling on 18 September. Measurements of hyphal lengths and frequency distributions of hyphal diameters among 7 classes were used to estimate fungal volumes per g of dry litter or soil, assuming that fungal hyphae are perfect cylinders. Hyphal diameter distributions were determined before drying the agar films to ensure that the hyphae were not collapsed. Conversion factors of 0·26 and 0·20 g cm^{-3} fungus were used to convert fungal volume into fungal biomass in litter and soil respectively (Lodge, 1987).

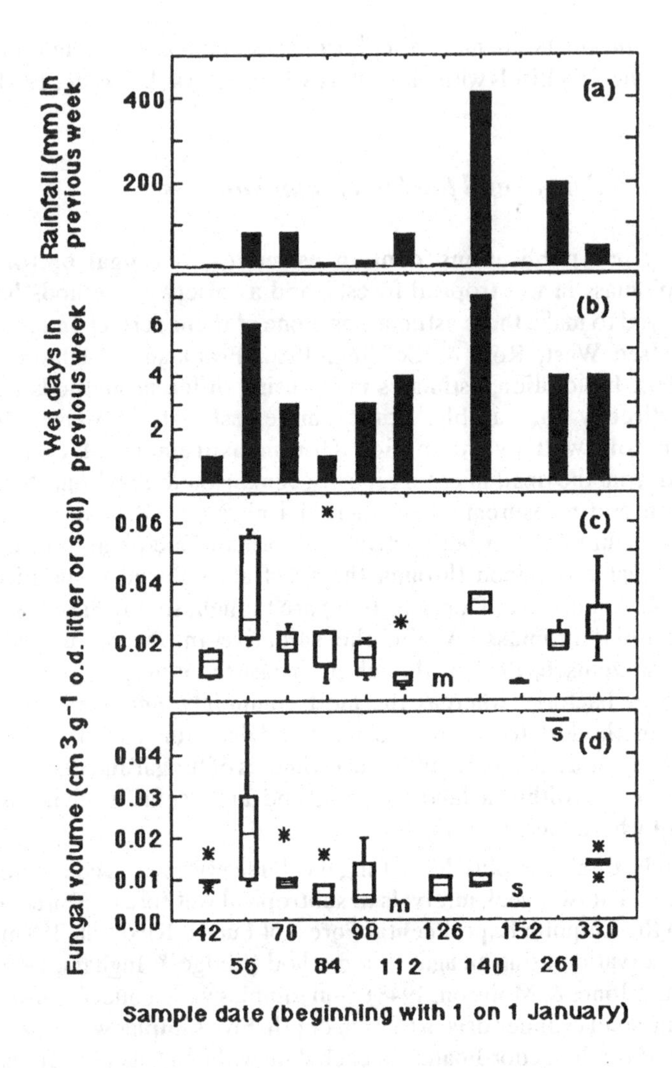

Fig. 1. Box plots showing cumulative weekly rainfall during the 7 days preceding sampling (a), number of days in the previous week with rainfall sufficient to reach the forest floor; >3 mm (b), and fluctuations in fungal volume in the litter layer (c) and upper 9 cm of soil (d) in a subtropical wet forest in El Verde, Puerto Rico. Fungal data are expressed per gram oven dry litter or soil. Median values are shown by horizontal bars, the boxes contain the 25% of the variation on either side of the median, the bars extending from the boxes indicate variation (see McNeil, 1977), and extreme values are shown with an asterisk. Data for soil on date 112 and for litter on date 126 are missing (m). Data represent five samples on each date except for soil samples on dates 42, 152 and 261 (s, where n = 3, 2, and 1 respectively).

Median fungal biovolume estimates for litter and soil on each date are shown by the bars within the box plots in Figs 1c and 1d respectively. Mean fungal biomass estimates for each sampling date were used to obtain overall mean fungal biomass estimates ± S.E.M. of 5·2 ± 2·4 mg fungus g^{-1} leaf litter and 2·7 ± 2·7 mg fungus g^{-1} soil in the upper 0-9 cm (Table 1). Fungal biomass was sometimes much higher in individual samples: up to 15·2 mg g^{-1} litter and 9·7 mg g^{-1} soil. All these extreme high values were from samples containing cords of decomposer basidiomycetes, and all but two were collected near decomposing wood.

Data on fungal and total microbial biomass in soil and litter from El Verde, Puerto Rico (described above) as well as from various other tropical and subtropical sites are compared in Table 1. The data cited from other sources are: Guzman, Puerto Rico, microbial N (L. M. Babilonia, pers, comm., 0-10 cm depth); El Verde, Puerto Rico, substrate-induced respiration (unpublished data of D. J. Lodge, C. E. Asbury, A. Masso & R. Pollit, 0-10 cm depth); Manãus, Brazil (Luizão, Bonde & Rosswall, 1992, 0-5 and 5-20 cm depths); Hainan Island, China (Chang & Insam, 1991, 0-12·5 cm depth); Orissa, India (Behera *et al.*, 1991, 0-3, 3-8, 8-13 and 13-18 cm depths, winter dry and summer monsoon seasons); pastures and grasslands in New Zealand, direct observations and fumigation incubation (from West *et al.*, 1986, 0-7·5 cm depth, low number from dry season and high from wet season, no direct observations in wet season), while the substrate-induced respiration data for New Zealand are from Sparling *et al.* (1985, 0-5 cm depth). Ranges given in Table 1 include variation among seasons and soil depths. Calculations were performed to convert microbial carbon into microbial biomass for data from Puerto Rico, Brazil and China by assuming that microbial C is 0·4× microbial biomass.

The estimates of total fungal biomass in soil at El Verde (means for each date ranging from 0·2 to 9·7 mg g^{-1} soil) were high when compared with other tropical forests and grasslands (0·2 to 1 mg g^{-1}; Table 1). One comparable study of fungal biomass in tropical forest using direct observation was carried out in a seasonally dry monsoon forest in Orissa, India, where total fungal biomass varied from 0·2 to 0·4 mg g^{-1} dry soil in the upper 0- 7·5 cm horizon (Behera *et al.*, 1991). Yang & Insam (1991) studied soil fungal biomass in another tropical monsoon forest in China using the selective inhibition method and obtained values for active fungal biomass (0·2 to 0·5 mg g^{-1}) that were comparable with the total fungal biomass estimates for monsoon forest in India (Behera *et al.*, 1991). The only other published study of fungal biomass or volume in tropical forest known was that on a 22 year old native second growth forest in Puerto Rico (Lodge & Ingham, 1991) at *ca* 350 m altitude in Guzman, near El Verde (Luquillo Experimental Forest Tropical Soil Biology and Fertility

Table 1. Comparison of fungal and total microbial biomass and microbial nitrogen in various tropical ecosystems

	Fungal biomass		Biomass of all microorganisms			Microbial N
	Live & dead	Active	Live & dead	Active & dormant	Active	Active & dormant
	Method					
	Direct observation	Selective inhibition	Direct observation	Fumigation-incubation	SIR	Fumigation-extraction
Leaf litter						
Subtropical wet forest El Verde, Puerto Rico	5·2±2·4 (1·7–9·5)	–	–	–	–	–
Upper soil horizon						
Subtropical wet forest El Verde, Puerto Rico	2·7±2·7 (0·2–4·8)	–	–	–	25·6±1·3	–
Secondary wet forest Guzman, Puerto Rico	6·1±1·2		3·1±0·2 (2·8–3·4)	–	–	0·123
Pine stand, wet forest Guzman, Puerto Rico	–	–	2·2±0·3 (1·9–2·4)	–	–	0·156
Tropical humid forest Manaus Amazonas Brazil	–	–	–	1·6–4·2	–	0·03–0·04
Tropical humid forest Hainan Island, China	–	0·2–0·5	–	–	0·6–1·6	–
Tropical monsoon forest Orissa, India	0·2–0·4	–	–	–	–	–
Pasture and grassland temperate and sub-tropical, New Zealand	0·2–1·0	–	1·2–ND	0·8–2·5	1·0–3·3	–

Entries are mean values ± S.E. mg g^{-1} oven dried litter or soil. ND = not done. SIR = substrate induced respiration. Figures in parentheses indicate range of values.

Programme plots). In this study using direct observation, Ingham examined soils from 0-5 cm depth and found hyphal lengths ten-fold greater than in temperate Douglas fir forest (Lodge & Ingham, 1991). These hyphal lengths were converted into hyphal volumes using mean hyphal diameter distributions for September and then to fungal mass, as above, to obtain an estimated 5·2 mg fungus cm^{-3} (6·1 mg g^{-1}) soil in the upper 5 cm (Table 1). These data suggest that fungal biomass is generally high in the subtropical wet forest life zone of Puerto Rico, and that the fungal biomass estimates for El Verde are probably not excessive.

Fungi comprise only a fraction of the soil microbial biomass. For example, fungi represented 21 to 41% of the active microbial biomass carbon in the monsoon tropical forest soils studied by Yang & Insam (1991). Thus, some check on the validity of fungal biomass estimates can be done by determining whether or not they are within an order of magnitude or less than the estimated biomass of all microorganisms. Data presented in Table 1 from an unpublished *in situ* substrate-induced respiration (SIR) experiment suggested that there were relatively large quantities of active microbial biomass in the upper 10 cm of soil at El Verde, which is consistent with the high values for fungal biomass. This experiment (D. J. Lodge, C. E. Asbury, R. Pollit & A. Masso, unpublished) was carried out using intact soil cores 25·4 cm diam. and 10 cm deep, rather than sieved soils, for two reasons. Firstly, previous attempts to use the SIR method on sieved soils resulted in declining levels of carbon dioxide evolution at all levels of glucose addition. This result occurred because of very high initial respiration induced as an artifact of sieving, possibly because of exposure of new substrate surfaces, or by disruption of fungal hyphae whose contents then fuelled respiration. Secondly, intact soil cores containing severed roots were used to mimic the conditions in soil nitrification tubes as well as in the forest following hurricane Hugo (after Hugo, mortality of fine roots was near 100% in the top 0 to 10 cm; Parrotta & Lodge, 1991). Each core was placed on a plastic funnel containing washed silica sand and replaced in its original hole. Respiration at 23°C was measured for 1 h before and 1 h after the addition of sugar by enclosing a trap of sodium hydroxide solution on the soil surface and then sealing the top of each tube with Mylar. The quantity of CO_2 evolved was determined by titration of the NaOH in the CO_2 traps. Weight of soil in the cores was estimated using a 5067 cm^3 volume and a mean bulk density for ridge tabonuco forest sites of 0·77 g cm^{-3}. Glucose (3·76 g) was added to each tube in 50 ml of water (1·51 mg C g^{-1} soil), and an equal volume of distilled water was added to the control cores (4 replicates per treatment).

Respiration rates of the soil microbial biomass of intact soil cores using the SIR method increased from 84·1 ± 33·9 to 610·0 ± 28·6 mg CO_2-C

$m^{-2} h^{-1}$ in response to the addition of glucose solution, whereas respiration in control tubes increased from 82·9 ± 19·8 to only 119·7 mg CO_2-C m^{-2} h^{-1} in response to addition of water. Active microbial biomass was estimated to be 25·6 ± 1·3 mg g^{-1} soil using the formula from Ocio & Brookes (1990), who added 1·6 mg glucose-C g^{-1} soil. This value is 10 times higher than soil microbial biomass estimated using the SIR method in temperate and subtropical pastures and grasslands in New Zealand (Sparling, West & Whale, 1985; Table 1) and 100 times higher than estimates from China in humid tropical forest on the island of Hainan (Yang & Insam, 1991; Table 1). The high soil organic carbon (3·9%) and the presence of dead roots and surface litter in the Puerto Rican study probably contributed to the high rates of microbial respiration, since microbial biomass and soil carbon are closely related (H. Insam, pers. comm.). The use of unsieved soils in Puerto Rico also differed from the other studies. However, Domsch *et al.* (1979) have also reported that the SIR method gave high values for active fungal biomass compared with direct observation of total fungal biomass.

The flush of CO_2 evolution following reinoculation of fumigated soils (fumigation-incubation; FI) has been the most widely used method to determine live (active plus dormant) soil microbial biomass in tropical soils. Estimates of live soil microbial biomass obtained using FI have ranged from 0·8 to 4·2 mg microbial biomass g^{-1} surface soil (assuming 0·45 g C g^{-1} microbial biomass; Table 1). The highest estimates of live microbial biomass were found in native second growth subtropical wet forest (2·8 to 3·4 mg g^{-1} soil) in Puerto Rico (personal observations using direct observation methods in Anderson & Ingram, 1989). Soil of a pine plantation in Puerto Rico (personal observations) and of a tropical humid forest in Manaus, Brazil (Luizão, Bonde & Rosswall, 1992) contained a similar or slightly lower amount of soil microbial biomass (1·9 to 2·4 and 1·6 to 4·2 mg g^{-1} soil). These data are consistent with the high estimates for total fungal biomass and active microbial biomass obtained for subtropical wet forest soils in Puerto Rico.

Total soil fungal biomass per square metre in wet and seasonally wet tropical forests was equal to or greater than fungal biomass in temperate forests. Among the tropical forest sites, Behera *et al.* (1991) found 116 g m^{-2} total soil fungal biomass from 0 to 23 cm depth in India; Yang & Insam (1991) found 8 to 80 g m^{-2} active fungal biomass from 0 to 12·5 cm depth in China, and this study found total fungal biomass of 14 to 333 g m^{-2} (mean 207 g m^{-2}) from 0 to 9 cm depth at El Verde and 260 gm^{-2} from 0 to 5 cm depth at Guzman in Puerto Rico. Among the studies summarized by Kjøller & Struwe (1982), estimates of total fungal biomass for temperate woodlands ranged from 16 to 51 g m^{-2} in the 0 to 10 cm horizon and 20 to 112 g m^{-2} in the 0 to 20 cm horizon. There are few studies of fungal biomass

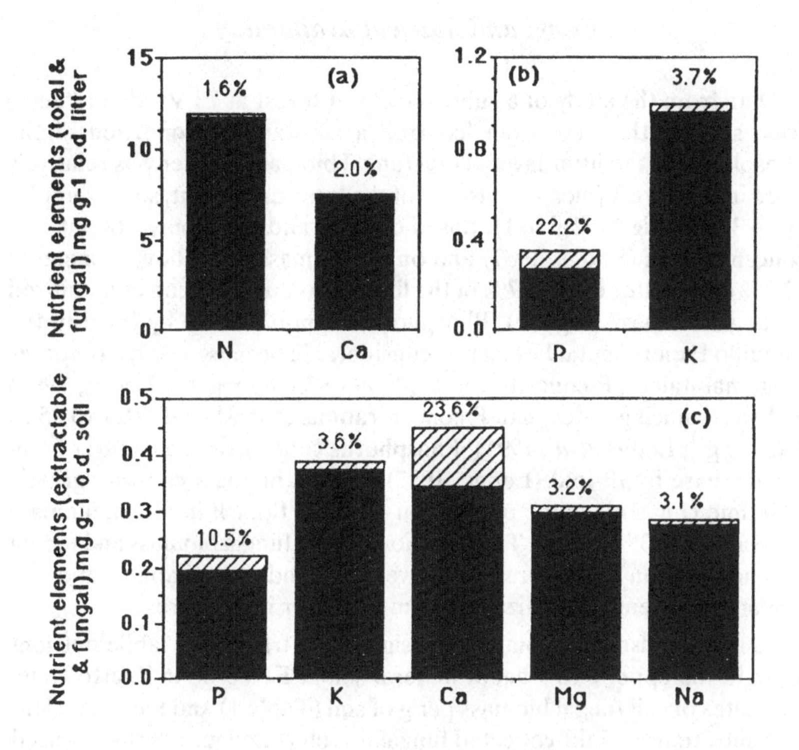

Fig. 2. Fungal elemental content (upper part of each bar) expressed as a proportion of the total nutrient content in litter (a & b), and of the extractable nutrient content in the surface soil horizon (c) in a subtropical wet forest in Puerto Rico. Elemental contents are expressed per gram oven dry litter or soil. Percentage values indicate the mean percentage of the total nutrient pool in litter or the labile nutrient pool in soil which was estimated to be immobilized in fungal biomass.

---

in litter, but this study found 1 to 5 g fungus m^{-2} at El Verde in Puerto Rico (9 yr mean litter mass of 525 g m^{-2}; Lugo, 1992) which was comparable to the values of 1·3 and 8·0 g m^{-2} determined for litter of deciduous woodlands in the U.K. (Meathop Wood; J. C. Frankland, cited in Kjøller & Struwe, 1982) and the U.S.A. (*Liriodendon* forest; Witkamp, 1974), respectively.

Fungi and nutrient availability

Data from the study of a subtropical wet forest at El Verde in Puerto Rico suggest that fungi can 'control' a substantial proportion of the phosphorus in the litter layer. Total fungal biomass in litter was relatively large in this forest (mean, 0·2 to 1% of the litter dry weight, some samples up to 3·6%; Fig.1c, Table 1). Based on litter and basidiomycete nutrient concentrations (Lodge, 1987) and on the biomass data above, a mean of 22·2% of the litter P and 3·7% of the litter K could have been immobilized in fungal biomass (Fig. 2c). Phosphorus immobilization by fungi in the Luquillo Experimental Forest was considerable because leaf decomposer fungi maintained P concentrations (4·99 to 35·66 mg g^{-1}; Lodge, 1987) that were much greater than P concentrations in fresh leaf litter (0·25 to 0·42 mg g^{-1}; Lodge *et al.*, 1991). Phosphorus concentrations in litter fungi can increase by 10-fold (Lodge, 1987) and the biomass can vary greatly with time (Fig. 1c), so the proportion of forest floor P in fungal biomass may vary from 3% to 85%. The variation in both fungal biomass and fungal P concentrations therefore may have regulated P availability and the timing of nutrient mineralization from leaf litter in this forest.

Fungi may also have controlled a significant fraction of labile nutrient pools in the upper 0 to 9 cm of mineral soil at El Verde in Puerto Rico. Estimates of soil fungal biomass per g of soil (Table 1) and mean nutrient concentrations in field-collected fungal mycelia (Lodge, 1987) were used to calculate the quantities of nutrients stored in the upper 9 cm at El Verde. Mean fungal biomass was 2·7 mg g^{-1} in the upper 9 cm of soil (range 0·2 to 4·8 mg g^{-1}). Assuming that the appropriate specific fungal density (0·2 g cm^{-3}) and nutrient concentrations were applied, fungal nutrient stores in the upper 9 cm of soil at El Verde were 7 to 178 μg N g^{-1}, 2 to 40 μg P g^{-1}, 1 to 24 μg K g^{-1}, 1 to 17 μg Mg g^{-1}, and 8 to 189 μg Ca g^{-1}. These fungal nutrient stores represent 0·8 to 20% of the Olson-extractable phosphorus from dried soils, and 24% of the total soil calcium but only 0·5% of the nitrogen, 3·5% of the extractable soil potassium, and 3% of the magnesium in the upper 10 cm of soil (soil nutrient data from Odum, 1970; Fig. 2). The percent extractable soil P held in fungal biomass at the Puerto Rican site (0·8 to 20%) was similar to the percent of soil organic P in all microbial biomass of a monsoon forest woodland and a teak plantation in India (9 to 12%; Srivastava & Singh, 1988). Fungal control of the labile pools of soil phosphorus is probably significant in regulating soil fertility in the Puerto Rican forest. Fluctuations in fungal and microbial biomass may well determine whether such biomass acts as a net sink for nutrients or as a potential source of nutrients for plants.

Timing of nutrient mineralisation from fungal and microbial biomass

Fluctuations in fungal and microbial biomass with moisture conditions are common, even in non-seasonal and slightly seasonal wet tropical forests. Fungal biomass was found to change significantly between sampling dates both in the litter and in the soil at El Verde in Puerto Rico (Fig. 1; $P = 0 \cdot 001$, and $P = 0 \cdot 008$ for litter and soil respectively; Kruskal-Wallis one-way non-parametric ANOVA). Similar fluctuations were observed in monsoon tropical forests by Yang & Insam (1991) in China and by Behera *et al.* (1991) in India. At El Verde in Puerto Rico, fungal biomass in both litter and soil were significantly correlated with percent moisture in the substratum ($P = 0 \cdot 038$ and $0 \cdot 001$, respectively; Fig. 1). Fungal biomass in litter was also strongly related to the number of days in the preceding week in which rainfall was sufficient to reach the forest floor (3 mm; Wilcoxon Signed Ranks test, $P < 0 \cdot 001$) but not with the cumulative amount of rainfall in the preceding week (linear regression; $P = 0 \cdot 27$). These fluctuations in fungal and microbial biomass can be very rapid, especially in the litter layer, and can result in pulses of nutrient mineralisation.

Raghubanshi *et al.* (1990) and Singh *et al.* (1989) showed that the pulsed release of nutrients by death and mineralisation of the microbial biomass during the first 4 weeks of the rainy season closely synchronised with rapid plant uptake and growth in Indian monsoon forest. Similarly, a study of fungal biomass in another Indian monsoon forest (Behera *et al.*, 1991) and in a slightly seasonal humid forest on Hainan Island in China (Yang & Insam, 1991) both showed that fungal biomass was directly related to soil moisture. In both studies cited above, the authors attributed the conservation of nutrients against losses during the rainy season to immobilisation of nutrients in fungal and other microbial biomass. Yang & Insam (1991) found that soil microbial biomass was related to the abundance of decomposable organic matter as well as soil moisture. Root activity, as measured by ^{32}P uptake, also reached an annual maximum soon after the first rain at the end of the dry season in a seasonal African forest (FAO/IAEA, 1975). Root production was frequently greatest at the beginning of the rainy season at La Selva (R. Sanford, pers. comm.), and at Barrow Colorado Island in Panama there were two peaks in root production which corresponded with peaks in nutrient leaching from litter during both transitions between wet and dry seasons (J. Wright, pers. comm.).

The proportion of nutrients available for plant growth may differ, depending upon timing and whether the nutrients are released in large bursts or gradually at low concentrations. The studies cited above suggest that massive bursts can occur, especially in transitions between wet and dry periods. According to the current paradigm, plants cannot compete effectively with saprotrophic microbes for limiting nutrients (Elliott *et al*., 1989). When nutrients are released in large pulses, they may saturate the immobilisation capacity of the existing soil microbial biomass. Therefore, proportionately more nutrients may be available to higher plants when mineralisation is a pulsed rather than being a gradual process of release. Alternatively, such pulsed nutrient release might also increase the susceptibility of nitrogen to leaching losses or the complexing of phosphorus with iron and aluminium in highly weathered soils.

Asynchrony of nutrient mineralisation with plant uptake may lead to net nutrient export from the ecosystem via leaching (Frankland, 1982) or loss from the biotic system via chemical fixation into aluminium and iron oxides (Sanchez, 1976). Consequently, both the INTECOL Tropical Biology Workshop (Hayes & Cooley 1987) and the Tropical Soil Biology and Fertility Programmes (Anderson & Ingram, 1989) have emphasised the importance of the timing of nutrient release. Lack of synchrony between nitrogen mineralisation and root uptake in tabonuco forests of northeastern Puerto Rico following hurricane Hugo may have resulted in significant nitrogen losses from these ecosystems (Lodge & McDowell, 1991). The virtual absence of shallow fine roots for three months following the hurricane (Parrotta & Lodge, 1991) at the time when nitrogen was mineralised from large quantities of hurricane litter may have accelerated export of stream nitrogen and denitrification (Steudler *et al*., 1991). After a temporary disappearance of nitrate from stream waters (probably by microbial immobilisation) nitrate concentrations increased 10-fold in streams draining the subtropical wet forest (C. Asbury, pers. comm.).

Nutrient conservation by fungi in wet tropical forests

Since fungi can physically translocate nutrients among resources that are separated in space (Thompson & Rayner, 1983; Thompson, 1984), they tend to conserve limiting nutrients more tightly than other microorganisms. Temperate wood-decomposing fungi have been found to incorporate most of the nitrogen and phosphorus of one food base and then translocate up to 81% of the original nitrogen and phosphorus into a new food base via hyphal cords (Watkinson, 1971; 1984), rather than releasing these nutrients upon death when the availability of carbon in their resource base becomes limiting. The translocation of limiting nutrients causes a short circuit in the nutrient cycle whereby nutrients are re-immobilised by fungi rather than being mineralised in a form available to plants. Thus, wood-decomposing fungi can influence the availability of nutrients to higher plants. Wells, Hughes & Boddy (1991) found that phosphorus was translocated from soil to wood by temperate wood decomposer fungi, and the same phenomenon apparently occurs in wet subtropical forest of Puerto Rico. Zimmerman *et al.* (1992) recorded significantly lower forest productivity as measured by rates of leaf litterfall 1·5 to 2 years after hurricane Hugo in control plots compared to plots in which all the fine and woody hurricane litter had been removed. Simulation modeling using the Century model (Parton, 1988; Sanford *et al.*, 1991) for tropical forests indicated that immobilisation of phosphorus by fungi in the 'hurricane-downed' wood (and presumably translocation of phosphorus from soil to wood) was responsible for less productivity in the control plots (W. Parton & R. Sanford, pers. comm.). It is probably not coincidental that in this study all but two of the extremely high values for fungal biomass were in samples of both litter and soil (Table 1) taken close to decomposing wood, and that cords and mycelia of wood-decomposing basidiomycetes were observed in these samples. Nutrient recycling between resource bases also occurs among fungi that decompose leaf litter in the tropics where the phosphorus content of the leaf litter is extremely low. The generally low availability of phosphorus in many wet tropical soils is associated with lower phosphorus concentrations in foliage and greater nutrient retranslocation during senescence (Vitousek & Sanford, 1986). For example, retranslocation of leaf phosphorus before abscission in subtropical wet forest of Puerto Rico was estimated as 72%. This was calculated from nutrient concentration data of Odum (1970) and Medina, Cuevas & Weaver (1981), using the formula of Vitousek & Sanford (1986) which includes ratios of phosphorus to calcium to control for declines in leaf specific gravity during leaf senescence, because calcium is not retranslocated and thus remains constant. As a consequence of differences in nutrient retention by plants, concentrations of phosphorus in the leaf litterfall of many wet tropical forests are ten times lower than concentrations in leaf litter of temperate forests (Vitousek, 1984). For example, mean concentrations of phosphorus in leaf litterfall are 0·24 to 0·42 mg g^{-1} in subtropical wet and lower montane rain forest of Puerto Rico (Lodge *et al.*, 1991; Lugo, 1992), while concentrations of phosphorus in individual litter species can be even lower: 0·16 mg g^{-1} in *Dacryodes excelsa* Vahl. leaves, the dominant tree of subtropical wet forest in Puerto Rico (Odum, 1970), and 0·27 mg g^{-1} in *Ficus fistulosa* Reinw.: Blume leaves in Hong Kong (Lam & Dudgeon, 1985).

Several species of agaric decomposers from Puerto Rico were found to translocate ^{32}P from partially decomposed leaf litter into freshly fallen

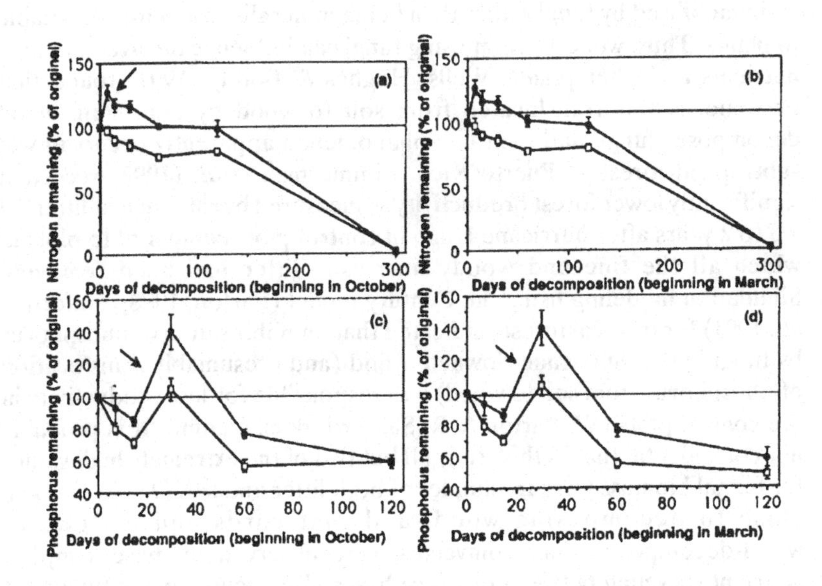

Fig. 3. Percentage of the original mass of nitrogen (a & b) and phosphorus remaining (c & d) in decomposing leaf litter enclosed in 1 mm-mesh bags placed in two plots (3 & 3x, shown with closed circles and open squares, respectively) at El Verde, Puerto Rico. Fresh litter from a 1 ha plot (plot 3) was placed in bags and was distributed in the same plot (3), as well as in a small, 10 x 10 m uniform plot nearby (plot 3x) during October (a & c) and March (b & d). (Carol P. Zucca, unpublished data). Arrows point to important accumulations of nitrogen or phosphorus in litter during the early stages of decomposition (a, c & d), although some nutrient accumulations are preceded by a leaching phase (c & d).

---

unlabelled leaves through rhizomorphs and hyphal cords across a 4 mm gap (personal observations). Translocation rates varied among species, and in some cases were quite high. The translocating structures of these litter fungi also contribute to nutrient conservation in mountainous wet tropical forests when they bind the thin litter layer into a mat, thereby protecting the forest floor and surface soil from loss of nutrients by erosion during heavy rains (Lodge & Asbury, 1988). Rainfall in these forests is typically 2500-3000 mm yr^{-1}.

Nutrient translocation by decomposer fungi probably accounts for much of the increase in nutrient content (% of initial nutrient mass remaining) observed during the first 3 to 8 weeks of leaf litter decomposition in wet tropical forests, although depositions from the canopy and N-fixation could be partly responsible. C.P. Zucca (pers. comm.; Fig. 3) found increases in phosphorus content of freshly fallen mixed leaf litter of 120 to 140% at El Verde in the Luquillo Experimental Forest of Puerto Rico during the first 6 weeks of decomposition. Similarly, a litterbag study of *F. fistulosa* leaves in Hong Kong indicated that phosphorus content increased 135% to 150% above the initial content after 12 to 18 days of decomposition (Lam & Dudgeon 1985; data reanalysed by Lodge (1989) to obtain % initial content remaining).

Nitrogen is apparently translocated by leaf-decomposing fungi in some wet tropical forests but not in others. For example, C. P. Zucca (pers. comm.; Fig. 3) found increases in nitrogen content of leaf litter up to 128% of initial nitrogen during decomposition in Puerto Rico, when freshly fallen mixed litter was placed in one area (plot 3x; Fig. 3) but not when the same litter was placed in another area (plot 3; Fig. 3). Cuevas & Medina (1988) showed an increase in nitrogen content of decomposing leaves of 110% to 160% during the first 10 weeks of decomposition in three types of Amazonian forest in the Rio Negro area of Venezuela (Tierra Firme, Caatinga, and especially in Low Bana forest types) but little change in phosphorus (probably a small net loss). Lam & Dudgeon's data (1985; Lodge, 1989) showed an initial leaching loss followed by a 150% increase in nitrogen content of *Ficus* leaves between the 8th and 18th day of decomposition.

That mycorrhizas associated with the litter layer in tropical forests take up phosphorus and other nutrients released from leaf litter has been well documented, but whether mycorrhizal fungi in association with plant roots are capable of decomposing leaf litter to obtain nutrients (i.e. direct nutrient cycling, Went & Stark, 1968*a*, *b*) is highly controversial (Janos, 1987). In subtropical wet forest of Puerto Rico, the appearance of mycorrhizas inside leaf litter bags after about 6-8 weeks of decomposition coincides with increased rates of nutrient loss from the litter (C. Zucca, pers. comm.). Cuevas & Medina (1988) showed in Venezuela that permanent or intermittent separation of leaf litter decomposition bags from the soil substratum to prevent root penetration or attachment greatly reduced the rate of disappearance of phosphorus, calcium and magnesium in Tierra Firme forest. This treatment lowered the mass loss rate only slightly in the Tall Caatinga forest type, an effect that may be unrelated to the activity of mycorrhizal fungi. Elliott *et al.* (1989) applied ^{32}P tracer to leaf litter in a vesicular-arbuscular mycorrhizal (VAM) coastal rainforest in northeastern Brazil, and found that 20% to 70% was apparently initially absorbed by saprotrophic microorganisms in humus, 0·1 to 57% was found in the upper 0 to 5 cm of mineral soil, and only 5% was removed by the mycorrhizas. They suggested that VAM fungi do not compete effectively for nutrients against saprotrophic microbes in the litter layer. In contrast, the Amazonian white sands vegetation studied by Went & Stark (1968*a*, *b*) was dominated by ectotrophic rather than VAM plants, but as yet there are no definitive studies of direct nutrient cycling by this very different group of fungi in tropical forests. Evidence from laboratory experiments shows thast some of the basidiomycetes that form ectomycorrhizas with temperate forest trees obtain organic forms of nitrogen directly from organic matter (Read, Leake & Langdale, 1989), lending credence to the Went & Stark (1968a, b) hypothesis of direct nutrient cycling in ectrotrophic tropical forests. However, most wet tropical forests are dominated by VAM plants (Janos, 1989). Nutrient cycling is nevertheless tight in these VAM forests where any superficial root mats are the first to take up nutrients mineralised upon death of saprotrophic microbes. For example, at El Verde in Puerto Rico, eventual uptake by mycorrhizas of radioactive phosphorus applied to the litter layer was 10 to 20 times greater than the amount transferred to soil (Luse, 1970).

Conclusions

From these studies it can be concluded that fungi are important in controlling the availability of nutrients to plants in wet and seasonally wet tropical forests. VAM root mats were very efficient in absorbing nutrients mineralized from the litter mat (Luse, 1970; Cuevas & Medina, 1988; Elliott *et al*., 1989), but mineralisation of nutrients from litter was apparently due to the activity of saprotrophic rather than VAM fungi (Elliott *et al*., 1989). Both litter- and wood-decomposer fungi in tropical forests translocated nutrients among resource bases, thereby accentuating their ability to immobilise nutrients important for plant growth. Fungi comprised 0·3 to 1·7% (mean 5·2 mg g^{-1}) of the mass of the litter layer of a wet forest in Puerto Rico, accounting for up to one third to two thirds of the phosphorus on the forest floor. Soil fungal biomass ranged from 0·02 to 1·3% of the surface soil dry weight (mean 2·7 mg g^{-1}) accounting for 24% of the extractable soil Ca and up to 20% of the labile soil phosphorus in Puerto Rico. Biomass of fungi in wet tropical forests in China (Yang & Insam, 1991), India (Behera *et al*., 1991), and Puerto Rico was similar or greater than fungal biomass in temperate forests (Witkamp, 1974; Kjøller & Struwe, 1982). Fungal and other microbial biomass fluctuated rapidly in response to wetting and drying cycles even in non-seasonal tropical forests. Such fluctuations in fungal biomass may have resulted in net nutrient immobilisation followed by pulses of mineralisation. Thus, fungi may influence nutrient uptake by plants or losses from the ecosystem by the timing of nutrient immobilization and mineralization. Immobilisation of nutrients by fungi was highest in the rain season, which may have helped to conserve nutrients against leaching (Behera *et al.*, 1991; Yang & Insam, 1991), whereas mineralization of nutrients from microbial biomass synchronized with fine root production in seasonally wet forests contributed to tight nutrient cycling.

Fungi were found to play other important roles in nutrient cycles of tropical forests in addition to immobilising and mineralising nutrients from organic matter and aided uptake of plant nutrients. For example, Lee & Jordan (1990) found that the activities and products of fungi and other microorganisms kept phosphorus in labile forms, thus conserving it against leaching and fixation onto weathered clay. In addition, Lodge & Asbury (1988) showed that rhizomorphs and cords of litter decomposers helped to retain litter and soil organic matter on steep slopes during heavy rains thereby reducing erosion. Thus, the importance of fungi in nutrient cycling in wet and seasonally wet tropical forests far exceeds that due to the quantity of nutrients held in fungal biomass.

Acknowledgments This research was performed under grant DE-AC05-760RO from the U.S. Department of Energy, Office of Health and Environmental Research, and grant BSR-8811902 from the National Science Foundation to the Center for Energy and Environment Research (University of Puerto Rico) and the Institute of Tropical Forestry (Southern Forest Experiment Station, USDA Forest Service). Advice and permission to cite unpublished data were kindly provided by Drs C. E. Asbury, T. A. Bonde, H. Insam and T. Rosswall, and by L. M. Babilonia, R. C. C. Luizão, C. P. Zucca and A. Masso. I am also grateful to Drs M. Larsen, H. Burdsall, A. E. Lugo, and especially J. C. Frankland for their helpful reviews.

References

Anderson, J. P. E. & Domsch, K. H. (1973). Quantification of bacterial and fungal contributions to soil respiration. *Archiv für Mikrobiologie* **93**, 113-127.

Anderson, J. P. E. & Domsch, K. H. (1978). A physiological method for the quantitative measurement of microbial biomass in soils. *Soil Biology and Biochemistry* **10**, 215-221.

Anderson, J. M. & Ingram, J. S. I. (1989). *Tropical Soil Biology and Fertility: a Handbook of Methods*. CAB International, Wallingford, UK.

Behera, N., Pati, D. P. & Basu, S. (1991). Ecological studies of soil microfungi in a tropical forest soil of Orissa, India. *Tropical Ecology* **32**, 136-143.

Brookes, P. C., Landman, A., Pruden, G. & Jenkinson, D. S. (1985). Chloroform fumigation and the release of soil nitrogen: a rapid direct extraction method to measure microbial biomass nitrogen in soil. *Soil Biology and Biochemistry* **17**, 837-842.

Cuevas, E. & Medina, E. (1988). Nutrient dynamics within amazonian forests. II. Fine root growth, nutrient availability and leaf litter decomposition. *Oecologia* **76**, 222-235.

Domsch, K. H., Beck, T., Anderson J. P. E., Söderström, B. Parkinson, D. & Trolldenier, G. (1979). A comparison of methods for soil microbial population and biomass studies. *Zeitschrift für Pflanzenernahrung und Bodenkunde* **142**, S20-S33.

Elliott, R. L., Salcedo, I. H., Sampaio, E. V. S. B. & Rose, S. (1989). Microbial biomass control of tracer phosphorus movement in a wet tropical forest floor. *Bulletin of the Ecological Society of America* **70** (2, Supplement, abstracts, 74th Annual ESA Meeting, University of Toronto, Ontario, Canada), 105-106.

FAO/IAEA (1985). *Root Activity Patterns of some Tree Crops*. Joint FAO/IAEA Division of Atomic Energy in Food and Agriculture. International Atomic Energy Agency, Technical Reports series No. 170.

Frankland, J. C. (1982). Biomass and nutrient cycling by decomposer basidiomycetes. In *Decomposer Basidiomycetes: their Biology and Ecology*, (ed. J. C. Frankland, J. N. Hedger & M. J. Swift), pp. 241-261. Cambridge University Press: Cambridge, U.K.

Frankland, J. C., Dighton, J. & Boddy, L. (1990). Methods for studying fungi in soil and forest litter. In *Methods in Microbiology*, **vol. 22**, (ed. R. Grigorova & J. R. Norris), pp. 343-404. Academic Press: London.

Hayes, M. J. & Cooley, J. H. (1987). Tropical soil biology: current status. *International Association for Ecology, Bulletin* **14**.

Hilton, G. (1987). Nutrient cycling in tropical rainforests: implications for management and sustained yield. *Forest Ecology and Management* **22**, 297-300.

Hunt, H. W., Elliott, E. T. & Walter, D. E. (1989). Inferring trophic transfers from pulse-dynamics in detrital food webs. *Plant and Soil* **115**, 247-259.

Janos, D. P. (1983). Tropical mycorrhizas, nutrient cycles and plant growth. In *Tropical Rain Forest: Ecology and Management*, (ed. S. L. Sutton, T. C. Whitmore & A. C. Chadwick), pp. 327-345. Blackwell: Oxford, U.K.

Janos, D. P. (1987). Roles of mycorrhizae in nutrient cycling and retention in tropical soils and organic matter. *International Association for Ecology, Bulletin* **14**, 41-44.

Jenkinson, D. S. & Powlson, D. S. (1976). The effects of biocidal treatments on metabolism in soil. V. A method for measuring soil biomass. *Soil Biology and Biochemistry* **8**, 209-213.

Jones, P. C. T. & Mollison, J. E. (1948). A technique for the quantitative estimation of soil micro-organisms. *Journal of General Microbiology* **2**, 54-69.

Jordan, C. F. (1985). *Nutrient Cycling in Tropical Forest Ecosystems: Principles and their Application in Management and Conservation*. John Wiley: New York.

Kjøller, A. & Struwe, S. (1982). Microfungi in ecosystems: fungal occurrence and activity in litter and soil. *Oikos* **39**, 389-422.

Lam, P. K. S. & Dudgeon, D. (1985). Breakdown of *Ficus fistulosa* (Moraceae) leaves in Hong Kong, with special reference to dynamics of elements and the effects of invertebrate consumers. *Journal of Tropical Ecology* **1**, 245-264.

Lee, D., Han, X. G. & Jordan, C. F. (1990). Soil phosphorus fractions, aluminum, and water retention as affected by microbial activity in an ultisol. *Plant and Soil* **121**, 125-136.

Lodge, D. J. (1985). Preliminary estimates of fungal biomass and nutrient stores in the litter and soil of a tropical rainforest. *Agronomy Abstracts, Soil Science Society of America*, Chicago, Illinois, USA. pp. 158-159.

Lodge, D. J. (1987). Nutrient concentrations, percentage moisture and density of field-collected fungal mycelia. *Soil Biology and Biochemistry* **19**, 727-733.

Lodge, D. J. & Asbury, C. E. (1988). Basidiomycetes reduce export of organic matter from forest slopes. *Mycologia* **80**, 888-890.

Lodge, D. J. & Ingham, E. R. (1991). A comparison of agar film techniques for estimating fungal biovolumes in litter and soil. *Agriculture, Ecosystems and Environment* **34**, 131-144.

Lodge, D. J., Scatena, F. N., Asbury, C. E. & Sanchez, M. J. (1991). Fine litterfall and related nutrient inputs resulting from Hurricane Hugo in subtropical wet and lower montane rain forests of Puerto Rico. *Biotropica* **23** (Suppl.), 336-342.

Lodge, D. J. & McDowell, W. H. (1991). Summary of ecosystem-level effects of Caribbean hurricanes. *Biotropica* **23** (Suppl.), 373-378.

Lugo, A. E. (1992). Comparison of four small tropical tree plantations (*Pinus caribaea* and *Swietenia macrophylla*) with secondary forests of similar age. *Ecological Monographs* **62**, 1-41.

Luizão, R. C. C., Bonde, T. A. & Rosswall, T. (1992). Seasonal variation of soil microbial biomass the effects of clearfelling a rainforest and establishment of pasture in the central Amazon. *Soil Biology and Biochemistry* **24**, 805-813.

Luse, R. A. (1970). The phosphorus cycle in a tropical rain forest. In *A Tropical Rain Forest*, (ed. H. T. Odum & R. F. Pigeon), pp. 161-166. U.S. Atomic Energy Commission: Washington, D.C.

Marumoto, T., Anderson, J. P. E. & Domsch, K. H. (1982). Mineralization of nutrients from soil microbial biomass. *Soil Biology and Biochemistry* **14**, 469-475.

McNeil, O. R. (1977). *Interactive Data Analysis. A Practical Primer*. John Wiley: New York.

Medina, E., Cuevas, E. & Weavel, P. L. (1981). Composicion foliar y transpiracion de especies lenosas de Pico del Este, Sierra de Luquillo, Puerto Rico. *Acta Cientifica Venezolana* **32**, 159-165.

Ocio, J. A. & Brookes, P. C. (1990). An evaluation of methods for measuring the microbial biomass in soils following recent additions of wheat straw and the characterization of the biomass that develops. *Soil Biology and Biochemistry* **22**, 685-694.

Odum, H. T. (1970). Summary: an emerging view of the ecological system at El Verde. In *A Tropical Rain Forest*, (ed. H. T. Odum & R. F. Pigeon), pp. 191-277. U.S. Atomic Energy Commission: Washington, D.C.

Parrotta, J. A. & Lodge, D. J. (1991). Fine root dynamics in a subtropical wet forest following hurricane disturbance in Puerto Rico. *Biotropica* **23** (Suppl.), 343-347.

Parton, W. J., Stewart, J. W. B. & Cole, C. V. (1988). Dynamics of C, N, P and S in grassland soils: a model. *Biogeochemistry* **5**, 105-131.

Raghubanshi, A. S., Srivastava, S. C., Singh, R. S. & Singh, J. S. (1990). Nutrient release in leaf litter. *Nature* **346**, 227.

Read, D. J., Leake, J. R. & Langdale, A. R. (1989). The nitrogen nutrition of mycorrhizal fungi and their host plants. In *Nitrogen, Phosphorus and Sulphur Utilization by Fungi* (ed. L. Boddy, R. Marchant & D. J. Read), pp. 181-204. Cambridge University Press: Cambridge, U. K.

Sanchez, P. A. (1976). *Properties and Management of Soils in the Tropics*. John Wiley & Sons: New York.

Sanford, R. L., Jr., Parton, W. J., Ohima, D. S. & Lodge, D. J. (1991). Hurricane effects on soil organic matter dynamics and forest production in the Luquillo Experimental Forest, Puerto Rico: results of simulation modeling. *Biotropica* **23** (Suppl.), 364-372.

Singh, J. S., Raghubanshi, A. S., Singh, R. S. & Srivastava, S. C. (1989). Microbial biomass acts as a source of plant nutrients in dry tropical forest and savanna. *Nature* **338**, 499-500.

Sparling, G. P., West, A. W. & Whale, K. N. (1985). Interference from plant roots in the estimation of soil microbial ATP, C, N and P. *Soil Biology and Biochemistry* **17**, 275-278.

Srivastava, S. C. & Singh, J. S. 1988. Carbon and phosphorus in the soil biomass of some tropical soils of India. *Soil Biology and Biochemistry* **20**, 743-747.

Steudler, P. A., Melillo, J. M., Bowden, R. D., Castro, M. S. & Lugo, A. E. (1991). The effects of natural and human disturbances on soil nitrogen dynamics and trace gas fluxes in a Puerto Rican wet forest. *Biotropica* **23** (Suppl.), 356-363.

Swift, M. J., Heal, O. W. & Anderson, J. M. (1979). *Decomposition in Terrestrial Ecosystems*. Blackwell Scientific Publications: Oxford, U.K.

Tanner, E. V. J., Kapos, V. & Franco, W. (1992). Nitrogen and phosphorus fertilization effects on Venezuelan montane forest trunk growth and litterfall. *Ecology* **73**, 78-86.

Thompson, W. (1984). Distribution, development and functioning of mycelial cord systems of decomposer basidiomycetes of the deciduous woodland floor. In *The Ecology and Physiology of the Fungal Mycelium*, (ed. D. H. Jennings & A. D. M. Rayner), pp. 185-214. Cambridge University Press: Cambridge, U.K.

Thompson, W. & Rayner, A. D. M. (1983). Extent, development and function of mycelial cord systems in soil. *Transactions of the British Mycological Society* **81**, 333-345.

Vitousek, P. M. (1984). Litterfall, nutrient cycling, and nutrient limitation in tropical forests. *Ecology* **65**, 285-298.

Vitousek, P. M. & Sanford, R. L. (1986). Nutrient cycling in moist tropical forest. *Annual Review of Ecology and Systematics* **17**, 137-167.

Watkinson, S. C. (1971). The mechanism of mycelial strand induction in *Serpula lacrimans*: a possible effect of nutrient distribution. *New Phytologist* **70**, 1079-1088.

Watkinson, S. C. (1984). Distribution, development and functioning of mycelial cord systems of decomposer basidiomycetes of the deciduous woodland floor. In *The Ecology and Physiology of the Fungal Mycelium*, (ed. D. H. Jennings & A. D. M. Rayner), pp. 165-185. Cambridge University Press: Cambridge, U.K.

Wells, J. M., Hughes, C. & Boddy, L, (1990). The fate of soil-derived phosphorus in mycelial cord systems of *Phanerochaete velutina* and *Phallus impudicus*. *New Phytologist* **114**, 595-606.

Went, F. W. & Stark, N. (1968a). The biological and mechanical role of soil fungi. *Proceedings of the National Academy of Sciences, U.S.A.* **60**, 497-504.

Went, F. W. & Stark, N. (1968b). Mycorrhiza. *Bioscience* **18**, 1035-1039.

West, A. W., Ross, D. J. & Cowling, J. C. (1986). Changes in microbial C, N, P and ATP contents, numbers and respiration on storage of soil. *Soil Biology and Biochemistry* **18**, 141-148.

Witkamp, M. (1974). Direct and indirect counts of fungi and bacteria as indexes of microbial mass and productivity. *Soil Science* **118**, 150-155.

Yang, J. C. & Insam, H. (1991). Microbial biomass and relative contributions of bacteria and fungi in soil beneath tropical rain forest, Hainan Island, China. *Journal of Tropical Ecology* **7**, 385-395.

Zimmerman, J. K., Lodge, D. J., Asbury, C. E. & Parrotta, J. A. (1992). Effect of fertilization and wood removal on litterfall in low montane and cloud forests of the Luquillo Experimental Forest, Puerto Rico. Abstracts, *Proceedings of the Ecological Society of America* **73**, 398-399.

Chapter 4

The enigma of desert ecosystems: the importance of interactions among the soil biota to fungal biodiversity

John C. Zak

Ecology Program, Department of Biological Sciences, Texas Tech University, Lubbock, Texas 79409, U.S.A.

Introduction

The tropics are defined as occurring between about 30° N and 30° S of the equator (Walter, 1973; Kirk, Chapter 6 this volume). This region includes components of the vast hot deserts, occurring between latitudes 20° and 40° N and S, and is the belt of the subtropical anticyclone, which creates high pressure systems reducing rainfall in these latitudes. Low and irregular rainfall is the most important condition defining a desert environment (McGinnies, 1979). Other features, such as orographic barriers, can also create deserts. For example, mountain rain shadows are responsible for deserts in North and South America, South Africa, and Madagascar. The outer Namib and the Peruvian-Chilean fog deserts, Baja California, and the Atlantic sea coast of the Sahara are due to cold ocean currents (Evenari, 1985). Together, arid and semi-arid regions occupy 13 - 14% of the total land surface of the globe (Evenari, 1985), and constitute nine phytogeographic regions (Schmida, 1985).

Much of the recent publicity over the loss of global biodiversity has focused on wet tropical ecosystems (e.g. Ehrlich & Wilson, 1991). In an attempt to focus attention on other equally threatened areas, Mares (1992) posed the question as to whether tropical rainforests are actually more diverse than other ecosystems. He indicated that this bias has not been adequately addressed in previous discussions on conservation in the tropics. Olson (1989) had stated previously that perhaps conservation attitudes had underestimated the potential diversity of mesic and arid regions in the tropics. Redford, Taber & Simonetti (1990) also questioned the dogma of tropical rainforests as the sole repositories of high species richness (diversity). They presented data for medium and large size mammals, which indicated that semi-arid and arid regions of South America are as diverse in species as the reported richest humid sites.

Mares (1992) estimated that in South America alone, while Amazonian lowlands and all other rainforests below 1500 m cover 30% of the land mass, arid and semi-arid regions occupy approximately 57% of the continent. Mares (1992) emphasized that when developing conservation plans to maintain neotropical biodiversity on a continental scale, the size and variety of ecosystems should be the principal consideration, and plans should not focus solely on rainforest habitats.

Mares (1992) included arid and semi-arid habitats in tropical conservation initiatives, based on two critical pieces of information regarding mammal biodiversity. First, drylands in South America support more endemic species (211 versus 138) and genera (509 versus 434) than Amazonian rainforest habitats. Greater than expected numbers of species in semi-arid ecosystems were also reported by Schmida, Evenari & Noy-Meir (1985). They showed that although plant biomass production was highest in subhumid and humid (moist) tropical forests, plant species richness was greater in shrub-deserts. The second argument of Mares (1992) addressed genetic concerns. He argued that while rainforests may contain more species per unit area, the species are more closely related genetically than species in semi-arid and arid regions. By conserving distantly related taxa from various habitats, genetic diversity, not merely species diversity, can be maximized.

Although the mycological literature cannot, as yet, provide a detailed analysis of species diversity for moist tropical versus desert ecosystems as conducted by Mares (1992), evidence from Gochenaur (1975) indicated, albeit from limited sampling, that the distribution of conidial fungi in moist versus arid and semi-arid tropical habitats may be similar to that of mammals. Sampling a moist *Casuarina* forest and a xeric coconut grove in the Bahamas, Gochenaur (1975) observed that taxonomic diversity was of the same order or even slightly higher in the xeric site (Table 1). Moreover, the *Casuarina* site was dominated by few genera, while a larger number of taxonomic groups were codominant in the xeric environment (Table 2). The more variable abiotic conditions in the coconut grove compared with those of the *Casuarina* forest may have contributed to the greater taxonomic diversity and larger number of moderately abundant species. Wicklow (1981) indicated that desert ecosystems have an even greater number of fungal species than would be predicted on the basis of abiotic conditions that exist in these habitats. While low moisture and high temperature certainly limit species activity, it is the high temporal and spatial heterogeneity in abiotic conditions and resource availability that may account for the unexpectedly high fungal species diversity in arid

Table 1. Fungal taxonomic diversity at a mesic and a xeric tropical site in the Bahamas

	Mesic *Casuarina* forest	Xeric Coconut grove
Number of genera	23	27
Number of species	47	51
Number of endemics	24	26
% Endemics	51	51

Data adapted from Gochenaur (1975); n = 10 samples per site

Table 2. Distribution (% of total isolates) of fungal taxa between a mesic and a xeric tropical ecosystem

	Mesic site	Xeric site
Ascomycetes	<1	11
Aspergillus sp.	<1	28
Penicillium sp.	46	20
Moniliaceae (excluding *Penicillium* & *Aspergillus*)	33	20
Dematiaceous–sphaeropsidaceous	2	32
Sterile isolates	<1	4

environments. Spatial and temporal heterogeneity in deserts occur on many different scales, affecting the composition and richness of communities (Polis, 1991a). Deserts have the greatest variability for a large number of parameters compared with other systems (Polis, 1991b).

In response to a highly variable and unpredictable environment, desert organisms tend to be functionally diverse, ensuring quick adaptation to favourable conditions (e.g. Crawford, 1991). Functional diversity for a given ecosystem may be as important a consideration in biodiversity issues as overall species richness. Gochenaur (1975) reported that the functional diversity of the fungi isolated from the xeric location was greater than that of fungi from the moist environment. She found that 75% of the isolates from the dry site exhibited cellulolytic activity, while less than 33% of the isolates from the moist site could hydrolyse cellulose, and she concluded that fungi in xeric environments should exhibit a greater nutritional versatility (functional diversity) and less nutrientspecialization. H. G. Wildman (personal communication) has also speculated that fungi from desert environments might be more functionally diverse (i.e. produce more secondary metabolites) than those occurring in moist tropical systems.

Interactions among root-region biota

Water availability is the key factor defining a desert. More specifically, potential evapotranspiration is greater than annual moisture inputs in desert environments, and is the major regulator of all biotic activity (e.g. Whitford, 1989). One cannot discuss any aspect of fungal dynamics, soil microfaunal activity, or primary production in arid ecosystems without acknowledging the fact that the activity of all these trophic groups is intimately regulated by moisture availability. Not only is total moisture important, but the temporal patterning (Fig. 1a) of these inputs is crucial, providing 'windows of opportunity' that concentrate biotic activity within these brief periods of favourable moisture (Fig. 1b). It is the high temporal heterogeneity in frequency and amplitude of the moisture 'windows' that provides the mechanism for increased biodiversity in deserts.

In arid and semi-arid ecosystems, the extremely short periods in which soil water potentials are conducive for bacterial growth suggest that soil food webs in these systems would be primarily fungal based (Whitford, 1989), since fungi are less dependent on free water for growth. The primary transfer of energy and nutrients from the fungal component to the upper trophic levels in the Chihuahuan Desert of Southwestern United States (Whitford, 1989) appears to be via grazing by fungivorous mites. Hunt, Elliott & Walter (1989) also showed that fungi accounted for the largest amount of material transferred from detritus (Table 3) in a semi-arid, shortgrass prairie, and that most fungal turnover could be attributed to normal evacuation of fungal hyphae, and to death by abiotic stress or from grazing by fungivorous microfauna. During high moisture periods, nematodes can be a major component of the soil food web in arid environments consuming bacteria, fungi, and soil algae, but like bacteria their activity is severely restricted in dry soils (Freckman, 1982; Freckman, Whitford & Steinberger, 1987). Nematodes in the Chihuahuan Desert were found to be active in surface litter for only a few hours after rain and for 48 to 72 h in the soil after artificial moisture inputs (Whitford *et al.*, 1981).

Even when moisture levels are adequate for plant growth, low rates of nitrogen mineralization may severely limit plant growth in deserts (Parker *et al.*, 1984). Species composition of floral communities as well as

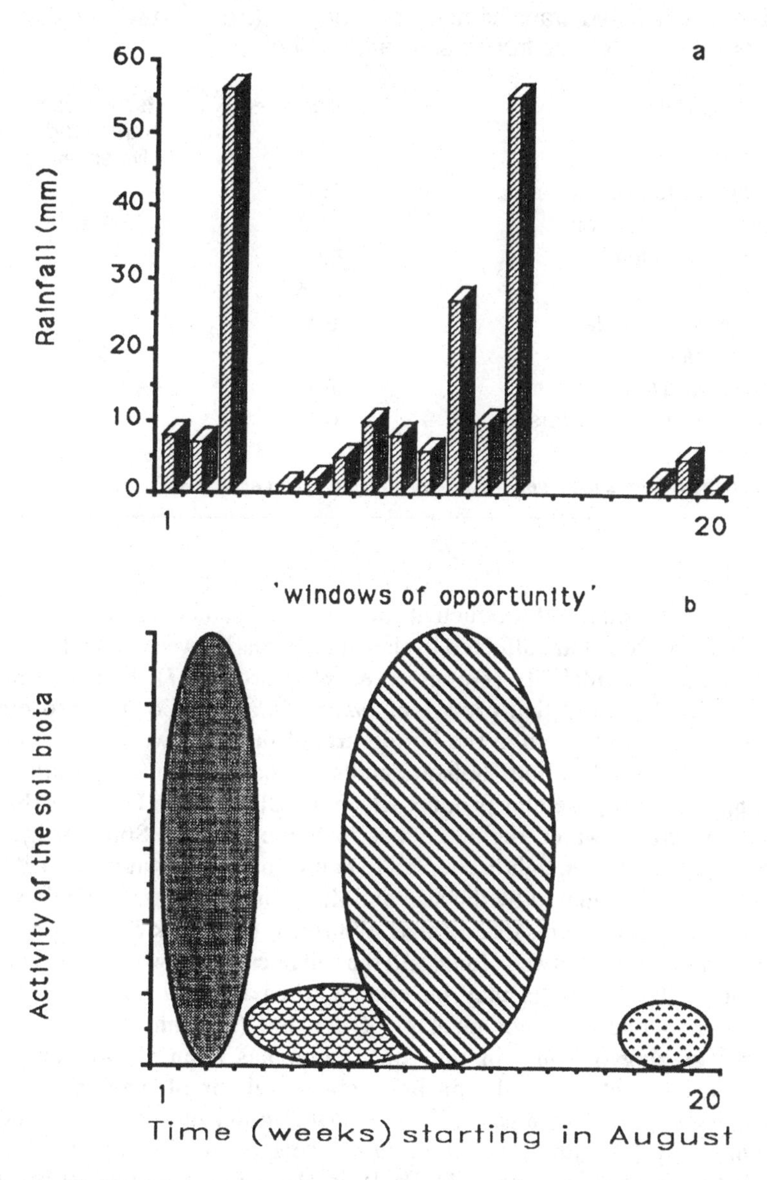

Fig. 1. (a) Short-term temporal pattern in rainfall for a 20 week period at the Jornada Long Term Ecological Research Site in the northern Chihuahuan Desert. (b) The theorized relationship between microfloral and microfaunal activities and the temporal patterning in 'windows' of optimum moisture. Each ellipse represents a specific 'window of opportunity' for biotic activity in desert soils, root-regions and litter as dictated by the size and duration of the moisture input. As shown, some windows may overlap while others are distinct in time.

Table 3. Predicted transfer rates of biomass ($\mu g\ g^{-1}\ day^{-1}$) within a decomposer food web from a semi-arid prairie

Trophic level	Transfer rates	% biomass transferred between levels
Substrate to bacteria	56·1	–
Bacteria to protozoa	11·1	19·8
Substrate to fungi	80·5	–
Fungi to microfauna	0·75	0·9
Fungi to nematodes	2·0	1·2
Fungal death	28·6	35·5
Microfauna to predators	0·07	–
Nematodes to predators	0·2	–

Data adapted from a food web model developed by Hunt *et al.* (1989)

individual life span, and productivity of spring and summer annuals in the Chihuahuan Desert are affected by the interactions between soil nitrogen and moisture availability (Gutierrez & Whitford, 1987). Fisher *et al.* (1988) also reported that growth of *Larrea tridentata* Coult. (creosote bush), a dominant shrub in the hot deserts of the Southwestern United States, was limited by both moisture and nitrogen. The central paradigm that has been proposed for productivity in the Chihuahuan Desert is that plant growth is water regulated but nitrogen limited. Soil nitrogen dynamics are, in turn, tightly coupled to microfloral-microfaunal activity, which is itself regulated by the size and duration of moisture 'windows' (Whitford & Freckman, 1988; Zak & Whitford, 1988; Zak & Freckman, 1991). Interactions between fungi and the soil microfauna within the root-region of desert plants may be crucial when mineralization by the microflora alone does not meet plant demand (Ingham *et al.*, 1985). Faunal grazing of fungi in the root-region has been shown, under controlled conditions, to be an important regulator of plant nutrient dynamics (e.g. Coleman *et al.*, 1988). Furthermore, fungal growth and activity may be stimulated at certain levels of grazing by the soil microfauna (e.g. Moore, 1988). It is the temporal and spatial heterogeneity in the 'windows of opportunity' that regulate the interactions among plants, fungi, and fungivorous soil microfauna, and which in turn may contribute to increased diversity of fungal species in arid ecosystems.

Interactions among root-region biota

To illustrate the effects of interactions among the root-region biota of desert plants on plant growth and nitrogen dynamics, data are presented from a study at the Jornada Long Term Ecological Research Site near Las Cruces, New Mexico, U.S.A. (personal observation). The study attempted to examine the response of the root-region biota to 'windows of opportunity' and the temporal linkages between 'windows' as they affect continued microfloral-microfaunal activity. Root-region is defined here as that volume of soil under the plant crown to a depth of 15 cm. *Erioneuron pulchellum* (H.B. & K.) Tateaka (fluffgrass), a common grass in the bajada areas of the Chihuahuan Desert, was chosen as the test plant. Microarthropod and nematode densities were manipulated over a two-year period, beginning in November 1986, by the application of Nemacur (nematocide) and Chlordane (to reduce microarthropod densities). The biocides were applied to randomly selected plots in November 1986, and April and September 1987 according to manufacturers' specified application rates. Control plots were also established. Sampling continued from January 1987 to January 1989. Root-region samples were obtained directly from the centre of a grass clump, and analyzed for microarthropod and nematode densities and species compositions, hyphal lengths, and soil nitrogen content. Plants also were sampled and analyzed for root production and total N content.

Microarthropods

Fungivores were the largest trophic group of microarthropods in the root-region for all biocide applications. Seasonal changes in densities and trophic structure were observed in all treatments, with the lowest densities occurring during April and July (a dry period in the Chihuahuan Desert). The effects of Chlordane were not evident until April 1987 at which time microarthropod densities were almost zero. In general, trophic structure in the Nemacur-treated plots was similar to that of the control but densities of the various microarthropod trophic groups differed between the control and Nemacur treatment at certain sampling times suggesting that modification of nematode numbers did alter microarthropod trophic dynamics to some extent.

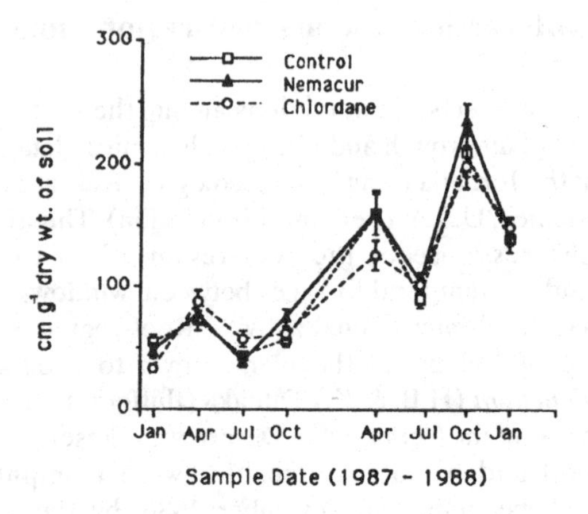

Fig. 2. Hyphal lengths (mean ± S.E.) associated within the root-region of *Erioneuron pulchellum* following manipulation of the root-region microfauna with biocides.

Nematodes

Bacterivorous nematodes were the largest trophic group in all treatments based on relative abundances. Nemacur appeared to reduce fungivorous nematodes selectively compared with other trophic groups. In the Chlordane plots, differences in trophic interactions were not apparent until the second year (1988) when fungivorous nematodes increased relative to control plots, possibly in response to the elimination of nematophagous microarthropods.

Fungi

Hyphal lengths were measured to assess the response of the fungi in the root-region to changes in the trophic structure of the soil fauna, using a membrane filter technique (see Bardgett, 1991 for a review of the procedure). Using a repeated measures MANOVA to analyse for statistical differences in hyphal lengths, changes in the trophic dynamics among the microfauna and fungal components (Fig. 2) were found not to be consistent between sampling periods during the first year (interaction between treatment and date; $P \leq 0{\cdot}001$). Changes in root-region hyphal

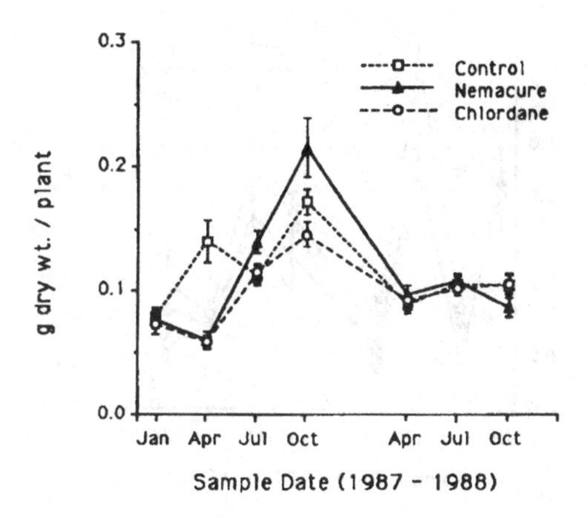

Fig. 3. Root growth (mean ± S.E.) of *Erioneuron pulchellum* in response to changes in the trophic interactions within the root-region through biocide applications.

lengths depended apparently upon the trophic condition of the microfauna prior to sampling. Only date was significant ($P \leq 0 \cdot 001$) for the second growing season (1988) indicating that changes in microfaunal trophic dynamics had no detectable effect on hyphal lengths (Fig. 2).

Plant response

Using root weights as in indicator of the plant's response to microfloral-microfaunal interactions within the root region, a significant interaction between treatment and date ($P \leq 0 \cdot 002$) was detected during the first growing season (Fig. 3). The apparent effects of trophic dynamics within the root-region varied through time possibly in response to the heterogeneity in the size and frequency of moisture 'windows'. In the second growing season (1988), no significant treatment by date ($P \leq 0 \cdot 325$) nor date effects ($P \leq 0 \cdot 171$) were detected for root weights (Fig. 3).

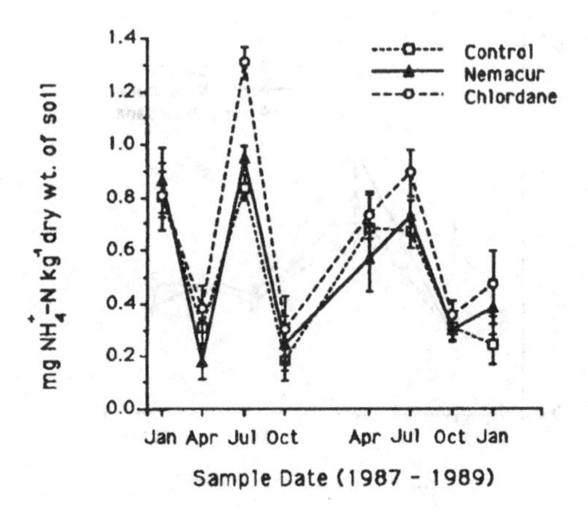

Fig. 4. Temporal changes in the amount of available ammonium nitrogen (mean ± S.E.) in the root-region of *Erioneuron pulchellum* following biocide applications to alter trophic interactions.

Plant and soil nitrogen

Although trophic interactions were apparently affected by reducing microarthropod and nematode densities in the root-region, followed by changes in amounts of mycelium, soil and plant nitrogen pools were not significantly altered (Figs 4 & 5). Small frequent rainfall events may normally stimulate mineralization of dead microbial biomass (Fisher *et al.*, 1987). Thus the pulse in available ammonium observed in July corresponded to the start of summer convection storms in the northern Chihuahuan Desert.

Discussion

Integration of these root-region trophic relationships with the 'windows of opportunity' concept suggests that the microflora and microfauna in the root-region of desert plants are highly dynamic with a short response time to favourable conditions. Grazing of fungi by root-region microarthropods and nematodes is highly dependent upon the temporal patterning of the 'windows of opportunity', and subsequent effects on nutrient dynamics do not appear to be cumulative.

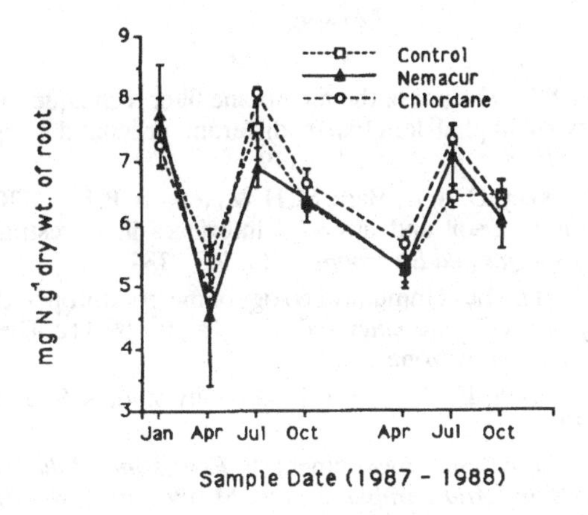

Fig. 5. Changes in total nitrogen content of *Erioneuron pulchellum* roots (mean ± S.E.) following biocide application.

Given the importance of abiotic constraints in structuring desert communities (e.g. Whitford, 1989), fungi from desert and semi-arid environments may have evolved a greater functional diversity to cope with the extreme spatial and temporal heterogeneity characteristic of these ecosystems than that of fungi from more mesic environments. With biodiversity issues receiving much recent attention, the evidence presented here suggests that desert and semi-arid regions of the tropics may be just as important for species conservation as the more mesic regions. Arid regions are far more threatened by human activities than most realise (Mares, 1992), and Schlesinger *et al.* (1990) emphasized that semi-arid and arid regions of the globe are more likely than mesic areas to be adversely affected by global climate change. With increasing human impact, arid and semi-arid regions are more likely to experience increased rates of desertification and species extinctions.

Acknowledgements Dr Solange De Silva and Dr Diana Freckman provided the skills for quantifying and identifying the microarthropods and nematodes respectively. Collection of data for portions of this manuscript were supported by a National Science Foundation Grant (BSR-8604716) to J. Zak. The comments and suggestions from Dr Shivcharn Dhillion, Dr Daryl Moorhead, and an anonymous reviewer are appreciated.

References

Bardgett, R. D. (1991). The use of the membrane filter technique for comparative measurements of hyphal lengths in different grassland sites. *Agriculture, Ecosystems, and Environment* **34**, 115 - 119.

Coleman, D. C., Crossley, D. A. Jr., Bare, M. H. & Hendrix, P. F. (1988). Interactions of organisms at root/soil and litter/soil interfaces in terrestrial ecosystems. *Agriculture, Ecosystems and Environment* **24**, 117 - 134.

Crawford, C. S. (1991). The community ecology of macroarthropod detritivores. In *The Ecology of Desert Communities*, (ed. G. Polis), pp. 89-112. The University of Arizona Press: Tucson, Arizona.

Ehrlich, P. R. & Wilson, E. O. (1991). Biodiversity studies: Science and policy. *Science* **253**, 758.

Evenari, M. (1985). The desert environment. In *Ecosystems of the World*, Volume 12A: *Hot Deserts and Arid Shrublands*, (ed. M. Evenari, I. Noy-Meir & D. W. Goodall), pp. 1-22. Elsevier: Amsterdam.

Fisher, F. M., Parker, L. W., Anderson, J. P. & Whitford, W. G. (1987). Nitrogen mineralization in a desert soil: interacting effects of soil moisture and N fertilizer. *Soil Science of America Journal* **51**, 1033-1041.

Fisher, F. M., Zak, J. C., Cunningham, G. L. & Whitford, W. G. (1988). Water and nitrogen effects on growth and allocation patterns of creosotebush in the northern Chihuahuan Desert. *Journal of Range Management* **31**, 387-391.

Freckman, D. W. (1982). Parameters of the nematode contribution to ecosystems. In *Nematodes in Soil Ecosystems*, (ed. D. W. Freckman), pp. 80-97. University of Texas Press: Austin, Texas.

Freckman, D. W., Whitford, W. G. & Steinberger, Y. (1987). Effect of irrigation on nematode population dynamics and activity in desert soils. *Biology and Fertility of Soils* **3**, 3-10.

Gochenaur, S. E. (1975). Distributional patterns of mesophilous and thermophilous microfungi in two Bahamian soils. *Mycopathologia and Mycoapplicata* **57**, 155-164.

Gutierrez, J. R. & Whitford, W. G. (1987). Chihuahuan Desert annuals: importance of water and nitrogen. *Ecology* **68**, 2032-2045.

Hunt, H. W., Elliott, E. T. & Walter, D. E. (1989). Inferring trophic transfers from pulse-dynamics in detrital food webs. In *Ecology of Arable Lands: Perspectives and Challenges*, (ed. M. Clarholm & L. Bergstrom), pp. 191-203. Kluwer Academic Publishers: Dordrecht.

Ingham, R. E., Trofymow, J. A., Ingham, E. R. & Coleman, D. C. (1985). Interactions of bacteria, fungi, and their nematode grazers: effects on nutrient cycling, and plant growth. *Ecological Monographs* **55**, 119-140.

Mares, M. A. (1992). Neotropical mammals and the myth of Amazonian biodiversity. *Science* **255**, 976-979.

McGinnies, W. G. (1979). Arid-land ecosystems - common features throughout the world. In *Arid-Land Ecosystems: Structure, Functioning and Management*, **Volume 1**, (ed. D. W. Goodall, R. A. Perry, K.M. & W. Howes), pp. 299 - 316. Cambridge University Press: Cambridge, U.K.

Moore, J. C. (1988). The influence of microarthropods on symbiotic and nonsymbiotic mutualisms in detrital-based belowground food webs. *Agriculture, Ecosystems and Environment* **24**, 117-134.

Olson, S. (1989). Extinctions on islands: man as a catastrophe. In *Conservation for the Twenty-First Century*, (ed. D. Western & M. C. Pearl), pp. 50-58. Oxford University Press, New York.

Parker, L. W., Santos, P. F., Phillips, J. & Whitford, W. G. (1984). Carbon and nitrogen dynamics during decomposition of litter and roots of a Chihuahuan Desert annual, *Lepidium lasiocarpum*. *Ecological Monographs* **54**, 339-360.

Polis, G. A. (1991a). Desert communities: an overview of patterns and processes. In *The Ecology of Desert Communities*, (ed. G. Polis), pp. 1-26. The University of Arizona Press: Tucson, Arizona.

Polis, G. (1991b). *The Ecology of Desert Communities*. The University of Arizona Press: Tucson, Arizona.

Redford, K. H., Taber, A. & Simonetti, J. A. (1990). There is more to biodiversity than the tropical rainforests. *Conservation Biology* **4**, 328-330.

Schlesinger, W. H., Reynolds, J. F., Cunningham, G. L., Huenneke, L. F., Jarrell, W. M., Virginia, R. A. & Whitford, W. G. (1990). Biological feedbacks in global desertification. *Science* **247**, 1043-1048.

Schmida, A. (1985). Biogeography of the desert flora. In *Ecosystems of the World*, Volume 12A: *Hot Deserts and Arid Shrublands*, (ed. M. Evenari, I. Noy-Meir & D. W. Goodall), pp. 23-77. Elsevier: Amsterdam.

Schmida, A., Evenari, M. & Noy-Meir., I. (1985). Hot desert ecosystems: an integrated view. In *Ecosystems of the World*, Volume 12B: *Hot Deserts and Arid Shrublands*, (ed. M. Evenari, I. Noy-Meir, & D. W. Goodall), pp. 379-387. Elsevier: Amsterdam.

Walter, H. (1973). *Vegetation of the Earth in Relation to Climate and the Eco-Physiological Conditions*, 2nd edn. The English Universities Press Ltd.: London.

Whitford, W. G. (1989). Abiotic controls on the functional structure of soil food webs. *Biology and Fertility of Soils* **8**, 1-6.

Whitford, W. G. & Freckman, D. W. (1988). The role of the soil biota in soil processes in the Chihuahuan Desert. In *Arid Lands: Today and Tomorrow*, (ed. E. E. Whitehead, C. F. Hutchinson, B. N. Timmerman & R. G. Varady), pp. 1063-1073. University of Arizona Press: Tucson, Arizona.

Whitford, W. G., Freckman, D. W., Elkins, N. Z., Parker, L., Parmalee, R., Phillips, J. & Tucker, S. (1981). Diurnal migration and responses to simulated rainfall in desert soil: microarthropods and nematodes. *Soil Biology and Biochemistry* **13**, 417-425.

Wicklow, D. T. (1981). Biogeography and conidial fungi. In *Biology of Conidial Fungi*, **Volume 1**, (ed. G. T. Cole & B. Kendrick), pp. 417-447, Academic Press: New York.

Zak, J. C. & Whitford, W. G. (1988). Interactions among soil biota in desert ecosystems. *Agriculture, Ecosystems and Environment* **24**, 87-100.

Zak, J. C. & Freckman, D. W. (1991). Soil communities in deserts: microarthropods and nematodes. In *The Ecology of Desert Communities*, (ed. G. A. Polis), pp. 55-88, The University of Arizona Press: Tucson, Arizona.

Chapter 5

Tropical marine fungi

E. B. Gareth Jones

School of Biological Sciences, University of Portsmouth, King Henry Building, King Henry I Street, Portsmouth PO1 2DY, U.K.

Introduction

While there is a significant volume of published information on the occurrence of marine fungi, most of this refers to collections or lists of species, predominantly in Europe, North America and the Far East (Jones, 1971; Henningson, 1974; Koch, 1982; Kohlmeyer, 1984; Vrijmoed, Hodgkiss & Thrower, 1986; Hyde & Jones, 1988). The biogeography of marine fungi has received scant attention. Early attempts were lists of species abstracted from the literature (Jones, 1971) with tentative assignment of fungi as tropical or temperate: *Cirrenalia macrocephala* (Kohlm.) Meyers & Moore, *Antennospora quadricornuta* (Cribb & Cribb) T.W. Johnson (previously *Halosphaeria quadricornuta* Cribb & Cribb), *Torpedospora radiata* Meyers and *Lindra thalassiae* Orpurt, Meyers, Boral & Simms were regarded as tropical species while *Ceriosporopsis* spp., *Corollospora* spp., *Halosphaeria* spp., *Lulworthia* spp. and *Lignincola laevis* Höhnk were regarded as cosmopolitan species.

Hughes (1974) was the first to consider seriously the geographical distribution of marine fungi when he proposed five zones based on seawater temperatures: arctic, temperate, subtropical, tropical and antarctic. Twenty lignicolous fungi were shown to be cosmopolitan in their distribution. The maps that Hughes (1974) produced have been widely used since to plot the geographic distribution of marine fungi (Kohlmeyer & Kohlmeyer, 1971; Booth, 1979; Boyd & Kohlmeyer, 1982; Kohlmeyer, 1983; Hughes, 1986).

The most detailed analyses of the distribution of marine fungi have been made by Hughes (1986) using coefficients of similarity, multivariate analysis and mapping techniques, and also Booth & Kenkel (1986) who used ordination techniques to evaluate the temperature-salinity regimes of 68 marine fungi. Booth & Kenkel recognised six regimes: group **I**, warm water homeothermic euryhaline; group **II**, warm water euryhalothermic; group **III**, cool water eurythermic homohaline; group **IV**, cool water euryhalothermic; group V, cool water homeothermic euryhaline; and group VI, eurytolerant species.

These studies confirmed temperature as the most important factor in controlling the geographic distribution of marine fungi. However, it is the temperature of the water that is important, not the air temperature of the nearest land mass. Water currents play a key role in determining seawater temperatures. Hughes (1986) defined the tropical zone as 'determined by the 20°C surface water isocryme for the coolest calendar month (February in the north, August in the south) and the subtropical zone between 17-20°C isocrymes'. *Antennospora quadricornuta* is a tropical ascomycete that occurs in the temperature range 24-28°C. Hughes (1986) reported that in Australia the southernmost collection was in a region where temperatures range from 21-24°C; however, on these occasions the water temperature was 24°C.

Approximately 500 marine fungi have been described, of which around 135 of the higher marine fungi are to be found in the tropics. Marine fungi are an ecological group adapted for life in the sea. Many possess appendaged propagules that aid in dispersal and attachment (Figs 1-3, 5; Rees & Jones, 1984; Hyde & Jones, 1989). Many of the marine ascomycetes have asci that deliquesce early in development and the ascospores are released passively into the surrounding water (Fazzani & Jones, 1977). However, not all marine fungi have appendaged ascospores (Figs 4, 6); some discharge their spores forcibly and the asci are persistent (Fig. 6).

Marine fungi can be grouped into categories according to their biogeographical distribution: those restricted to the tropics and subtropics and cosmopolitan species (Table 1), and others which are restricted to temperate or arctic waters. However, many of the species listed in Table 1 have only recently been described or/and are known from one locality. Further data are required to confirm their placement into these geographic zones.

Figs 1–6 (facing page). Ascospores of tropical/cosmopolitan marine fungi. Figures 1–3. Scanning electron micrographs. Figures 4-6. Light micrographs. Fig. 1. *Nimbospora effusa* (tropical) with sheath forming a pad-like attachment (arrowed) to the polycarbonate membrane, and drop of mucilage arrowed m. Fig. 2. *Antennospora quadricornuta* (tropical), ascospore with sub-polar appendages. Fig. 3. *Halosarpheia retorquens* (cosmopolitan), ascospores with long, polar, viscous, thread-like appendages wrapped around debris (arrowed). Fig. 4. *Nais glitra* (tropical), ascospores lack appendages. Fig. 5. *Halosarpheia abonnis* (tropical), polar hamate appendages (arrowed) closely adpressed to spore wall. Fig. 6. *Savoryella lignicola* (cosmopolitan), asci with ascospores; dehisced ascus showing apical pore (arrowed). Scale bars = 5 μm.

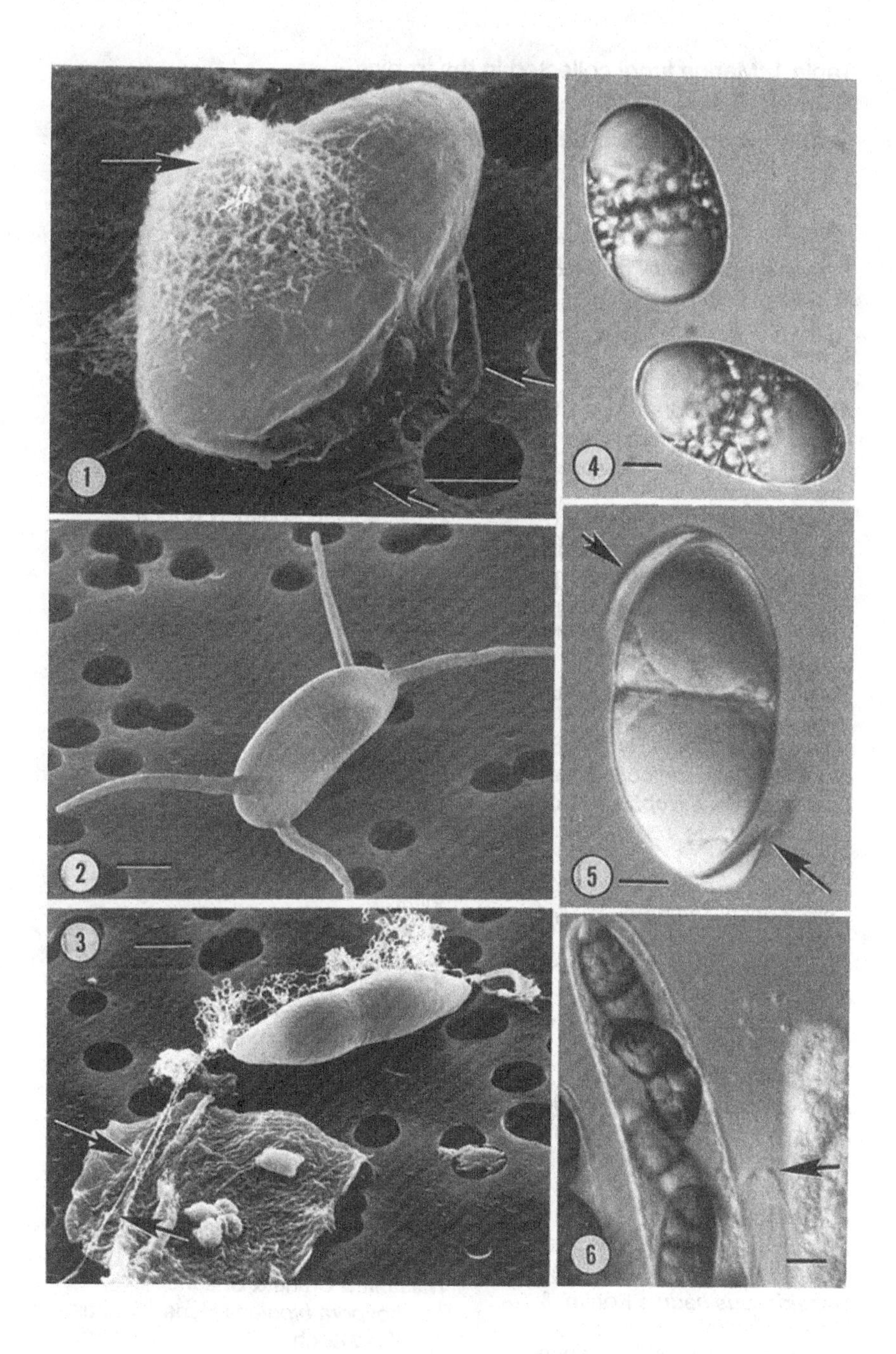
1
2
3
4
5
6

Table 1. Marine fungi collected in the tropics

Group I: Fungi restricted to subtropic and tropics

Ascomycotina

Acrocordiopsis patilii Borse & Hyde
Adomia avicenniae Schatz
Aigialus grandis Kohlm. & Schatz
A. mangrovei Borse
A. parvus Schatz & Kohlm.
A. rhizophorae Borse
Aniptodera longispora Hyde
Antennospora quadricornuta (Cribb & Cribb) Meyers
Arenariomyces parvulus Koch
Ascocratera manglicola Kohlm.
Bathyascus avicennniae Kohlm.
B. grandisporus Hyde & Jones
B. mangrovei Ravikuma & Vittal
B. tropicalis Kohlm.
Belizeana tuberculata Kohlm. & Volkm.-Kohlm.
Biatriospora marina Hyde & Borse
Capillataspora corticola Hyde
Caryosporella rhizophorae Kohlm.
Corallicola nana Volkm.-Kohlm. & Kohlm.
Corollospora cinnamomea Koch
Coronopapilla mangrovei (Hyde) Kohlm. & Volkm.-Kohlm.
Cucullosporella mangrovei (Hyde & Jones) Hyde & Jones
Dactylospora haliotrepha (Kohlm. & Kohlm.) Hafellner
Didymella avicenniae Patil & Borse
Falciformispora lignatilis Hyde
Halosarpheia abonnis Kohlm.
H. bentotensis Koch
H. fibrosa Kohlm. & Kohlm.
H. minuta Leong
H. ratnagiriensis Patil & Borse
Halosphaeria salina (Meyers) Kohlm.
Hapsidascus hadrus Kohlm. & Volkm.-Kohlm.
Helicascus kanaloanus Kohlm.
H. nypae Hyde
Hydronectria tethys Kohlm. & Kohlm.
Hypophloeds rhizophorae Hyde & Jones
Hypoyxlon oceanicum Schatz
Julella avicenniae (Borse) Hyde
Lanspora coronata Hyde & Jones
Lautospora gigantea Hyde & Jones
Leptosphaeria australiensis (Cribb & Cribb) G C Hughes
L. avicenniae Kohlm. & Kohlm.
Lignincola longirostris (Cribb & Cribb) Kohlm.
L. tropica Kohlm.
Lindra hawaiiensis Kohlm. & Volkm.-Kohlm.
L. marinera Meyers
L. thalassiae Orpurt, Meyers, Boral & Simms
Lineolata rhizophorae (Kohlm. & Kohlm.) Kohlm. & Volkm.-Kohlm.
Linocarpon appendiculatum Hyde
L. nypae (P. Henn.) Hyde
Lophiostoma mangrovei Kohlm. & Vittal
Lulworthia grandispora Meyers
Manglicola guatemalensis Kohlm. & Kohlm.
Marinosphaera mangrovei Hyde
Massarina acrostichi Hyde
M. ramunculicola Hyde
M. thalassiae Kohlm. & Volkm.-Kohlm.
M. velatospora Hyde & Borse
Moana turbinulata Kohlm. & Volkm.-Kohlm.
Mycosphaerella pneumatophorae Kohlm.
Nais glitra Crane & Shearer
Nimbospora bipolaris Hyde & Jones
N. effusa Koch
N. octonae Kohlm.

Ophiodeira monosemeia Kohlm. & Volkm.-Kohlm.
Paraliomyces lentiferus Kohlm.
Passeriniella savoryellopsis Hyde & Mouzouras
Payosphaeria minuta Leong
Pedumispora rhizophorae Hyde & Jones
Phomatospora acrostichi Hyde
Quintaria lignatilis (Kohlm.) Kohlm. & Volkm.-Kohlm.
Remispora crispa Kohlm.
Rhizophila marina Hyde & Jones
Salsuginea ramicola Hyde
Savoryella paucispora (Cribb & Cribb) Koch
S. appendiculata Hyde & Jones
Swampomyces armeniacus Kohlm. & Volkm.-Kohlm.
Trematosphaeria lineolatispora Hyde
T. mangrovei Kohlm.
T. striatispora Hyde
Verruculina enalia (Kohlm). Kohlm. & Volkm.-Kohlm.
Xenus lithophylli Kohlm. & Volkm.-Kohlm.

Basidiomycotina

Halocyphina villosa Kohlm. & Kohlm.
Calathella mangrovei Jones & Agerer

Deuteromycotina

Cirrenalia pygmea Kohlm.
C. pseudomacrocephala Kohlm.
C. tropicalis Kohlm.
C. basiminuta Raghu-Kumar & Zainal
Nypaella frondicola Hyde & Sutton
Periconia prolifica Anastasiou
Phomopsis mangrovei Hyde
Plectophomella nypae Hyde & Sutton
Rhabdospora avicenniae Kohlm. & Kohlm.
Varicosporina ramulosa Meyers & Kohlm.

Group II: Fungi with a cosmopolitan distribution

Ascomycotina

Aniptodera chesapeakensis Shearer & Miller
Arenariomyces trifurcatus Höhnk
Ceriosporopsis caduca Jones & Zainal
C. halima Linder
Chaetosphaera chaetosa Kohlm.
Corollospora maritima (Linder) Kohlm.
Corollospora pulchella Kohlm., Schmidt & Nair
Dryosphaera navigans Koch & Jones
Halosarpheia retorquens Shearer & Crane
H. viscosa (I. Schmidt) Shearer & Crane ex Kohlm. & Volkm.-Kohlm.
Halosphaeriopsis mediosetigera (Cribb & Cribb) T.W. Johnson
Lignincola laevis Höhnk
Nautosphaeria cristaminuta Jones
Remispora galerita Tubaki*
R. maritima Linder*
R. stellata Kohlm.
Savoryella ligniciola Jones & Eaton
Torpedospora radiata Meyers

Basidiomycotina

Nia vibrissa Moore & Meyers

Deuteromycotina

Camarosporium roumeguerii Sacc.
Cirrenalia macrocephala (Kohlm.) Myers & Moore
Clavatospora bulbosa (Anast.) Nakagiri & Tubaki
Humicola alopallonella Meyers & Moore
Monodictys pelagica (Johnson) Jones
Phoma sp.
Trichocladium achrasporum (Meyers & Moore) Dixon in Shearer & Crane
Zalerion maritimum (Linder) Anastasiou
Z. varium Anastasiou

Table 2. Percentage frequency of occurrence of fungi on submerged mangrove woods (After Tan *et al.*, 1989)

	Avicennia alba	*Avicennia lanata*	*Bruguiera cylindrica*	*Rhizophora apiculata*
Verruculina enalia	66	62	63	73
Lulworthia sp.	62	71	58	30
Aigialus parvus	27	12	38	36
Lignincola laevis	19	21	50	47
Halosarpheia marina	22	47	18	10
Ascomycete 25	22	39	33	INF
Halocyphina villosa	22	10	INF	INF
Nais inornata-cf.	INF	INF	18	21
Aigialus mangrovei	INF	INF	INF	23
Deuteromycete 456	–	–	15	INF
Hypoxylon oceanicum	–	–	18	INF
Halosarpheia lotica	INF	–	10	–

Species which appeared only once have been omitted.
INF = infrequent, i.e. less than 10%; – = absent.

Until recently, little was known of marine fungi from tropical locations (Kohlmeyer, 1968, 1969, 1980, 1981). However, with the examination of fungi growing on mangrove wood, leaves and fruits, many new taxa have been described (Kohlmeyer, 1984; Kohlmeyer & Volkmann-Kohlmeyer, 1987a, b, 1991, 1992; Hyde, 1991, 1992; Hyde & Jones, 1986, 1987, 1992; Jones & Hyde, 1990, 1992; Hyde & Sutton, 1992; Volkmann-Kohlmeyer & Kohlmeyer, 1992). Mangroves are an ecological group of plants restricted largely to the tropics and subtropics and number 34 major taxa, 20 minor taxa and 60 associate taxa (these are usually herbaceous or climbing plants or those that can be considered as minor components of the mangrove ecosystem) (Tomlinson, 1986). Hyde & Jones (1988) list 90

Legend to Table 3 (facing page):

a, b = **Seychelles** (Hyde & Jones, 1988); **c** = **Philippines** (Jones, Uyenco & Follosco, 1988); **d** = **Thailand** (Hyde, Chalermpongse & Boonthravikoon, 1990), **d1** = total % occurrence, **d2** = on *Rhizophora apiculata*, **d3** = on *Sonneratia griffithii*; **e** = **Brunei** (Hyde, 1989); f = **Sumatra** (Hyde, 1989a); **g** = **Thailand** (Hyde, 1989b); **h** = **Brunei** (Hyde, 1990); **I** = **Mexico** (Hyde, 1992).
* believed to have cosmopolitan distribution.

Table 3. Frequent fungi on mangrove driftwood in Brunei, Mexico, Seychelles, Sumatra, the Philippines and Thailand

	a	b	c	d1	d2	d3	e	f	g	h	i
Halocyphina villosa	27	13	–	5	7	–	14	–	–	–	–
Lulworthia grandispora	23	–	2	4	9	–	8	–	16	–	31
Ascomycete sp. 4	16	15	–	–	–	–	–	–	–	–	–
Antennospora quadricornuta	12	6	–	–	–	–	–	–	–	–	–
Dactylospora haliotrepha	7	3	3	8	7	8	–	12	–	–	–
Aniptodera mangrovei	7	4	–	–	–	–	–	–	–	–	–
Caryosporella rhizophorae	5	–	–	–	–	–	–	–	12	7	–
Lulworthia sp.	5	6	–	–	–	–	6	9(3)	–	–	–
Hydronectria tethys	–	–	–	–	8	–	–	–	–	–	–
Massarina velatospora	5	–	19	7	–	10	–	–	–	–	–
Aigialus grandis	5	–	–	9	10	9	–	–	–	–	–
Halosarpheia marina	3	19	5	–	–	–	10	22	–	6	20
Humicola alopallonella	–	6	–	–	–	–	–	–	–	–	–
Biatriospora marina	–	3	–	–	–	–	–	–	–	–	–
Verruculina enalia	–	3	–	–	8	–	–	–	7	–	46
Zalerion varium	–	–	8	–	–	–	–	–	–	–	–
Savoryella lignicola	–	–	7	10	14	7	–	–	8	3	–
Leptosphaeria australiensis	–	–	–	5	7	–	–	–	11	–	–
Hypoxylon oceanicum	–	–	6	–	–	–	10	12	–	13	–
Helicascus kanaloanus	–	–	–	–	–	9	–	–	–	–	–
Trichocladium achrasporum	–	–	3	–	–	–	6	–	–	–	–
Lignincola laevis	–	–	–	–	–	–	7	11	–	–	14
Marinosphaera mangrovei	–	–	–	–	–	–	–	–	15	4	–
Halosarpheia abonnis	–	–	–	8	–	12	–	–	–	–	–
Cirrenalia pygmea	–	–	–	–	–	–	12	11	10	8	–
Halosarpheia ratnagiriensis	–	–	–	–	–	–	–	–	–	7	–
Aniptodera chesapeakensis	–	–	–	–	–	–	–	9	–	–	17
Rhizophila marina	–	–	–	–	–	–	–	19	6	–	–
Phoma sp.	–	–	–	–	–	–	–	15	–	3	–
Leptosphaeria cf. *avicenniae*	–	–	–	–	–	–	–	–	9	–	–
Lophiostoma cf. *mangrovis*	–	–	–	–	–	–	–	–	9	–	–
Xylomyces sp.	–	–	–	–	–	–	–	–	–	4	–
Trematosphaeria lineolatispora	–	–	–	–	–	–	–	–	–	–	20
Dictyosporium pelagica	–	–	–	–	–	–	–	–	–	–	20

higher marine fungi growing on 26 species of mangroves. The percentage frequency of occurrence of 12 common mangrove fungi are presented in Table 2 (Tan *et al.*, 1989). Two fungi are regarded as cosmopolitan: *Lignincola laevis* and *Nais inornata*-cf., the remainder have only been reported from the tropics.

Table 3 lists the frequent fungi recorded in various tropical mangroves. A number of fungi were frequent at more than three sites: *Cirrenalia pygmea* Kohlm., *Dactylospora haliotrepha* (Kohlm. & Kohlm.) Hafellner, *Halocyphina villosa* Kohlm. & Kohlm., *Hypoxylon oceanicum* Schatz, *Lulworthia grandispora* Meyers, *Savoryella lignicola* Jones & Eaton, while others were only recorded once, e.g. *Biatriospora marina* Hyde & Borse, *Helicascus kanaloanus* Kohlm. and *Halosarpheia ratnagiriensis* Patil & Borse. The factors that control the frequency of occurrence of the latter remain to be determined, e.g. the timber species on which they grow (e.g. *Verruculina enalia* (Kohlm.) Kohlm. & Volkm.-Kohlm. is found on *Rhizophora apiculata* but not on *Sonneratia griffithii*); and whether or not the fungus is an early colonizer (e.g. *Lignincola laevis, Lulworthia* sp.) or a late colonizer (e.g. *Dactylospora haliotrepha*). Fig. 7 presents the geographic distribution of two marine ascomycetes: *Corollospora maritima* Werdermann and *Hypoxylon oceanicum*. The latter species is particularly common on mangrove wood (especially *Bruguiera* spp.) and is restricted to the tropics/subtropics. *Corollospora maritima* is an arenicolous fungus and cosmopolitan in its distribution. In what way does temperature control the distribution of marine fungi? This will be considered under the following headings, effect on: vegetative growth, anamorph/teleomorph distribution, reproduction, weight loss of wood, and spore dimensions.

Temperature and vegetative growth

The effect of temperature on the growth of three tropical fungi and two isolates of a cosmopolitan species is shown in Fig. 8. For *Nia vibrissa*, optimum growth was at 20°C with little growth at 10 and 30°C. One isolate was obtained from a temperate habitat while the other was a tropical isolate, yet both behaved in the same manner with respect to temperature. For the three tropical species *Hypoxylon oceanicum, Hydronectria tethys* Kohlm. & Kohlm., *Periconia prolifica* Anastasiou, no growth occurred at 10°C and optimum growth was at 30°C. Boyd & Kohlmeyer (1982) investigated the growth of three marine hyphomycetes on a glucose-peptone yeast extract seawater medium. *Asteromyces cruciatus* Moreau & Moreau ex Hennebert (a temperate water species) grew at

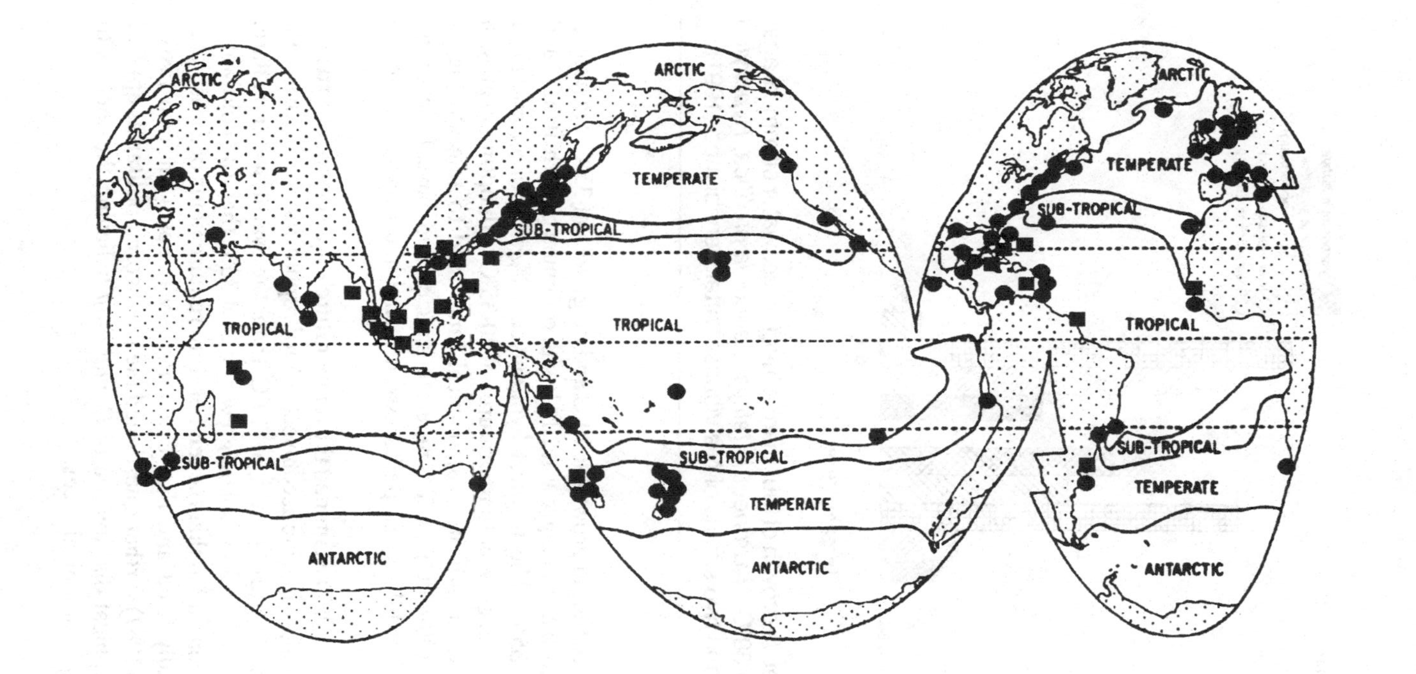

Fig. 7. Geographic distribution of the marine Ascomycotina *Corollospora maritima* (closed circles) and *Hypoxylon oceanicum* (closed squares) based on records in the literature. World map redrawn after Hughes (1974) and Bebout *et al*. (1987) with permission.

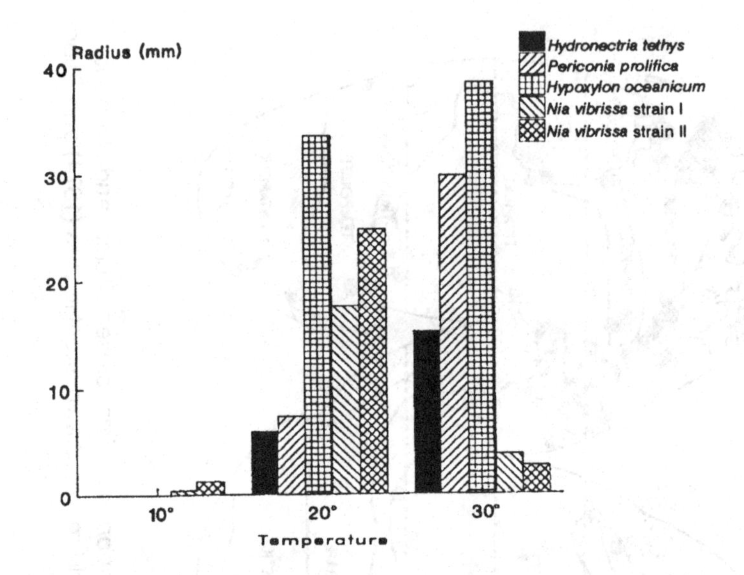

Fig. 8. Vegetative growth of four marine fungi on seawater cornmeal agar at 10, 20 and 30°C: *Hydronectria tethys, Periconia prolifica, Hypoxylon oceanicum,* and two strains of *Nia vibrissa*. After Panebianco (1991).

10°C but with optimum growth at 20°C and no growth at 35°C. *Sigmoidea marina* Haythorn & Jones grew at 20°C, with optimum growth at 30°C and no growth at 35°C. The tropical hyphomycete *Varicosporina ramulosa* Meyers & Kohlm. grew at temperatures of 20-35°C with optimum growth at 20°C. The data indicate that tropical fungi tend to tolerate higher temperatures (30-40°C) than temperate species and have little or no growth at 10°C, while cosmopolitan strains exhibit a broad tolerance to temperature.

Bebout *et al.* (1987) examined the temperature requirement of 5 strains of the cosmopolitan marine ascomycete *Corollospora maritima*. An isolate from cold water grew better at 10°C than the warm water isolate. Conversely, a warm water isolate grew better at 30°C than the cold water isolate. They concluded that the temperature requirements of the isolates were genetically fixed and that *C. maritima* was not a uniform taxon. Bebout *et al.* (1987) further questioned the label cosmopolitan as applied to the geographical distribution of *C. maritima* believing the species to have distinct physiological races.

Temperature and teleomorph/anamorph distribution

Nakagiri & Tubaki (1985) plotted the geographic distribution of the teleomorphs and anamorphs of *Corollospora pulchella* Kohlm., Schmidt & Nair, *Corollospora intermedia* Schmidt and *Halosphaeriopsis mediosetigera* (Cribb & Cribb) T.W. Johnson. From the limited data available, they were able to demonstrate that teleomorphs occurred in regions with lower temperatures while the anamorphs were found in areas with higher temperatures. Both teleomorphs and anamorphs were found in the intermediate zones. They further reported that the anamorph of *H. mediosetigera (Trichocladium achrasporum)* does not occur on the west coast of North America. However since the present author found both the anamorph and teleomorph on driftwood in Taiwan, Republic of China, there may be a more complex explanation for the distribution of *H. mediosetigera* which merits further investigation.

Temperature and reproduction

Although marine fungi can be grown successfully in culture, their vegetative growth is often very slow (e.g. 1 mm radial growth in 3 weeks for *Nais glitra* Crane & Shearer (Panebianco, 1991) while some fungi do not sporulate under laboratory conditions (e.g. *Nereiospora comata* (Kohlm.) Jones, Johnson & Moss, *Hydronectria tethys*), or only after prolonged incubation (e.g. *Corollospora* spp.) Hyde, Farrant & Jones, 1987). Many lose their ability to sporulate after a period of time in culture. Hyde *et al.* (1987) tested 67 marine fungi for their ability to sporulate under laboratory conditions: 41 produced ascomata, while a further 13 formed conidia, pycnidia or chlamydospores. Most studies have concentrated on the use of natural substrata to induce sporulation, e.g. balsa wood (Hyde *et al.,* 1987; Lutley & Wilson, 1972), birch sticks (Kirk, 1969), sterilized leaves of *Thalassia* (Meyers & Simms, 1965) or *Spartina* (Lloyd & Wilson, 1962).

This lack of information on the conditions required for sporulation in marine fungi has greatly hindered experimental studies and explains why so few have been undertaken on the effect of temperature on ascomata production. Nakagiri & Tubaki (1985) examined the effect of temperature on sporulation in *Corollospora pulchella, C. intermedia* and *Halosphaeriopsis mediosetigera,* and the data are summarised in Table 4. *Corollospora pulchella* produced no ascomata at any temperature, while in *C. intermedia* (temperate?) ascomata were formed at 15-20°C (teleomorph strain) and 25-35°C for the holomorphic strains. For *H. mediosetigera* (cosmopolitan

Table. 4. The effect of temperature on ascomata and conidial production in *Corollospora pulchella, C. intermedia* and *Halosphaeriopsis mediosetigera*. After Nakagiri & Tubaki (1985).

	Sporulation at the indicated temperature (°C)						
	10	15	20	25	30	35	40
Corollospora pulchella							
Holomorphic strains (conidia)	W	G	G	G	G	W	W
Anamorphic strains (conidia)		W	G	G	G	W	W
Corollospora intermedia							
Teleomorphic strain (ascospores)		G	G	W			
Anamorphic strain (ascospores)			W	G	G	G	W
Anamorphic strain (conidia)	W	G	G	G	G	G	G
Halosphaeriopsis mediosetigera							
Teleomorphic strains (ascospores)		W	G	G	W		
Anamorphic strains (conidia)			W	W	G	G	W

G = good sporulation, W = weak sporulation.

species) ascomata were formed at 20-25°C (teleomorph strains) and none by the anamorphic strains.

The results obtained are in agreement with the known geographic distribution of these fungi. It is apparent that optimum conditions for production of ascomata is at a lower temperature than for conidial production, yet fruiting structures are readily formed in the field at temperatures of 25-35°C. Clearly, further studies are required to elucidate the conditions required for ascomata production in tropical marine ascomycetes.

Ability to decay wood

Fifty-nine (of a total of 81) marine fungi have been shown to cause decay of wood (Mouzouras, 1989), soft rot decay being the most prevalent. Mouzouras (1986) also compared the ability of marine fungi to decay wood at 10 and 22°C (Table 5). He showed that a temperate water species (*Digitatispora marina* Doguet) produced the greatest weight loss at 10°C while the tropical *Halocyphina villosa* only caused decay at 22°C. The cosmopolitan species *Nia vibrissa* and *Monodictys pelagica* were able to

Table 5. Mean percentage loss in dry weight of *Ochroma lagopus* (balsa) by selected marine fungi at 24 weeks. After Mouzouras (1986).

	10°C	22°C
Control	3·6	4·8
Monodictys pelagica (Johnson) Jones	28·3	40·8
Nia vibrissa Moore & Meyers	13·7	27·9
Digitatispora marina Doguet	14·3	4·9
Halocyphina villosa Kohlm. & Kohlm.	−0·6	22·9

cause significant weight loss at both 10 and 20°C, confirming their greater versatility for growth over a range of temperatures.

Length of ascospores/propagules

Kirk (1986) has suggested that 'the length of ascospores or of polar appendages, or both, is often greatest in tropical, intermediate in cosmopolitan, and least in temperate species of the same genus, as in other plankton'. *Lulworthia grandispora* (tropical) ascospores measure 500-756 x 3-5 μm, while in the temperate water species *Lulworthia fucicola* Suth., they measure 70-110(-126) x 4-6 μm. Cosmopolitan species (*Lulworthia floridana* Meyers, *L. purpurea* (Wilson) T.W. Johnson have spores in the intermediate range 150-350 μm. While this is not correct for all marine genera, the fact remains that many (over 60) species of marine fungi have spores greater than 50 μm.

Conclusion

In general, marine fungi appear to be distributed in relation to seawater temperature: arctic, temperate, tropical; although others grow equally well over a range of temperatures (the cosmopolitan species). It remains to be seen if strains, from temperate and tropical locations, of species regarded as cosmopolitan are identical. Confirmation depends on investigations with a much wider range of taxa. Tropical fungi appear to require or tolerate higher temperatures for growth than temperate water species whereas cosmopolitan marine fungi have the ability to grow over a wider range of temperatures than species restricted solely to the tropics.

Marine fungi, because of their limited number and unique adaptation to the aquatic habitat, make an ideal group to study for their biogeographic distribution.

Acknowledgements I am grateful to a number of colleagues and friends who have helped with my field work on tropical fungi: K. D. Hyde (Australia), L. L. P. Vrijmoed (Hong Kong), T. K. Tan (Singapore), A. J. Kuthubutheen, T. Singh, Phang Siew Moi (Malaysia), H. S. Chang, S.-Y. Hsieh (Taiwan), T. D. Steinke (South Africa), V. Gacutan (Philippines) and A. Zainal (Kuwait). The financial support of The British Council, Natural Environment Research Council, Science Research Council of the Republic of China, the City Polytechnic of Hong Kong and the University of Malaya is gratefully acknowledged. Thanks also go to C. Derrick and G. Bremer for photographic and technical assistance.

References

Bebout, B., Schatz, S., Kohlmeyer, J. & Haibach, M. (1987). Temperature dependent growth in isolates of *Corollospora maritima* Werdermann (Ascomycetes) from different geographical regions. *Journal of Experimental Marine Biology and Ecology* **106**, 203-210.

Booth, T. (1979). Strategies for study of fungi in marine and marine influenced ecosystems. *Review of Microbiology* (S. Paulo) **10**, 123-138.

Booth, T. & Kenkel, N. (1986). Ecological studies of lignicolous marine fungi: A distribution model based on ordination and classification. In *The Biology of Marine Fungi* (ed. S.T. Moss), pp. 297-310. Cambridge University Press: Cambridge, U.K.

Boyd, P. E. & Kohlmeyer, J. (1982). The influence of temperature on the seasonal and geographic distribution of three marine fungi. *Mycologia* **74**, 894-902.

Fazzani, K. & Jones, E. B. G. (1977). Spore release and dispersal in marine and brackish water fungi. *Material und Organismen* **12**, 235-248.

Henningson, M. (1974). Aquatic lignicolous fungi in the Baltic and along the west coast of Sweden. *Svensk Botanisk Tidskrift* **68**, 401-425.

Hughes, G. C. (1974). Geographical distribution of the higher marine fungi. *Veroffentlichungen des Instituts für Meeresforschung* Bremerhaven Supplement **5**, 419-441.

Hughes, G. C. (1986). Biogeography and the marine fungi. In *The Biology of Marine Fungi* (ed. S. T. Moss), pp. 275-295. Cambridge University Press: Cambridge, U.K.

Hyde, K. D. (1988). Studies on the tropical marine fungi of Brunei. *Botanical Journal of the Linnean Society* **98**, 135-151.

Hyde, K.D. (1989a). Intertidal mangrove fungi from north Sumatra. *Canadian Journal of Botany* **67**, 3078-3082.

Hyde, K. D. (1989b). *Caryospora mangrovei* sp. nov. and notes on marine fungi from Thailand. *Transactions of the Mycological Society of Japan* **30**, 333-341.

Hyde, K. D. (1990). A comparison of the intertidal mycota of five mangrove tree species. *Asian Marine Biology* **7**, 93-107.

Hyde, K. D. (1991). *Helicascus kanaloanus, Helicascus nypae* sp. nov. and *Salsuginea ramicola* gen. et sp. nov. from intertidal wood. *Botanica Marina* **34**, 311-318.

Hyde, K. D. (1992). *Julella avicenniae* (Borse) comb. nov. (Thelenellaceae) from intertidal mangrove wood and miscellaneous fungi from the NE coast of Queensland. *Mycological Research* **96**, 939-942.

Hyde, K. D. (1992). Intertidal mangrove fungi from the west coast of Mexico including one new genus and two new species. *Mycological Research* **96**, 25-30.

Hyde, K. D. & Jones, E. B. G. (1986). Marine fungi from Seychelles. IV. *Cucullospora mangrovei* gen. et sp. nov. from dead mangrove. *Botanica Marina* **29**, 491-495.

Hyde, K.D. & Jones, E.B.G. (1987). Marine fungi from Seychelles. VIII. *Bathyascus grandisporus* sp. nov. from mangrove wood. *Botanica Marina* **30**, 413-416.

Hyde, K. D. & Jones, E. B. G. (1988). Marine mangrove fungi. *P.S.Z.N.I.: Marine Ecology* **9**, 15-33.

Hyde, K. D. & Jones, E. B. G. (1989). Observations on ascospore morphology in marine fungi and their attachment to surfaces. *Botanica Marina* **32**, 205-218.

Hyde, K.D. & Jones, E. B. G. (1992). Intertidal mangrove fungi: *Pedumispora* gen. nov. (Diaporthales). *Mycological Research* **96**, 78-80.

Hyde, K. D. & Sutton, B. C. (1992). *Nypaella frondicola* gen. et sp. nov., and *Pleurophomopsis nypae* sp. nov. (Coelomycetes) from intertidal fronds of *Nypa fructicans. Mycological Research* **96**, 210-214.

Hyde, K. D., Farrant, C. A. & Jones, E. B. G. (1987). Isolation and culture of marine fungi. *Botanica Marina* **30**, 291-303.

Hyde, K. D., Chalermpongse, A. & Boonthavikoon, T. (1990). Ecology of intertidal fungi at Ranong mangrove, Thailand. *Transactions of the Mycological Society of Japan* **31**, 17-27.

Jones, E. B. G. (1971). The ecology and rotting ability of marine fungi. In *Marine Borers, Fungi and Fouling Organisms of Wood* (ed. E. B. G. Jones & S. K. Eltringham), pp. 237-258, Organisation for Economic Cooperation and Development: Paris.

Jones, E. B. G. & Hyde, K. D. (1990). Observations on poorly known mangrove fungi and a nomenclatural correction. *Mycotaxon* **37**, 197-201.

Jones, E. B. G. & Hyde, K. D. (1992). Taxonomic studies on *Savoryella* Jones *et* Eaton (Ascomycotina). *Botanica Marina* **35**, 83-91.

Jones, E. B. G., Uyenco, F. R. & Follosco, M. P. (1988). Fungi on driftwood collected in the intertidal zone from the Philippines. *Asian Marine Biology* **5**, 103-106.

Kirk, P. W. (1969). Isolation and culture of lignicolous marine fungi. *Mycologia* **61**, 174-177.

Kirk, P. W. (1986). Evolutionary trends within the Halosphaeriaceae. In *The Biology of Marine Fungi* (ed. S. T. Moss), pp. 263-273, Cambridge University Press: Cambridge, U.K.

Koch, J. (1982). Some lignicolous fungi from Sri Lanka. *Nordic Journal of Botany* **2**, 163-169.

Kohlmeyer, J. (1968). Marine fungi from the tropics. *Mycologia* **60**, 252-270.

Kohlmeyer, J. (1969). Ecological notes on fungi in mangrove forests. *Transactions of the British Mycological Society* **53**, 237-250.

Kohlmeyer, J. (1980). Tropical and subtropical filamentous fungi of the Western Atlantic Ocean. *Botanica Marina* **23**, 529-544.

Kohlmeyer, J. (1981). Marine fungi from Martinique. *Canadian Journal of Botany* **59**, 1314-1321.

Kohlmeyer, J. (1983). Geography of marine fungi. *Australian Journal of Botany* Suppl. Ser. **10**, 67-76.

Kohlmeyer, J. (1984). Tropical marine fungi. *P.S.Z.N.I.: Marine Ecology* **5**, 329-378.

Kohlmeyer, J. & Kohlmeyer, E. (1971). Marine fungi from tropical America and Africa. *Mycologia* **63**, 831-861.

Kohlmeyer, J. & Volkmann-Kohlmeyer, B. (1987a). Marine fungi from Aldabra, the Galapagos, and other tropical islands. *Canadian Journal of Botany* **65**, 571-582.

Kohlmeyer, J. & Volkmann-Kohlmeyer, B. (1987b). Marine fungi from Belize with a description of two genera of ascomycetes. *Botanica Marina* **30**, 195-204.

Kohlmeyer, J. & Volkmann-Kohlmeyer, B. (1991). *Hapsidascus hadrus* gen. et sp. nov. (Ascomycotina) from mangroves in the Caribbean. *Systema Ascomycetum* **10**, 113-120.

Kohlmeyer, J. & Volkmann-Kohlmeyer, B. (1992). Two Ascomycotina from coral reefs in the Caribbean and Australia. *Cryptogamic Botany* **2**, 367-374.

Lloyd, L. S. & Wilson, I. M. (1962). Development of the perithecium in *Lulworthia medusa* (Ellis & Everh.) Cribb & Cribb, a saprophyte on *Spartina townsendii*. *Transactions of the British Mycological Society* **45**, 359-372.

Lutley, M. & Wilson, I. M. (1972). Development and fine structure of ascospores in the marine fungus *Ceriosporopsis halima*. *Transactions of the British Mycological Society* **58**, 393-402.

Meyers, S. P. & Simms, J. (1965). Thalassiomycetes VI. Comparative growth studies of *Lindra thalassiae* and lignicolous ascomycete species. *Canadian Journal of Botany* **43**, 379-392.

Mouzouras, R. (1986). Patterns of timber decay caused by marine fungi. In *The Biology of Marine Fungi* (ed. S. T. Moss), pp. 341-353. Cambridge University Press: Cambridge, U.K.

Mouzouras, R. (1989). Soft rot decay of wood by marine fungi. *Journal of the Institute of Wood Science* **11**, 193-201.

Nakagiri, A. & Tubaki, K. (1985). Teleomorph and anamorph relationships in marine Ascomycetes (Halosphaeriaceae). *Botanica Marina* **28**, 485-500.

Panebianco, C. (1991). Indagine sulla crescita di funghi liginicoli marini con particolare riferimento alla competizione per interferenza *in vitro* ed *in situ*. Dottorato di Ricerca in Scienze Ambientali: Ambiente Marino e Risorse, University of Messina, Italy.

Rees, G. & Jones, E. B. G. (1984). Observations on the attachment of spores of marine fungi. *Botanica Marina* **27**, 145-160.

Tan, T. K., Leong, W. F., Mouzouras, R. & Jones, E. B. G. (1989). Occurrence of fungi on mangrove wood and its decomposition. In *Recent Advances in Microbial*

Ecology, (ed. T. Hattori, Y. Ishida, Y. Maruyama, R. Y. Morita & A. Uchida), pp. 307-310. Japan Scientific Societies Press: Tokyo.

Tomlinson, B. (1986). *The Botany of Mangroves*. Cambridge University Press: Cambridge, U.K.

Volkmann-Kohlmeyer, B. & Kohlmeyer, J. (1992). *Corallicola nana* gen. et sp. nov. and other ascomycetes from coral reefs. *Mycotaxon* **44**, 417-424.

Vrijmoed, L. L. P., Hodgkiss, I. J. & Thrower, L. P. (1986). Factors affecting the distribution of lignicolous marine fungi in Hong Kong. *Hydrobiologia* **87**, 143-160.

Chapter 6

Distribution of Zygomycetes - the tropical connection

P. M. Kirk

International Mycological Institute, Bakeham Lane, Egham, Surrey TW20 9TY, U.K.

Introduction

Our state of knowledge of the Zygomycetes in temperate countries is relatively well documented and many modern diagnostic keys are available. However, members of the Zygomycetes from the tropics have been poorly researched because of inadequate collecting. Examples of taxa from the tropics which exhibit clear endemism are known at the family level, at the genus level within single families, and at the species level within a single genus. Since there are a significant number of genera and species having a widespread distribution there are no appreciable difficulties in studying these fungi in tropical countries. A few examples, chosen from this class of fungi, which exhibit interesting distribution patterns with reference to the tropics will be discussed.

The tropics may be definedby reference to the lines at 23°27' North and 23°27' South of the equator. In discussing the occurrence and distribution of fungi in the tropics the temperature requirements of the fungi in relation to the ambient temperatures is the most important consideration. Specifically, are the temperatures in a given area within the range which allows the fungus to survive from one year to another? The tropics, therefore, can be defined only with reference to annual temperature maxima and minima. A convenient baseline on which to define the tropics is the so called 'palm line', which delimits the natural range of Palmae north and south of the equator.

Of course, through the involvement of man, many palms are now growing outside their natural range and these areas are often outside the palm line. An example of this is found in the palm *Trachycarpus fortunei* H. Wendl., which, because of the warming effect of the gulf stream, can be seen growing on the west coast of Scotland.

Altitude is another important factor in defining the tropics, since average temperatures fall as altitude increases (usually considered to be 5°C for every 1000 m), thus allowing species which are unable to survive at the higher temperatures found in the tropics to grow. The altitude factor is not particularly relevant here as this chapter is concerned with fungi of tropical conditions but it can explain why some species which are otherwise temperate in distribution may be found in the tropics at higher elevation (see later discussion). It is, therefore, important that altitude is considered when drawing conclusions from distribution data.

Unlike the recording of higher plants or, indeed, the majority of animals, the recording of fungi is beset with difficulties. It requires either special techniques of observation, such as cultures or the use of microscopes, or, if their fruit bodies, on which identification often depends, are ephemeral, the right person to be there to make the observation. The distribution is, therefore, often a reflection of mycological expertise in the field and based only on confirmed records. With few exceptions the maxim 'absence of evidence is not evidence of absence' holds very strongly for the fungi. With higher plants, a survey of a particular area for a particular plant which yields no records can, with confidence, be recorded as a negative record. This is not the case with fungi although it goes without saying that species which are obligate parasites of higher plants follow the distribution of their hosts to the extent that they do not occur where there hosts are absent. The distributions which are referred to in the following pages are often, therefore, incomplete and reflect only positive records.

Zygomycetes

The Zygomycetes are a relatively small group of microfungi divided into six or seven orders. They are one of the two classes of the Zygomycotina, the other class being the strange Trichomycetes (Table 1). The Zygomycetes are characterized by aplanate, anamorphic spores borne in sporangia and thick-walled, teleomorphic spores termed zygospores, which appear to be survival propagules.

The Zygomycetes number some 120 genera containing about 700 species (Hawksworth, Sutton & Ainsworth, 1983). Most are saprotrophs although some are parasites of protozoa, arthropods and fungi or are secondary invaders of higher plants and animals including man. This Chapter concentrates on members of just four of these orders: the Mucorales, Dimargaritales, Kickxellales and Zoopagales. In the Zoopagales only the mycoparasites (Piptocephalidaceae) will be

Table 1. Outline classification of the Zygomycotina

Subdivision	Class	Order
ZYGOMYCOTINA	Zygomycetes	Mucorales*
		Zoopagales*
		Entomophthorales
		Endogonales
		Dimargaritales*
		Kickxellales*
		Glomales
	Trichomycetes	Harpellales
		Asellariales
		Amoebidiales
		Eccrinales

*Groups discussed in the text.

considered since our knowledge of the distribution of those species parasitic on other organisms is very limited (Duddington, 1973).

Mucorales

The Mucorales is the largest of the orders, including as it does over 50 genera containing nearly 200 species, and it is certainly one of the better researched (Hesseltine & Ellis, 1973; Zycha, Siepmann & Linnemann, 1969). Some members of the Mucorales are amongst the most common species of fungi with world-wide distribution patterns. Some of the rarer species are often known from only a single collection (specimen or isolate) orappear to be of restricted range. Only three of the ten or eleven families, the Choanephoraceae, Pilobolaceae and Thamnidiaceae, will be used to demonstrate specific distribution patterns.

Firstly the Choanephoraceae, an example of a family in the Mucorales in which all members appear to be virtually restricted to the tropics. They are soil fungi which can be facultative parasites of a wide range of host plants, particularly members of the Cucurbitaceae, on which they cause fruit rots and blossom end-rots (Cunningham, 1895; Sinha, 1940; Hesseltine, 1953).

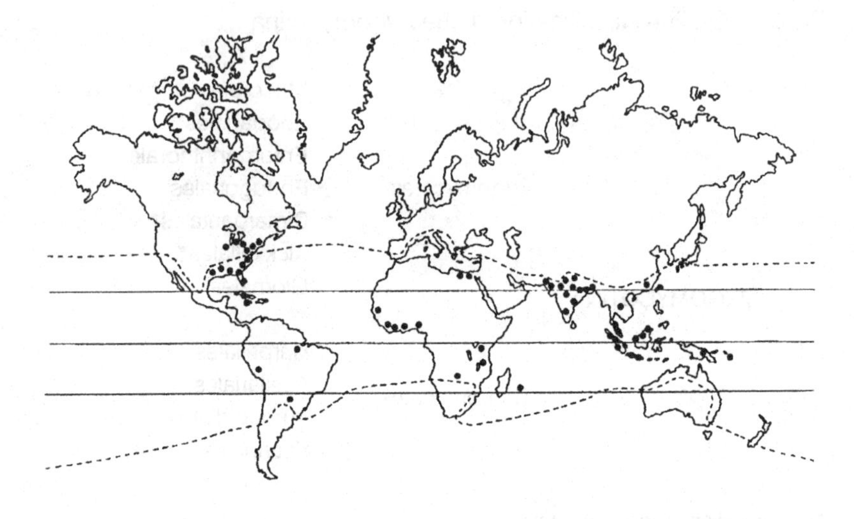

Fig. 1. Distribution of all taxa of *Choanephoraceae* (dashed line = palm line).

All collections of Choanephoraceae, with a few notable exceptions, fall within the boundaries set by the palm line (Fig. 1). The main exceptions are those which have been recorded from the east coast of the USA, as far north as New York State (Herb. IMI, unpubl.). The likely explanation of this is the occurrence of warm summer temperatures and the existence of suitable host plants which are, or were, a common summer crop plant in these areas. It is unlikely that these fungi can survive the cold temperatures of the winter since in culture the spores rapidly become non-viable when placed in a refrigerator at 2-5°C. Whether they can survive in the temperate regions in man-made environments such as farm buildings is unknown. It could be speculated that they re-invade these areas from the warmer southern regions during each growing season.

Members of the Choanephoraceae exibit morphological specialization in the anamorph which probably has evolved because of their tropical distribution and occurrence on the aerial parts of plants. The sporangia have a persistent wall, a characteristic which is uncommon in the Mucorales where in the majority of species the sporangium wall, at maturity, is diffluent and with the sporangiospores forms a translucent, liquid droplet (Zycha, Siepmann & Linnemann, 1969). The wall has a dark brown pigment and it is speculated that this deposition of pigment has

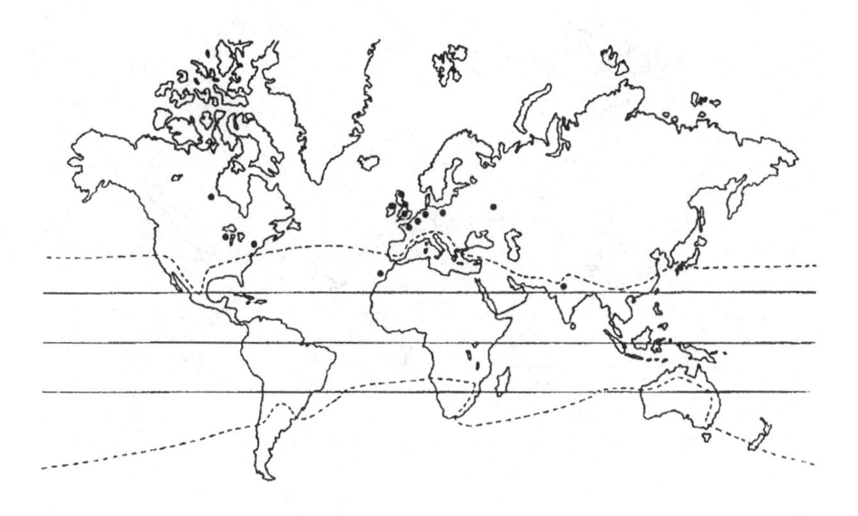

Fig. 2. Distribution of *Pilaira* spp. (dashed line = palm line).

evolved as a mechanism for protecting the genome in the sporangiospores from mutagenic UV light. Further, and again rather uncommonly in the Mucorales, the sporangiospore wall is pigmented, presumably providing further protection to the genome. With such distinct morphological characteristics from which to identify these taxa it is evident, due to the lack of reports of their occurrence, that members of the Choanephoraceae are specifically tropical.

The Pilobolaceae contains the three genera, *Pilobolus*, *Utharomyces* and *Pilaira*. These genera are either exclusively or mainly coprophilous in their habitat and exhibit a distinct distribution pattern in the tropics. Morphological specialization towards their particular habitat appears to have evolved to a greater or lesser extent in all three genera, but this apparently has no link with their distribution. The most simple, morphologically, is the genus *Pilaira*, but this is restricted to temperate regions (Zycha *et al*., 1969).

Pilaira has been searched for on suitable substrata by several workers in the tropics but has not been found. This is perhaps one of the few circumstances in which one can be reasonably confident that absence of records for a particular area visited by mycologists does mean that the

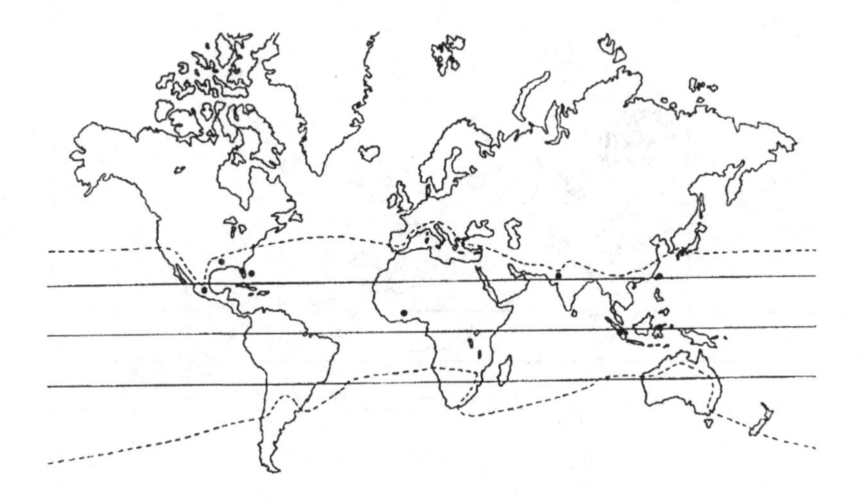

Fig. 3. Distribution of *Utharomyces eppalocaulus* Boedijn ex P. M. Kirk & Benny (dashed line = palm line).

fungus is absent. Those records from within the palm line are from high altitude areas in the Himalaya and Tenerife in the Canary Islands (Fig. 2).

The monotypic genus *Utharomyces* is intermediate in terms of morphological specialization but, unlike *Pilaira*, the single species is probably restricted to the tropics (Fig. 3). Again, that such a morphologically distinct species could occur in temperate regions and not have been found is unlikely.

The third genus of the Pilobolaceae is the type genus, *Pilobolus*, and the one that is familiar to most students of mycology. This genus is considered to be morphologically the most specialized because a mechanism for the forcible discharge of the sporangium has evolved (Buller, 1934). Species of *Pilobolus* are found in both temperate and tropical areas (Fig. 4) although there are insufficient data to determine whether individual species are restricted to only one of these areas.

The final family of Mucorales to consider is the Thamnidiaceae or perhaps, more correctly, the Chaetocladiaceae and Thamnidiaceae, two families which are closely related to the Mucoraceae, and thus unlike the Choanephoraceae and Pilobolaceae which appear to be only distantly related to the type family. They include about 12 genera. Some genera, for example *Chaetocladium*, are psychrophilic and, although they have been recorded from the tropics, are rarely isolated and when they are the localities are invariably at higher altitudes (Herb. IMI, unpubl.). One

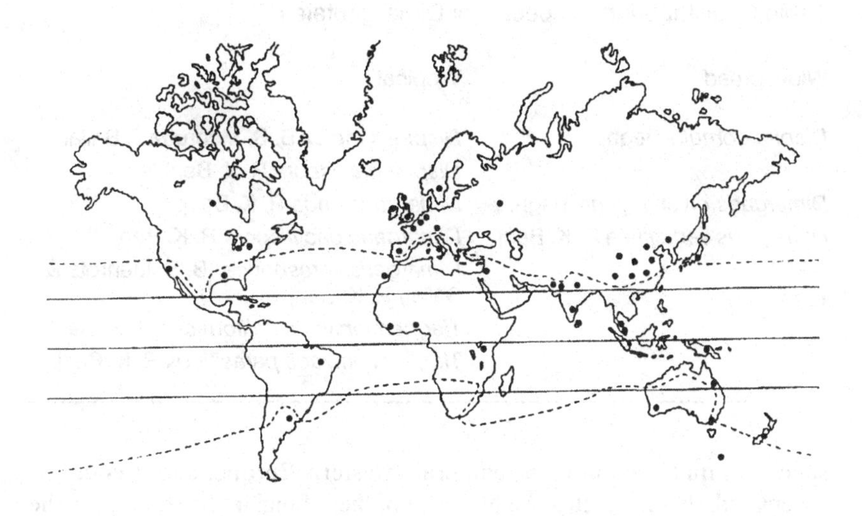

Fig. 4. Distribution of *Pilobolus* spp. Dashed line = palm line.

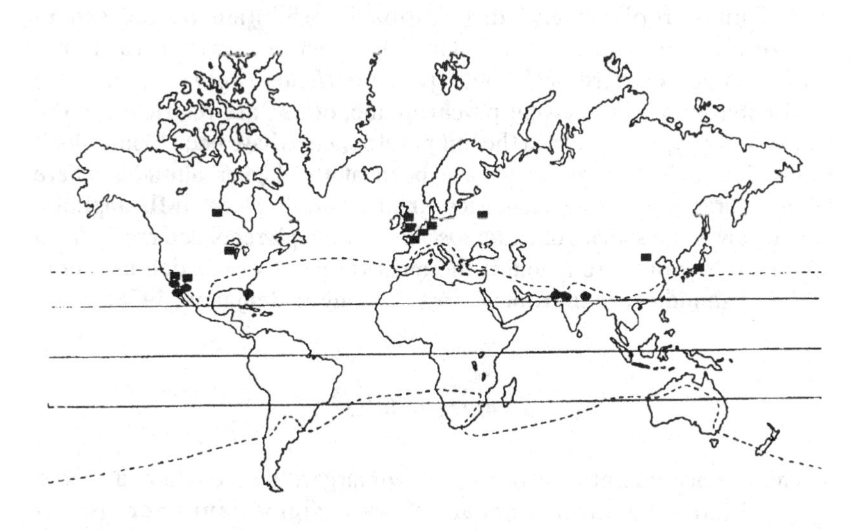

Fig. 5. Distribution of *Chaetocladium* spp. (squares) and *Dichotomocladium* spp. (circles). Dashed line = palm line.

Table 2. Distribution of species of Dimargaritales

Widespread	Tropical
Dispira cornuta Tiegh.	*Dispira simplex* B. S. Mehrotra & Baijal
	Dispira parvispora R. K. Benj.
Dimargaris crystalligena Tiegh.	*Dimargaris arida* R. K. Benj.
Dimargaris verticillata R. K. Benj.	*Dimargaris bacillispora* R. K. Benj.
	Dimargaris xerosporica (B. S. Mehrota & Baijal) R. K. Benj.
	Tieghemiomyces californicus R. K. Benj.
	Tieghemiomyces parasiticus R. K. Benj.

species is quite common in temperate western Europe, and it could be speculated, therefore, that the presence of these fungi in the tropics are the remains of more extensive populations that existed when the temperatures were lower. The closely related although morphologically distinct genus *Dichotomocladium* (Benny, 1973) appears to replace *Chaetocladium* in the tropics since all known species of *Dichotomocladium* were recorded from these areas (Fig. 5).

A similar 'replacement' distribution is exhibitied by the genera *Helicostylum* and *Thamnostylum*. Here, however, the two genera are not wholly temperate or tropical. Some species of *Helicostylum* are temperate in distribution and somewhat psychrophilic, others are restricted to the tropics (Benny, 1973). When the temperate species have been found, albeit rarely, in tropical areas, they have been at the higher altitudes where temperatures were, on average, less than tropical (Herb. IMI, unpubl.). Conversely, some species of *Thamnostylum*, although recorded rarely from temperate regions, are amongst the most frequently recorded members of the Thamnidiaceae in tropical areas (Benny & Benjamin, 1975).

Dimargaritales

The species comprising the order Dimargaritales, contained in the single family Dimargaritaceae, show a significant increase in morphological diversity in the tropics as compared with the species which occur in temperate regions (Table 2). The three genera which comprise this order also show some distinct patterns of distribution in the tropics. A similar pattern can be found in the Kickxellales.

Some members of the Dimargaritales are known to be restricted to temperate regions, and where they have been found in north temperate regions they appear to be at the northern limits of their range (Kirk & Kirk, 1984). The most frequently encountered member of the order is *Dispira cornuta*, which is known to occur in Western Europe, North America, the middle East, and East and South-east Asia. Other species in the genus are restricted to the tropics; they have, as yet, been found only in California. Species of *Dimargaris* are very rarely found outside the tropics, although the genus is based on one species, *D. crystalligena*, originally found in France (van Tieghem, 1876), and a second species, *D. verticillata*, has subsequently been found in England (Kirk & Kirk, 1984). Description of this latter species was originally based on a collection from California and it may also occur in Java according to an unconfirmed record. Distribution in three such disparate locations as Southern North America, West Europe and South-east Asia would be difficult to explain except in relation to the distribution of mycologists!

Species of the genus *Tieghemiomyces*, like *Dispira parvispora* and *D. simplex*, have been found only in California (Benjamin, 1959, 1963). Whether they are restricted to this single area cannot be confirmed since they have not been searched for elsewhere with anything like the dedication which led to their original discovery (R. K. Benjamin, pers. comm.). For the present, it is somewhat easier to accept that eventually they will be found in other suitable habitats around the world than that they are endemic to California.

Zoopagales

The Zoopagales is an order of parasitic fungi and includes four families. Only one, the Piptocephalidaceae, a family of mycoparasitic fungi, will be considered further. The other three, including species which are parasites of nematodes, amoebae and other protozoa, are so very poorly researched on a world-wide basis that no useful information on their distributions can be offered. The Piptocephalidaceae is better researched and some interesting patterns of distribution are emerging as more collecting and recording is carried out. The family contains two genera, *Piptocephalis* and *Syncephalis*, of which the former is the better known and contains the species whose distribution will be discussed further. Species of *Piptocephalis* are parasites of Mucorales although one species, appropriately named *P. xenophila* Dobbs & M.P. English, is a parasite of species of *Penicillium* Link. They occur, therefore, wherever their hosts

Table 3. Distribution of seven species of *Piptocephalis*

Temperate	*P. fimbriata* M. J. Richardson & Leadb. *P. freseniana* de Bary
Widespread	*P. arrhiza* Tiegh. & G. Le Monn. *P. cylindrospora* Bainier
Tropical	*P. curvata* Baijal & B. S. Mehrotra *P. indica* B.S. Mehrotra & Baijal

Table 4. Distribution of genera of Kickxellaceae

Temperate	Widespread	Tropical
Kickxella	*Coemansia*	*Dipsacomyces*
Martensella		*Linderina*
		Martensiomyces
		Spirodactylon
		Spiromyces

occur, and are thus frequently encountered in the soil (Richardson & Leadbeater, 1972).

The seven species of *Piptocephalis*, out of a total of about 20 accepted species, referred to in Table 3 show quite distinct distribution patterns. *Piptocephalis fimbriata* is frequent as a mycoparasite of soil mucors in temperate areas but is unkown in the tropics. Had it been present it would surely have been found since several studies, identical to those which produced the data for temperate regions, have now been carried out in tropical areas. Whether it is present in the tropics at the higher altitudes is not known. Of the species which are listed as widespread in distribution, one, *Piptocephalis arrhiza*, is more common in temperate areas whereas *Piptocephalis cylindrospora* is more common in tropical areas, being recorded from Southern North America, Northern South America, Asia and Australasia (Herb. IMI, unpubl.). Those species which appear to be strictly tropical, *Piptocephalis curvata* and *Piptocephalis indica*, can be stated, with more confidence, to be restricted to the tropics since they have not been found in any of the extensive studies of these fungi carried out in temperate areas (Richardson & Leadbeater, 1972). *Piptocephalis indica* is frequent in soil and has been recorded from Southern North America, Asia and South-east Asia (Herb. IMI, unpubl.).

There is no clear evidence of an increase in diversity within this group of fungi in the tropics although they are still relatively poorly researched in this area.

Kickxellales

The species comprising the order Kickxellales, contained in the single family Kickxellaceae, show a significant increase in morphological diversity in the tropics as compared to the species which occur in temperate regions. The eight genera which comprise this order also show some distinct patterns of distribution with regard to the tropics (Table 4). Species in the genus *Coemansia*, which is numerically the largest genus, appear to be widespread in distribution. They are known from Western Europe, North, Central and South America, tropical West Africa, India and South-east Asia (Herb. IMI, unpubl.). The seven remaining genera, including five which are monotypic, have, as far as is known, restricted distributions. *Kickxella*, the type genus of the order containing a single species, is restricted to temperate regions and is unknown in the tropics. Again, if it were there it would almost certainly have been recorded, since it has been looked for and is relatively easy to recognize. The genus *Martensella*, containing two species, apparently is also restricted to temperate regions.

All of the remaining five genera are, however, restricted to the tropics. The genera *Dipsacomyces*, *Martensiomyces* and *Spiromyces* are only known from the type collections of the monotype whereas more than one collection of *Spirodactylon* (with one species) and *Linderina* (with two species) are known. The Kickxellales in temperate regions is, when compared with those from the tropics, are relatively well researched (Zycha *et al.*, 1969), and although these fungi are relatively rare the above profile suggests that there is greater morphological diversity amongst tropical members of this order.

Conclusions

From the data available there is no evidence that fungi in the Zygomycetes are more numerous in the tropics than in temperate regions. There is, however, some indication that morphological diversity is greater in the tropics. With a background of continuing destruction of natural habitats in the tropics, work on collecting and describing members of the Zygomycetes from this area has been minimal and urgently needs to be increased.

References

Benjamin, R. K. (1959). The merosporangiferous Mucorales. *Aliso* **4**, 321-433.

Benjamin, R. K. (1963). Addenda to the "merosporangiferous Mucorales". *Aliso* **5**, 11-19.

Benny, G. L. (1973). *A taxonomic revision of the Thamnidiaceae (Mucorales)*. Ph.D. thesis, Claremont Graduate School, California, U.S.A.

Benny, G. L. & Benjamin, R. K. (1975). Observations on Thamnidiaceae (Mucorales). New taxa, new combinations, and notes on selected species. *Aliso* **8**, 301-351.

Buller, A. H. R. (1934). The biology and taxonomy of *Pilobolus*. In *Researches on Fungi*, vol. 6. Longman Green & Co.: London.

Cunningham, D. D. (1895). A new and parasitic species of *Choanephora*. *Annals of the Royal Botanical Garden, Calcutta* **6**, 163-174.

Duddington, C. L. (1973). Zoopagales. In *The Fungi, An Advanced Treatise*, vol. **IVB**, (ed. G. C. Ainsworth, F. K. Sparrow & A. S. Sussman), pp. 231-234. Academic Press: London.

Hawksworth, D. L., Sutton, B. C. & Ainsworth, G. C. (1983). *Ainsworth & Bisby's Dictionary of the Fungi, 7th Ed.* Commonwealth Mycological Institute: Kew.

Hesseltine, C. W. (1953). A revision of the Choanephoraceae. *The American Midland Naturalist* **50**, 248-256.

Hesseltine, C. W. & Ellis, J. J. (1973). Mucorales. In *The Fungi, An Advanced Treatise*, vol. **IVB**, (ed. G. C. Ainsworth, F. K. Sparrow & A. S. Sussman), pp. 187-217. Academic Press: London.

Kirk, P. M. & Kirk, J. P. (1984). *Dimargaris*, a genus new to the British Isles. *Transactions of the British Mycological Society* **82**, 551-53.

Richardson, M. J. & Leadbeater, G. (1972). *Piptocephalis fimbriata* sp. nov., and observations on the occurrence of *Piptocephalis* and *Syncephalis*. *Transactions of the British Mycological Society* **58**, 205-215.

Sinha, S. (1940). A wet rot of leaves of *Colocasia antiquorum* due to secondary infection by *Choanephora cucurbitarum* Thaxter and *Choanephora trispora* Thaxter sp. (=*Blakeslea trispora* Thaxter). *Proceedings of the Indian Academy of Science*, section B **11**, 167-176

Tieghem, P. H. van (1876). Troisième memoire sur les Mucorinées. *Annales des sciences naturelles, Botanique, série 6* **4**, 312-398.

Zycha, H., Siepmann, R. & Linnemann, G. (1969). *Mucorales*. Cramer: Lehre.

Chapter 7

Tropical Xylariaceae: their distribution and ecological characteristics

A. J. S. Whalley

Department of Microbiology, School of Biomolecular Sciences, Liverpool John Moores University, Byrom Street, Liverpool L3 3AF, U.K.

Introduction

In any investigation concerning the general distribution or frequency of occurrence of fungi, the need to be in the right place at the right time is well recognised as a necessary prerequisite for success (e.g. Watling, 1978; Kirk, Chapter 6 this volume). Charting the Xylariaceae is no exception and a further complication, by no means confined to this family, that records are almost exclusively based on their teleomorphic state, requires careful consideration.

The Xylariaceae is a large family of at least 36 genera and of these over 75% have representatives in the tropics (Eriksson & Hawksworth, 1991). Most Xylariaceae produce distinctive anamorphs assignable, for example, to the form genera *Nodulisporium*, *Geniculosporium* and *Xylocladium* (Greenhalgh & Chesters, 1968; Jong & Rogers, 1972) and considerable success has been achieved in recognising endophytic species by their anamorphic form (Petrini & Petrini, 1985). However, even for the well known European taxa the anamorphs are not always sufficiently exclusive to guarantee identity. When investigating tropical members of the family the situation is much more complex since on the basis of current information the majority of taxa can only be recognised by their teleomorphic features and it is doubtful if many identifications can be made confidently by consideration of anamorphic data alone.

The question of occurrence, in the absence of a teleomorph, exposes what is potentially a major problem in any critical study of Xylariaceae since they have been found to be common and widely occurring endophytes of healthy hosts (Petrini & Petrini, 1985; Rodrigues & Samuels, 1990). In a study of *Biscogniauxia* and *Hypoxylon* in the Nordic countries, Granmo *et al.* (1989) found *Hypoxylon fragiforme* (Pers.: Fr.) Kickx to be restricted to the temperate zone with distribution apparently exclusively linked to the distribution of its preferred host, *Fagus*. In contrast, *H. multiforme* (Fr.:Fr.) Fr. was found to be ubiquitous and occurred in all wooded areas throughout the region although it was noted that at its upper limit of latitude mature stromata often failed to develop. Furthermore *H. multiforme* had a broad host range being recorded on 13 of the 16 host genera of the region whilst *H. fragiforme* was virtually confined to *Fagus* (Granmo *et al.*, 1989). There are now indications that *H. fragiforme* and other species of *Hypoxylon* which are recognised as having a narrow host range on the basis of their teleomorph presence occur as endophytes on a steadily increasing host list (Petrini, Stone & Carroll, 1982; Petrini & Petrini, 1985; Rodrigues & Samuels, 1990) and therefore have the potential to extend their range well outside their recognised distribution. To what extent, if any, this phenomenon also occurs in the tropics is unknown. Undoubtedly the Xylariaceae are frequent endophytes of many tropical plants (Dreyfuss & Petrini, 1984; Petrini & Dreyfuss, 1981; Rodrigues & Samuels, 1990), but it might also be argued that with the enormous diversity of potential hosts, coupled with a climate which is often more conducive to fungal growth, the inability to produce a teleomorph will be a much rarer event in the tropics than in most temperate situations. It should be emphasised that apart from this recently recognised endophytic existence the Xylariaceae are best known as saprotrophic wood rotting fungi, as inhabitants of dung or litter or, as is becoming increasingly recognised, as pathogens of a range of plants (Rogers, 1979a; Whalley, 1985).

Another important point to be addressed concerns the widely disjunctive occurrence of what is apparently either the same taxon or at least a very closely related one. Thus *H. chestersii* Rogers & Whalley was based on a collection from North Wales and is notable for its distinctive and unusual longitudinal striations of the ascospore wall (Fig. 1; Rogers & Whalley, 1978). Further collections are now known from Britain, Switzerland and Germany (Petrini & Muller, 1986) whilst a small-spored variety, *H. chestersii* var. *microsporum* J.D. Rogers & Samuels has been described from Brazil (Rogers & Samuels, 1985). As pointed out by Rogers & Samuels (1985), *H. chestersii* was the second species of *Hypoxylon* with distinctly ornamented ascospores known to have varieties on different continental land masses. Prior to this, *Hypoxylon weldenii* J.D. Rogers was described from Louisiana (Rogers, 1977) and its small-spored counterpart, *H. weldenii* var. *microsporum* J.D. Rogers, was later described from Brazil (Rogers, 1980). Rogers & Samuels (1985) also reported the large-spored *H. aeruginosum* var. *macrosporum* J. D. Rogers and they stated: 'It is perhaps noteworthy that in the cases of *H. chestersii* and *H. aeruginosum* as well as *H. weldenii*, the small-spored subtaxa have been described from tropical regions whereas the large-spored subtaxa

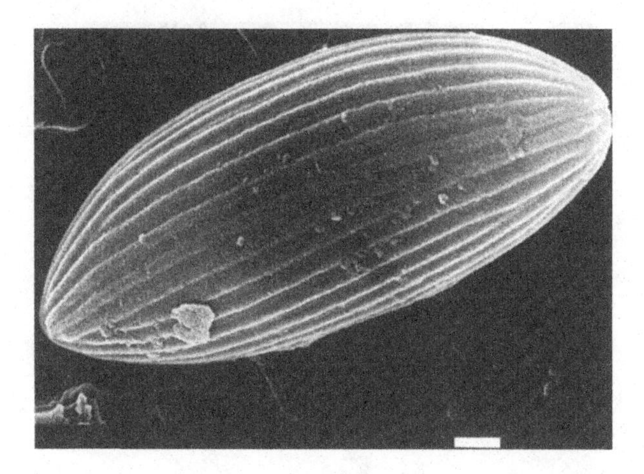

Fig. 1. Ascospore of *Hypoxylon chestersii*. Bar = 1 μm.

have been described from more temperate localities' and they speculated that 'This might indicate that, for a given taxon, larger spores are advantageous in cooler, or seasonally cool, environments and are selectively favoured. Any advantage of larger spores probably involves a greater capacity to store substrates for eventual germination over cool or cold seasons'. This tendency towards smaller-spored varieties in the tropics is difficult to confirm without many more examples and, as Rogers & Samuels (1985) conceded, some of the Xylariaceous taxa with the largest ascospores known to the family are tropical!

Another problem to overcome results from the occurrence of taxa which exhibit only very minor differences between the tropical and temperate forms. Sometimes these differences are so small that for accurate separation the quality of specimens is critical. In *Daldinia*, a genus containing species which possess some of the largest and most conspicuous stromata in the family, *Daldinia concentrica* (Bolt.: Fr.) Ces. & De Not. was considered by Child (1932) to be cosmopolitan whilst the closely related *D. eschscholzii* (Ehrenb.) Rehm inhabited the tropics and subtropics (Child, 1932; Dennis, 1963). Unless stromal material is in good condition the separation of these two species can be uncertain and although cultures of *D. eschscholzii* are characteristic (Petrini & Müller, 1986) its identification is, usually through necessity, based on examination of stromal characteristics, often with material of poor quality. Dennis (1963) considered *D. eschscholzii* to be a tropical analogue of *D. concentrica* which 'seems to differ from it only in its slightly narrower

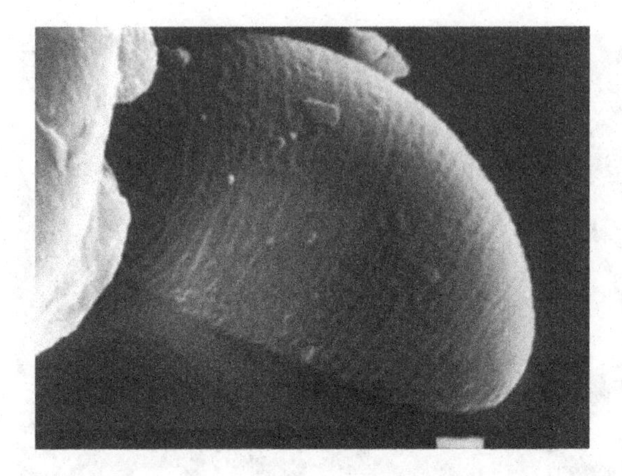

Fig. 2. Ascospores of *Daldinia eschscholtzii*. Bar = 1 μm.

ascospores and I do not think it worthy of specific rank'. The discovery that ascospores of *D. eschscholzii* are striate whilst those of *D. concentrica* are smooth provides an easy means of separation (van der Gucht; see abstract on page 309 and Fig. 2).

There is also the universal question, rarely answered satisfactorily, as to what constitutes a tropical fungus? Responses to this vary according to our state of knowledge of the group being considered, but the firm message that some taxa are truly cosmopolitan whilst others appear to be genuinely restricted to the tropics is gathering momentum. It is against this background of incomplete data and unanswered questions that this Chapter reviews the distribution of selected members of the Xylariaceae in the tropics and elsewhere, and speculates on the reasons for some of the patterns which are now emerging.

Overall geographical considerations

Perhaps because of their size and conspicuous nature the Xylariaceae have been well represented in collections from, and floras of, tropical regions. Studies by Dennis on South American Xylariaceae (Dennis, 1956; 1957; 1970) and more recently by Rogers, Callan & Rossman (1988) and Laessøe, Rogers & Whalley (1989) provide a good platform for comparative studies on distribution. Dennis also published accounts of the Xylariaceae in Africa (Dennis, 1958; 1961; 1963) which together with the papers of Martin (1967; 1968a, b; 1969a, b; 1970); Taligoola & Whalley (1976; 1977) and Whalley & Taligoola (1978) provide a broad background to the family in Africa. Unfortunately the majority of the information is entirely taxonomic and little information about host or habitat is given. However, there are the extensive collections of the late F. C. Deighton from Sierra Leone (housed at International Mycological Institute) and the recent collections of Watling from Korup, Cameroon (housed at Edinburgh) which together with the study of three forests in Uganda by Taligoola & Whalley (1976) provide some of this missing information and allow a picture of the 'likes' and 'dislikes' of different taxa to be built up. Modern studies in South East Asia (Rogers, Callan & Samuels, 1987; Whalley & Jones, unpublished), Papua New Guinea (van der Gucht & van der Veken, 1992; van der Gucht, 1992) and Australia (Whalley & Watling, 1989; Cribb, 1990a, b) together with numerous papers from the Indian subcontinent (e.g. Thind & Waraitch, 1969, Dargon, 1980) all contribute to a worldwide overview of tropical Xylariaceae.

There appears to be little doubt that the Xylariaceae are numerically important inhabitants of many tropical forests. Taligoola & Whalley (1976) in a three year study of the Mabira, Mpanga and Zika forests in Uganda monitored the occurrence of species belonging to *Xylaria* and *Hypoxylon*, including species now contained in *Biscogniauxia*, as a percentage of the total number of ascomycetes collected. They reported values of 31% to 47% for *Hypoxylon* (Taligoola & Whalley, 1976) and 5% to 16% for *Xylaria* (unpublished data) for the three sites surveyed. Seventeen species of *Hypoxylon sensu lato* were reported and differences in species distribution between the three forests were considered to be mainly a consequence of the different host tree species in each habitat. Thus, the Zika forest exhibited a narrower range of *Hypoxylon* species, dominated by *H. nummularium* var. *merrillii* (Bres.) Mill., which occurred mainly on *Piptadeniastrum africanum* Hook. and together with *H. rubiginosum* constituted about 70% of the species of *Hypoxylon* present.

The studies of Dennis (1956; 1957; 1970) also demonstrated that the Xylariaceae are well represented in tropical South America. In his flora of Venezuela and adjacent countries he recorded 14 species of *Camillea*, 47 species of *Xylaria* and 36 species of *Hypoxylon*, including 7 species of *Biscogniauxia*, in addition to representatives of some 10 other xylariaceous genera (Dennis, 1970). In a survey of Cerro de la Neblina, Venezuela, 41 species of *Xylaria* were recognised of which six were described as new (Rogers, Callan & Rossman, 1988). A preliminary comparison with African data indicates that *Xylaria* and many of the other Xylariaceae are better represented in tropical America. In studies of Xylariaceae from the Congo basin 25 species and varieties of *Xylaria* and 20 of *Hypoxylon* were

Table 1. Occurrence of taxa of Xylariaceae in different regions

Region	Number of species recorded		
	Hypoxylon	*Xylaria*	*Camillea*
Venezuela (Dennis, 1970)	36	47	14
Venezuela (Rogers *et al*., 1988)	–	41	–
Sierra Leone	21	18	2
Congo	20	25	–
Uganda	17	–	1
Papua New Guinea	24	–	1
Peninsular Malaysia	16	–	1
North Sulawesi	18	27	–

recorded (Dennis, 1961; 1963) whilst from Uganda a total of 17 *Hypoxylon* taxa was reported by Taligoola & Whalley (1976). Examination of the Deighton collections from Sierra Leone provided a comparative figure of 21 different *Hypoxylon* species (Whalley, unpublished). Thus, comparison of data from three African areas implies an approximate ratio of 2:1 in favour of South America at least for *Hypoxylon*. Comparing *Xylaria* in Venezuela with those from the Congo figures of 45:25 recorded species maintain this ratio (Dennis, 1961; 1970; Rogers, Callan & Rossman, 1988). In an earlier and broader based revision of *Xylaria* from tropical Africa, Dennis (1958) recognised 31 taxa and pointed out that the African species formed a less varied assemblage than those from South America. Extending this comparison to South East Asia a total of 60 Xylariaceous taxa were reported from North Sulawesi, Indonesia; these included 27 of *Xylaria* and 18 of *Hypoxylon* together with representatives of five other genera (Rogers, Callan & Samuels, 1987). In studies of the biota of Papua New Guinea, 24 species and varieties of *Hypoxylon* were recently recorded (van der Gucht & van der Veken, 1992, van der Gucht, 1992). However these accounts from Papua New Guinea do not contain information on members of the primo-cinerea subsection of *Hypoxylon* (Miller, 1961) and the figure given for comparison with other regions is, therefore, artificially low. In a preliminary study of Xylariaceae in peninsular Malaysia 16 species of *Hypoxylon* have been recognised (Whalley & Jones, unpublished). These figures are therefore in line with those for Africa but are considerably less than the impressive lists reported from South

America. A summary of the comparative data for these regions is given in Table 1.

It is important to note that many of the taxa reported from each of the regions compared were based on single collections suggesting that most individual taxa are not really abundant. Furthermore, the analysis of results of a study in Cerro de la Neblina, Venezuela is relevant. Rogers *et al.* (1988) reported that at least half of the 47 taxa of *Xylaria* recorded by Dennis (1970) in his study in Venezuela were not represented in their collections and likewise about half the taxa recognised in their study were not listed by Dennis (1970). Rogers *et al.* (1988) concluded that 'only additional intensive collecting and taxonomic studies will clarify the numbers of *Xylaria* taxa and their distributions based upon host, climatic, and other variables'. This comment clearly relates to other Xylariaceae as well.

Specific geographical considerations

Examination of the distribution of selected species or genera indicates that while many taxa are global in their distribution others genuinely seem to be restricted. Thus, *Camillea* is almost exclusively confined to the Americas with the largest concentration of species occurring in the Amazon region, including the Guianas, although the genus is quite well represented in Central America (Laessøe *et al.*, 1989). *Camillea broomeiana* (Berk. & Curt.) Laessøe, J. D. Rogers & Whalley, *C. punctulata* (Berk. & Rav.) Laessøe, J. D. Rogers & Whalley, *C. signata* (Jong & Benjamin) Laessøe, J. D. Rogers & Whalley and *C. tinctor* (Berk.) Laessøe, J. D. Rogers & Whalley have been recorded from the Southern United States with *C. broomeiana* also in Africa and *C. tinctor* in Africa and Asia. Miller (1961) was relucant to accept the occurrence of *C. tinctor* (as *Hypoxylon*) in Singapore on consideration of its known distribution at that time and Rogers *et al.* (1987) failed to record this or any other *Camillea* species from Sulawesi. Its presence in Singapore has however been confirmed (Laessøe *et al.*, 1989) and in addition several collections have recently been made in peninsular Malaysia (Whalley & Jones, unpublished) whilst van der Gucht (1992) recorded its occurrence on several occasions in Papua New Guinea. *Camillea tinctor* is therefore widely distributed throughout the tropics.

Thamnomyces is a strange genus known only from lowland tropical America except for the single African species *T. camerunensis* Henn. which is distributed from Sierra Leone to Uganda (Dennis, 1961). In this genus the perithecia are characteristically embedded singly or in clusters

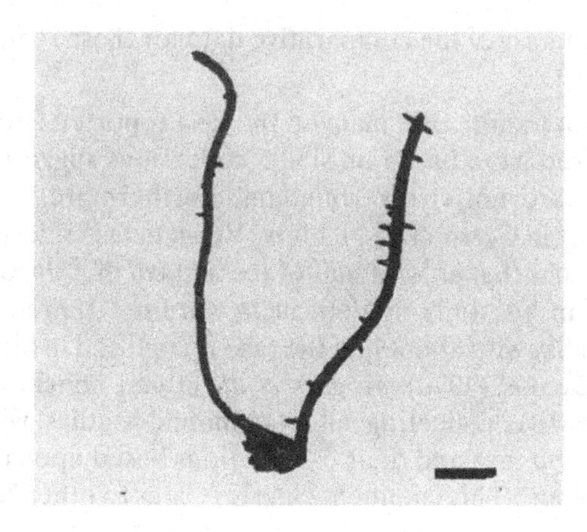

Fig. 3. Stromata of *Thamnomyces chordalis* Frr., Ecuador. Bar = 2 cm.

in the tips of a dendroid stroma or are scattered singly along a wiry axis (Fig. 3). The asci break down within the perithecium and liberate a spore mass which is exuded from the ostiole. Dennis (1957; 1970) suggested that the spores may be unfit for air-borne dispersal and that this might explain this very limited distribution. Whatever the reason, *Thamnomyces* is very much a South American genus.

A similar restriction, in this case in Africa, is exhibited by *Engleromyces* where the single species *E. goetzii* Henn. occurs in Uganda, Kenya and Tanzania (Dennis, 1961). Since the stromata are usually very large, up to 30 cm diameter with a light coloured surface it seems unlikely that this taxon is readily overlooked and its present recorded distribution is probably a fairly accurate assessment of its range. Since bamboos are apparently a specific host their availability will be a major factor influencing distribution. The fact that bamboos occur outside Africa but *Engleromyces* apparently does not is highly suggestive.

If *Thamnomyces* is predominently South American and *Engleromyces* solely African then *Rhopalostroma*, a genus erected to accommodate a curious group of Xylariaceous fungi having stipitate dark brown or black stromata with often abruptly expanded convex heads, belongs to India and South East Asia. Seven of the nine known species occur there whilst the other two are African (Hawksworth, 1977; Hawksworth & Whalley, 1985). *Leprieuria bacillum* (Mont.) Laessøe, Rogers & Whalley is another unusual and distinctive taxon. It is widely distributed in South and Central

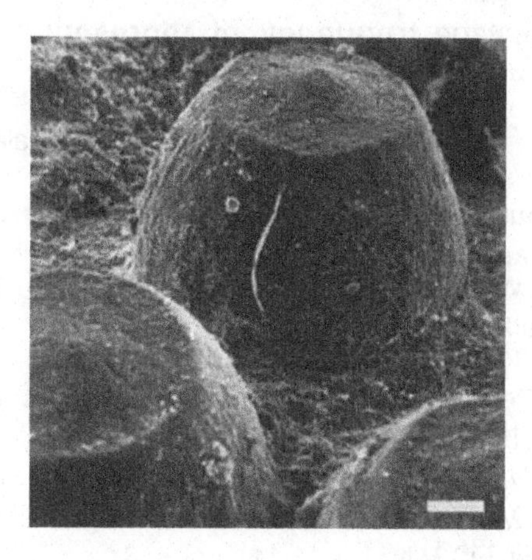

Fig. 4. Stromata of *Hypoxylon megannulata*, Uganda. Bar = 250 μm.

America and in spite of its distinctive form has not yet been reported from outside that region (Laessøe *et al.*, 1989).

Whilst there are several genera which apparently have specific but relatively broad distribution patterns, it is noteworthy that there are a considerable number of species or monotypic genera known only from their type localities or from the immediate vicinities. Thus *Versiomyces cahuchocosus* Whalley & Watling is from Queensland, Australia, *Hypoxylon megannulatum* Taligoola & Whalley from Uganda (Fig. 4) and *Camillea stellata* Laessøe, J. D. Rogers & Whalley from two localities in Amazonian Peru (Taligoola & Whalley, 1977; Laessøe *et al.*, 1989; Whalley & Watling, 1989) so it is unlikely that all of these taxa are that restricted and a combination of intensive collecting and taxonomic studies will be required to confirm or rectify individual situations. This is well illustrated by *Biscogniauxia citriforme* (Whalley, Hammelev & Taligoola) van der Gucht & Whalley, originally described from a single locality in Nigeria (Whalley, Hammelev & Taligoola, 1988), then found on several occasions in the Korup Rain Forest, Cameroon by Watling (pers. comm.) and very recently recorded from Malaysia (Whalley & Jones, unpublished). Now a large spored variety has been reported from Papua New Guinea (van der Gucht & Whalley, 1992) and its occurrence elsewhere should be anticipated.

If the emphasis so far has concentrated on taxa with restricted distributions it is equally important to appreciate that many members of the Xylariaceae are widely distributed in the tropics, the tropics and

Table 2. Percentage frequency of representative species of *Biscogniauxia*, *Camillea* and *Hypoxylon*

Species	Sierra Leone	Uganda	Malaysia	Papua New Guinea
*B. nummularia** (Bull.:Fr.) O. Kuntze	8	25	10	14
C. tinctor (Berk.) Laessøe, J. D. Rogers & Whalley	7	+	+	8
C. broomeiana (Berk. & Curt.) Laessøe, J. D. Rogers & Whalley	2	–	–	–
*H. rubiginosum** (Pers.:Fr.) Fr.	36	23	6	8
*H. sclerophaeum** Berk. & Curt.	7	1	–	2
H. investiens (Schw.) Curt.	7	18	10	9
H. haematostroma Mont.	4	–	–	+
H. truncatum (Schw.:Fr.) Mill.	11	3	+	15
H. stygium (Lév.) Sacc.	9	6	21	19
H. thouarsianum (Lév.) Lloyd	–	8	–	–
H. hypomiltum Mont.	2	1	–	2
H. jecorinum Berk. & Rav.	1	1	–	–
H. crocopeplum Berk. & Curt.	–	–	–	+
H. nectrioideum Trot. & Sacc.	1	–	3	+
H. oodes Berk. & Br.	–	–	–	9
H. bovei Speg.	–	–	9	–
H. bovei var. *microspora* Mill.	–	+	+	+
H. archeri Berk.	–	–	–	3
H. subannulatum P. Henn. & E. Nym.	–	–	3	2
H. macroannulatum Ito & Imai	–	–	–	+

* includes varieties

– =not recorded; + = present at less than 1%

subtropics or universally. *Hypoxylon rubiginosum* was stated by Miller (1961) to be cosmopolitan, occurring in all climatic zones and on a wide host range. Although there is some discussion on the true circumscription of the species with the probable recognition of a number of varieties (Petrini & Müller, 1986; Granmo *et al*., 1989). *H. rubiginosum* appears equally at home in the tropics, subtropics and temperate zones. However, *H. rubiginosum* var. *dieckmannii* (Theiss.) Mill. is a truly tropical fungus (Miller, 1961). *Hypoxylon stygium*, and *H. investiens* are further examples of taxa which are very common and widespread species throughout the tropics and subtropics and probably occur whereever there is a suitable substratum. Comparison of percentage occurrence of species of *Hypoxylon*, *Camillea* and *Biscogniauxia* recorded from a number of geographical areas indicates that the most abundant taxa are, perhaps not surprisingly, the most widespread (Table 2). Thus *H. rubiginosum* and its varieties, *Biscogniauxia, H. stygium* and *H. investiens* are the most common taxa overall. *H. rubiginosum* can clearly be seen to be much more common in Africa than either Malaysia and Papua New Guinea whilst for *H. stygium* the reverse is true.

Ecological aspects

Although knowledge of the ecological activities of the Xylariaceae is limited (Whalley, 1985) there are sufficient data to indicate trends amongst certain species or genera. Laessøe *et al.* (1989) reported that in Western Amazonia species of *Camillea* occurred mainly in lowland localities and only extended into the low subtropics of the Andes. Furthermore, they were found to be inhabitants of sun-exposed sites such as natural clearings, man-made fellings or on recently dead trunks and branches. Often the substratum was observed to have been attacked by termites. It was also noted that *Camillea* had a tendency to occur on branches still hanging in the tangle above ground. *Biscogniauxia*, *Hypoxylon* and *Daldinia* also appear to prefer a dry situation unlike *Xylaria* and *Kretzschmaria* which are rarely found in these conditions. Taligoola & Whalley (1977) found *Biscogniauxia* and *Hypoxylon* to be fungi of drier sites in Ugandan forests and reported that '*H.* (*Biscogniauxia*) *nummularium* var. *merrilli* (Bres.) Mill. and *H. thouarsianum* are primary colonizers being found mainly on fairly sound, clean material either on the forest floor or on branches still attached to trees. *Hypoxylon quisquiliarum* Mont. and *H. investiens* most frequently occur on highly decomposed wood which is still in direct contact with soil.' Watling (pers. comm.) in his studies in Korup, Cameroon also found that varieties of *Biscogniauxia nummularia*, *Biscogniauxia citriforme* and *C. tinctor* occurred on sound wood in dry situations and were usually associated with poles cut from forest vegetation for use in shelters or as handrails over difficult water. *Kretzschmaria clavus* (Fr.) Sacc., the most common and widespread member of the Xylariaceae in Korup, was usually found to be associated with large fallen branches or logs (R. Watling, pers. comm.). Recent studies in Malaysia have shown *H. nectriodeum* to be a frequent inhabitant of mangroves (A. J. S. Whalley & E. B. G. Jones, unpublished) and it has since been reported from similar situations in South Africa and Hongkong (E. B. G. Jones, unpublished).

Fig. 5. Stromata of *Xylaria ianthino-velutina* on fallen fruit of *Flindersia australis* R. Br., Queensland, Australia. Bar = 1 cm.

In Queensland, Australia *H. archeri* typically occurs in dry conditions and was not found deep in the rain forest and in general species of *Hypoxylon* and *Biscogniauxia* occurred in clearings or by trails. In contrast species of *Xylaria* mainly inhabited more inaccessible places with dense vegetation and high moisture conditions.

Although the majority of Xylariaceae inhabit wood there are a number of genera such as *Hypocopra* and *Podosordaria* which are dung inhabitants and they do not appear to include species which are especially tropical; those found in tropical zones exist at high elevation (Krug & Cain, 1974a, b). There are also several interesting seed inhabiting species of *Xylaria*, of which a number grow on fallen fruits and pods and are usually well represented in the tropics and subtropics. *Xylaria ianthino-velutina* Mont. (Fig. 5) is extremely widespread on a wide range of fruits in both tropical and subtropical situations (Rogers, 1979b).

In considering the activities of members of the Xylariaceae in nature it should be stressed that the vast majority are wood inhabitants and cause saprotrophic decomposition. The stromata of many of them appear very soon after the death of the branch or tree, often occurring on branches which are still attached. They are believed to appear very rapidly as a result of the latent invasion of the host followed by a change in conditions which now favours the fungus. Many species of *Biscogniauxia* and *Hypoxylon sensu stricto* can behave in this manner (Whalley, 1985). Not all Xylariaceae are saprotrophs and there are a number of important

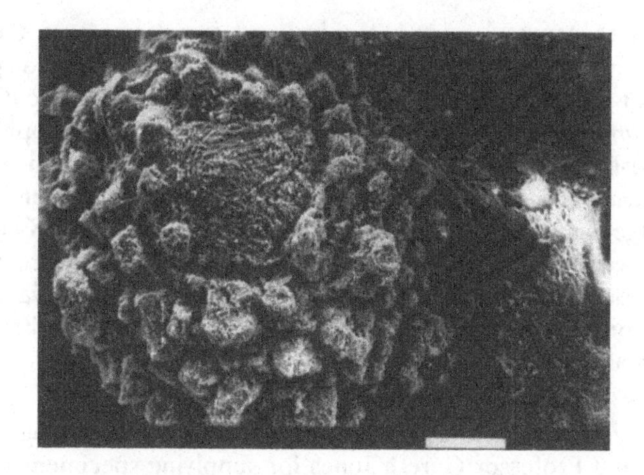

Fig. 6. Stromata of *Rosellinia bunodes* on root of *Hibiscus* sp., Puerto Rico. Bar = 1 mm.

pathogenic species in the family which mainly cause canker or root rot diseases (Whalley, 1985). *Rosellinia necatrix* Prill., *R. bothrina* (Berk. & Br.) Sacc. and *R. bunodes* (Berk. & Br.) Sacc. (Fig. 6) are important root infecting species in the tropics causing diseases of a variety of commercially important plants (Whalley, 1985; Teixeira de Sousa & Whalley, 1991). Interestingly, species of *Rosellinia* are rarely found in tropical rainforests (Rogers *et al*., 1987), but this might in part be a result of the apparent reluctance of some species to produce perithecia in nature (Teixeira de Sousa & Whalley, 1991). Two other species which could be important tropical pathogens are *Ustulina zonata* (Lé.) Sacc., which is responsible for a serious collar rot of *Hevea* in Malaysia (Varghese, 1971), and *Kretzschmaria clavus* now recognised as an important pathogen of macadamia (*Macadamia integrifolia* Maiden & Betche) in Hawaii (Ko, Kunimoto & Maedo, 1977). The latter is a widespread fungus throughout the tropics growing on many different forest trees and its potential as a significant pathogen of commercial crops in the future should not be underestimated (Ko, Tomita & Short, 1986).

Conclusions

Predictably, some species and genera of the Xylariaceae are widespread throughout the tropics and often the wide distribution displayed reflects an ability to occur also in subtropical and sometimes even temperate regions. There are also reasonably certain, restricted distribution patterns for some taxa which might be associated with inefficient spore dispersal, as in *Thamnomyces*. The vast majority of taxa are saprotrophic wood decomposers but species such as *Kretzschmaria clavus* and *Rosellina necatrix* can cause serious diseases of tropical crops. To extend current knowledge on the distribution and activities of tropical Xylariaceae intensive collecting at the right time of year is essential. Corner (1935) demonstrated that agarics exhibit seasonal fruiting in Malaysia and field observations also indicate that the fruiting of *Xylaria* and other related genera is affected by season.

Acknowledgements It is a pleasure to thank Dr Roy Watling, Thomas Laessøe and Professor Gareth Jones for supplying specimens and data and for many lively discussions about this topic. I also wish to thank Katleen van der Gucht for allowing me to use her important data from Papua New Guinea. Finally I am very grateful to Dr Geoff Hadley for his valuable help in the preparation of this chapter.

References

Child, M. (1932). The genus *Daldinia*. *Annals of the Missouri Botanical Garden* **19**, 429-496.

Corner, E. J. H. (1935). The seasonal fruiting of agarics in Malaya. *Gardens' Bulletin, Straits Settlements* **9**, 79-88.

Cribb, A. B. (1990a). Species of the fungus genus *Xylaria* at Iron Range, Queensland. *Queensland Naturalist* **30**, 17-23.

Cribb, A. B. (1990b). *Xylaria aristata* in Queensland. *Queensland Naturalist* **30**, 87-88.

Dargon, J. S. (1980). The family Xylariaceae in India–a review. *Journal of Indian Botany* **59**, 53-59.

Dennis, R. W. G. (1956). Some Xylarias of tropical America. *Kew Bulletin* **1956**, 401-444.

Dennis, R. W. G. (1957). Further notes on tropical American Xylariaceae. *Kew Bulletin* **1957**, 297-332.

Dennis, R. W. G. (1958). Some Xylosphaeras of tropical Africa. *Revista de Biologia* **1**, 175-208.

Dennis, R. W. G. (1961). Xylarioideae and Thamnomycetoideae of Congo. *Bulletin Jardin Botanique de l'Etat* (Bruxelles) **31**, 109-154.

Dennis, R. W. G. (1963). Hypoxyloideae of Congo. *Bulletin Jardin Botanique de l'Etat* (Bruxelles) **33**, 317-343.

Dennis, R. W. G. (1970). *Fungus flora of Venezuela and adjacent countries*. Kew Bulletin Additional series 3. HMSO: London.

Dreyfuss, M. & Petrini, O. (1984). Further investigations on the occurrence and distribution of endophytic fungi in tropical plants. *Botanica Helvetica* **94**, 33-40.

Eriksson, O.E. & Hawksworth, D.L. (1991). Outline of the ascomycetes–1990. *Systema Ascomycetum* **9**, 39-271.

Granmo, A., Hammelev, D., Knudsen, H., Laessøe, T., Sasa, M., & Whalley, A. J. S. (1989). The genus *Biscogniauxia* and *Hypoxylon* (Sphaeriales) in the Nordic countries. *Opera Botanica* **100**, 59-84.

Greenhalgh, G. N. & Chesters, C. G. C. (1968). Conidiophore morphology in some British members of the Xylariaceae. *Transactions of the British Mycological Society* **51**, 57-82.

Hawksworth, D. L. (1977). *Rhopalostroma*, a new genus in the Xylariaceae s.l. *Kew Bulletin* **31**, 421-431.

Hawksworth, D. L. & Whalley, A. J. S. (1985). A new species of *Rhopalostroma* with a *Nodulisporium* anamorph from Thailand. *Transactions of the British Mycological Society* **84**, 560-562.

Jong, S. C. & Rogers, J. D. (1972). Illustrations and descriptions of conidial states of some *Hypoxylon* species. *Washington Agricultural Experiment Station Technical Bulletin* **71**, 1-51.

Ko, W. H., Kunimoto, R. K. & Maedo, I. (1977). Root decay caused by *Kretzschmaria clavus*: its relation to macadamia decline. *Phytopathology* **67**, 18-21.

Ko, W. H., Tomita, J. & Short, R. L. (1986). Two natural hosts of *Kretzschmaria clavus* in Hawaiian forests. *Plant Pathology* **35**, 254-255.

Krug, J. C. & Cain, R. F. (1974a). A preliminary treatment of the genus *Podosordaria*. *Canadian Journal of Botany* **52**, 589-605.

Krug, J. C. & Cain, R. F. (1974b). New species of *Hypocopra* (Xylariaceae). *Canadian Journal of Botany* **52**, 809-843.

Laessøe, T., Rogers, J. D. & Whalley, A. J. S. (1989). *Camillea*, *Jongiella* and light-spored species of *Hypoxylon*. *Mycological Research* **93**, 121-155.

Martin, P. (1967). Studies in the Xylariaceae: I. New and old concepts. *Journal of South African Botany* **33**, 205-240.

Martin, P. (1968a). Studies in the Xylariaceae: III. South African and foreign species of *Hypoxylon* sect. Entoleuca. *Journal of South African Botany* **34**, 153-199.

Martin, P. (1968b). Studies in the Xylariaceae: IV. *Hypoxylon*, sections Papillata and Annulata. *Journal of South African Botany* **34**, 303-330.

Martin, P. (1969a). Studies in the Xylariaceae: V. *Euhypoxylon*. *Journal of South African Botany* **35**, 149-206.

Martin, P. (1969b). Studies in the Xylariaceae: VI. *Daldinia*, *Numulariola* and their allies. *Journal of South African Botany* **35**, 267-320.

Martin, P. (1970). Studies in the Xylariaceae: VIII. *Xylaria* and its allies. *Journal of South African Botany* **36**, 73-138.

Miller, J. H. (1961). *A monograph of the world species of* Hypoxylon. University of Georgia Press: Athens, GA, U.S.A.

Petrini, L. E. & Müller, E. (1986). Untersuchungen über die Gattung *Hypoxylon* (Ascomycetes, Xylariaceae) und verwandte Pilze. *Mycologia Helvetica* **1**, 501-639.

Petrini, L. E. & Petrini, O. (1985). Xylariaceous fungi as endophytes. *Sydowia* **38**, 216-234.

Petrini, O. & Dreyfuss, M. (1981). Endophytische Pilze in epiphytischen Araceae, Bromeliaceae und Orchidaceae. *Sydowia* **34**, 135-148.

Petrini, O., Stone, J. K. & Carroll, F. E. (1982). Endophytic fungi in evergreen shrubs in western Oregon: a preliminary study. *Canadian Journal of Botany* **60**, 789-796.

Rodrigues, K. F. & Samuels, G. J. (1990). Preliminary study of endophytic fungi in a tropical palm. *Mycological Research* **94**, 827-830.

Rogers, J. D. (1977). A new *Hypoxylon species* with appendaged, ornamented ascospores. *Canadian Journal of Botany* **55**, 372-375.

Rogers, J. D. (1979a). The Xylariaceae: systematic, biological and evolutionary aspects. *Mycologia* **71**, 1-42.

Rogers, J. D. (1979b). *Xylaria magnoliae* sp. nov. and comments on several other fruit-inhabiting species. *Canadian Journal of Botany* **57**, 941-945.

Rogers, J. D. (1980). *Hypoxylon weldenii* var. *microsporum* and *H. punctidiscum*. *Mycologia* **72**, 829-832.

Rogers, J. D., Callan, B. E. & Samuels, G. J. (1987). The Xylariaceae of the rain forests of North Sulawesi (Indonesia). *Mycotaxon* **29**, 113-172.

Rogers, J. D., Callan, B. E. & Rossman, A. Y. (1988). *Xylaria* (Sphaeriales, Xylariaxeae) from Cerro de la Neblina, Venezuela. *Mycotaxon* **31**, 103-153.

Rogers, J. D. & Samuels, G. J. (1985). New taxa of *Hypoxylon*. *Mycotaxon* **22**, 367-373.

Rogers, J. D. & Whalley, A. J. S. (1978). A new *Hypoxylon* species from Wales. *Canadian Journal of Botany* **56**, 1346-1348.

Taligoola, H. K. & Whalley, A. J. S. (1976). The genus *Hypoxylon* in Uganda forests. *Transactions of the British Mycological Society* **67**, 517-519.

Taligoola, H. K. & Whalley, A. J. S. (1977). *Hypoxylon megannulatum* sp. nov. A new species of *Hypoxylon* from Uganda. *Transactions of the British Mycological Society* **68**, 298-300.

Teixeira de Sousa, A. J. & Whalley, A. J. S. (1991). Induction of mature stromata in *Rosellinia necatrix* and its taxonomic implications. *Sydowia* **43**, 281-290.

Thind, K. S. & Waraitch, K. S. (1969). Xylariaceae of India–I. *Proceedings of the Indian Academy of Science B* **70**, 131-138.

van der Gucht, K. (1992). Contribution towards a revision of the genera *Camillea* and *Biscogniauxia* (Xylariaceae, Ascomycetes) from Papua New Guinea. *Mycotaxon* **44**, 275-299.

van der Gucht, K. & van der Veken, P. (1992). Contribution towards a revision of the genus *Hypoxylon* s. str. (Xylariaceae, Ascomycetes) from Papua New Guinea. *Mycotaxon* **44**, 275-299.

van der Gucht, K. & Whalley, A.J.S. (1992). A new variety and combination for *Hypoxylon citriforme*. *Mycological Research* **96**, 895-896.

Varghese, G. (1971). Infection of *Hevea brasiliensis* by *Ustulina zonata* (Lev.) Sacc. *Journal of the Rubber Research Institute of Malaya* **23**, 157-163.

Watling, R. (1978). The distribution of larger fungi in Yorkshire. *Naturalist* **103**, 39-57.

Whalley, A. J. S. (1985). The Xylariaceae: some ecological considerations. *Sydowia* **38**, 369-382.

Whalley, A. J. S. & Taligoola, H. K. (1978). Species of *Hypoxylon* from Uganda. *Transactions of the Royal Botanical Society of Edinburgh (Cryptogamic Centenary Symposium)* **42**, 93-98.

Whalley, A. J. S., Hammelev, D. & Taligoola, H. K. (1988). Two new species of *Hypoxylon* from Nigeria. *Transactions of the British Mycological Society* **90**, 139-141.

Whalley, A. J. S. & Watling, R. (1989). *Versiomyces cahuchucosus* gen. et sp. nov. from Queensland, Australia. *Notes, Royal Botanic Garden, Edinburgh* **45**, 401-404.

Chapter 8

Diversity and phylogenetic importance of tropical heterobasidiomycetes

F. Oberwinkler

Lehrstuhl Spezielle Botanik und Mykologie, Universität Tübingen, Auf der Morgenstelle, 1, D-7400 Tübingen 1, Germany

Introduction

Heterobasidiomycetous fungi from the tropics are especially important for the interpretation of natural relationships and evolutionary trends in basidiomycetes. Most heterobasidiomycetous groups, especially the simple-pored auricularioid taxa, inclusive of rusts and Septobasidia, the tremelloid and tulasnelloid fungi, are highly diverse in tropical areas. Many rust genera and the majority of Septobasidia associated with scale insects are only known from the tropics; several genera of auricularioid, tremelloid and tulasnelloid fungi appear to be restricted to tropical regions, and all species of the Cryptobasidiales occur in tropical and subtropical areas. In addition several stilboid fungi from the tropics could be reinterpreted and assigned to basidiomycetes, and the recently described *Cryptomycocolax abnorme* Oberwinkler & Bauer from Costa Rica is a most important taxon for a better understanding of the evolution of basidiomycetes in general.

Heterobasidiomycetous fungi have a surprisingly high percentage of diverse and important characteristics. (1) There are transversely, longitudinally and obliquely septate basidia and also holobasidia. In all groups active spore discharge and passive spore release are known. (2) Basidiospores can germinate by budding, production of ballistospores (secondary spores), microconidia or hyphae. (3) Life cycles can follow the normal basidiomycetous behaviour of a short haplophase and an extensive dikaryophase. However, in rusts the haplontic stage is extended considerably. (4) Anamorph stages are also rather diverse in Heterobasidiomycetes, especially in Uredinales. (5) Evolution of the basidiome produced most of the principal types of structure with corticioid, pulvinate, clavarioid, hydnoid or porioid, and surprisingly also lamellate, hymenia. (6) Gastroid taxa are found in most heterobasidiomycetous groups. (7) Main types of septal pore, a variation of simple pores, dolipores without parenthesomes, and others with cup-like, continuous or perforated ones occur in Heterobasidiomycetes. (8) Spindle pole body (SPB) structures and cycles are also remarkably diverse. These nuclear associated organelles are essential for spindle formation and nuclear division. (9) Recently, several unique cell organelles have been discovered in Heterobasidiomycetes, viz. colacosomes and symplechosomes. (10) There is a high diversity of haustorial types in both mycoparasites and plant pathogens. (11) 5S rRNA primary and secondary structures are so heterogeneous that more than 50% nucleotide exchanges occur in 'distant' Heterobasidiomycetes.

Such diversity of principal characteristics are unknown in Homobasidiomycetes. Consequently, Heterobasidiomycetes have to be considered as (a) heterogeneous and (b) not as 'derived' as the Homobasidiomycetes. There are several very important character transformations which are sometimes correlated with each other, indicating a general, i.e. organismic evolutionary strategy. These are: (i) repeated derivation of holobasidia from phragmobasidia; (ii) loss of flexibility in spore germination and loss of an ontogenetic yeast stage; (iii) a gradual increase in complexity of septal pore types, and (iv) considerable conformity of these character transformations with 5S rRNA patterns. Therefore, heterobasidiomycetous fungi play an essential role in the understanding of evolutionary processes of basidiomycetes in general.

Brief comments follow on Cryptomycocolacales, Uredinales, Septobasidiales, Agaricostilbales, Atractiellales, Ustilaginales *sensu lato*, Graphiolales, Exobasidiales, Cryptobasidiales, Tremellales *sensu stricto*, Auriculariales, Tulasnellales *sensu lato*, and Dacrymycetales.

Cryptomycocolacales

The mycoparasitic *Cryptomycocolax abnorme*, known only from Costa Rica, combines important basidiomycetous and ascomycetous features (Oberwinkler & Bauer, 1990). Septal pores are simple and associated with Woronin bodies, a unique example in basidiomycetes. SPBs appear ascomycete-like in interphase with one or two discs in a flat position, attached to the nuclear envelope and duplication of the disc during interphase occurs by splitting. Bar-like initial stages of SPBs, typical of basidiomycetes, are lacking. However, in late interphase and prophase, conspicuous midpieces, typical of basidiomycetes, occur. Hyphae bear clamp connections, undoubtedly a basidiomycetous feature. The mycoparasitic interaction is established via colacosomes (Oberwinkler & Bauer, 1989; Bauer & Oberwinkler, 1991), cell organelles which are so far only known from basidiomycetes. Basidial ontogeny of *C. abnorme* is unique. A subulate basidium develops a terminal spore-like body, finally abstricted by a transverse septum. This developmental stage represents a two-celled phragmobasidium. The first meiotic division provides nuclei for both basidial cells. The second meiotic divisions begin shortly before the apical cell is released. The attached, basal basidial cell then develops a gasteroid holobasidium with successively formed, terminal, sessile basidiospores. Further development of the second basidial cell which has been released before sporulation of the basal cell is unknown. Systematically *C. abnorme* has an interesting position, with implications for other species. Septal pore and SPB features indicate a very early phylogenetic position, the most primitive one so far discovered among basidiomycetes. The presence of clamps also appears to be an original characteristic. Consequently, basidiomycetes without clamps, e.g. rusts, Septobasidia, and many other hetero- and homobasidiomycetous taxa, may be regarded as derived. Obviously, mycoparasitic interaction with colacosomes is also an ancient character, found only in a few other heterobasidiomycetes, for example in the genus *Colacogloea* (Oberwinkler *et al.*, 1991).

Finally, *C. abnorme* confirms that phragmobasidia have to be interpreted as phylogenetically old, and holobasidia as derived. An unsolved problem, however, is whether passive spore release in *C. abnorme* is already derived or represents an evolutionary stage before the unique basidiomycetous mechanism of ballistospore abstriction has been evolved. Since this mechanism is already present in rusts, all other basidiomycetes with passive spore release have to be derived. There are many well documented examples of convergent evolution of gasteroid taxa in both hetero- and homobasidiomycetous fungi.

Uredinales

The inventory of tropical rusts is still fragmentary. The known tropical taxa, however, display enormous diversity in morphology and host selectivity. There are good reasons to assume that clues for understanding the origin of Uredinales and of systematic relationships within the rusts essentially depend on the knowledge of representatives from the tropics. For example, there are many autoecious tropical rusts and a considerable number of species lack teliospores or have thin-walled probasidia.

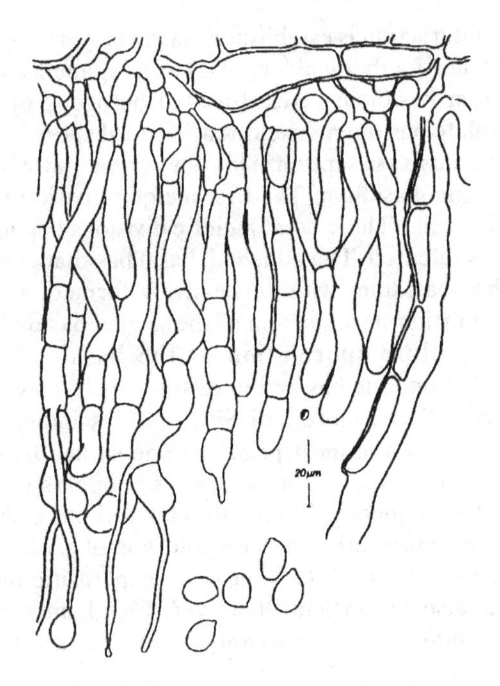

Fig. 1. *Goplana micheliae* Racib.: basidia in different stages of development. Note that teliospores are lacking.

The following discussion aims to demonstrate the diversities and phylogenetics of tropical rusts. The information has been extracted from Cummins (1959) and Cummins & Hiratsuka (1983).

Rusts constitute a unique group within the basidiomycetes. Typical Uredinales have a macrocyclic and heteroecious ontogeny. All species are obligate parasites of cormophytes, i.e. ferns, gymnosperms, and angiosperms. Some 10 species are known in the genus *Goplana*, rusts without teliospores (Fig. 1), i.e. auricularioid basidia lacking probasidia, from tropical and subtropical regions. The type species, *G. micheliae* Racib., occurs on *Michelia*, a genus of the Magnoliaceae. On *Mikania hirsutissima* DC. in Costa Rica there are additional tropical rusts without teliospores, e.g. *Chrysella mikaniae* H. Sydow, the only species of its genus with pedicellate basidia. Similar basidial structures are found in the five tropical species of *Achrotelium*. Three to four neotropical species constitute the genus *Chrysocyclus* with unusual basidia, developed from apical and subapical cells on a common pedicel but without probasidial telia.

Rusts with *Cystobasidium*-type meiosporangia are potentially interesting for phylogenetic considerations in the Uredinales. There are several tropical representatives, such as *Botryorhiza hippocrateae* Whetzel & Olive, the single species of its genus, occuring in the neotropics and known only in its telial and basidial stages. More than 20 species from tropical regions may constitute *Maravalia* (including *Scopella*, according to Cummins & Hiratsuka, 1983) with similar meiosporangia.

Remarkably diverse rusts with telial columns occur in the tropics and subtropics, e.g. *Cerotelium* with the pantropical *C. fici* (Butl.) Arthur. Predominantly restricted to tropical and subtropical Bignoniaceae are the five species of the genus *Phragmidiella*. Some 15 tropical species constitute *Crossopsora*, and approximately 10 telial column-forming species of tropical origin are included in *Skierka*. Species of the genus *Masseeella* appear to be restricted to tropical southern and southeastern parts of Asia, and the Philippines. *Trichopsora tournefortiae* Lagerh. on neotropical *Tournefortia* (Heliotropiaceae) represents a monotypic genus. Only three species are included in the neotropical *Cionothrix*, parasitizing Asteraceae. The endophylloid *Dietelia* with aecidium-like telia (Buriticá & Hennen, 1980) comprises five species from the neotropics and one from the Philippines. Two-celled, catenulate teliospores, separated by intercalary cells and forming telial columns are known from *Pucciniosira*, *Didymopsora*, and *Gambleola*, the latter considered to be congeneric with *Pucciniosira* by Buriticá (1974), thus comprising 17 species, seven of which occur in the neotropics on Tiliaceae, Solanaceae, and Asteraceae. In *Didymopsora* six species, growing on Melastomataceae, Tiliaceae, Solanaceae, and Asteraceae in Africa and the Neotropics are recognized. Telial columns are due to the pedicel-like, long intercalary cells in the four species of *Chardoniella*, all pathogenic on Asteraceae in South America. Columns of *Kernella lauricola* (Thirum.) Thirum. on *Litsea* (Lauraceae) from tropical Asia, are composed of puccinioid teliospores, probably indicating a closer relationship to *Gymnosporangium*.

Cummins & Hiratsuka (1983) proposed the Uropyxidaceae to accommodate several genera, most of tropical origin, e.g. *Dasyspora*, a monotypic genus on neotropical Annonaceae. *Didymopsorella*, comprising two species on *Toddalia* (Rutaceae), is restricted to southern India. *Dipyxis* is represented by two species on Bignoniaceae from Mexico and Brazil. Also *Newinia* is recorded exclusively from bignoniaceous hosts in Africa and Burma. Four species comprise *Porotenus*, the type species, *P. concavus* Viégas on Bignoniaceae in Brazil. Some 50 species are included in *Prospodium* which predominantly occur on tropical and subtropical Bignoniaceae and Verbenaceae. Also most species of the genus *Uropyxis* are restricted to warmer regions.

The majority of rusts assigned to the Raveneliaceae occur on Fabales in tropical and subtropical regions and all are autoecious. Most of these genera are defined by teliospore characteristics. There are several monotypic genera, e.g. *Apra*, *Diabole*, *Diorchidiella*, and *Lipocystis*, all occurring on *Mimosa* in various parts of the neotropics. Additional monotypic genera, *Anthomyces* from Brazil, and *Cystomyces* from Costa Rica are recorded from unidentified legumes. The majority of more than 150 species of *Ravenelia* occur also on Fabales in the tropics and subtropics. The eight species of the related genus *Kernkampella* occur preferentially on Euphorbiaceae in warmer regions.

Haplophragmium is another genus with autoecious species on legumes. Some 15 species are known from tropical Africa and Asia. Most of the 18 species of *Sphaerophragmium* also occur on Fabales in the tropics and subtropics.

Several rusts with unique characteristics and isolated taxonomic positions are known from warmer regions. Some 50 species included in *Hemileia* occur mainly in Africa and Asia, pathogenic on di- and monocotyledoneous hosts, including the economically most important coffee rust and generic type, *Hemileia vastatrix* Berk. & Br. Two species are recognized in the genus *Alveolaria*, both occurring on *Cordia* in the neotropics. Their teliospores form short columns, composed of radially arranged, plate-like cells. *Chrysopsora gynoxidis* Lagerh. on *Gynoxis* (Asteraceae) from South America and the single species of the genus is unique in basidial arrangement: terminal and subterminal cells are probasidia and develop into basidia. The four recognized superstomatal species of the genus *Desmella* parasitic on tropical ferns are of considerable importance for the interpretation of coevolutionary processes. Similar uredinial and telial stages are found in the three species of the genus *Edythea*, which grow on *Berberis* in South America. Cummins & Hiratsuka (1983) included also *Cerradoa palmaea* Hennen & Ono, a palm rust on *Attalea ceraensis* Barbosa Rodrigues from Brazil, in *Edythea*. The monotypic *Cumminsia* known from *Grewia* (Tiliaceae) in Africa has multicellular teliospores.

Species of the genera *Olivea* (*ca* 10 species on Euphorbiaceae, Lamiaceae and Verbenaceae), *Phacopsora* (*ca* 50 on diverse angiosperms), and *Physopella* (*ca* 20 on di- and monocotyledons) occur predominantly in warmer regions.

Kweilingia bambusae (Teng) Teng, the single representative of the genus, parasitic on bamboos in China, has not been studied in sufficient detail for proper systematic placement. The genus was assigned to the Ustilaginales by Teng (1940), who, however, referred to a close relationship with the rust genus *Chrysomyxa*. Cummins & Hiratsuka (1983) treated *Kweilingia* in the Uredinales, in contrast to Thirumalachar & Narasimhan (1951) who placed it in the Auriculariales. The genus was tentatively included again in the Ustilaginales *sensu lato* by Vánky (1987).

Septobasidiales

The unique biology of species of *Septobasidium* arises from their obligatory association with scale insects, and their growth together on living spermatophytes. The majority of species has been described from the tropics and warmer regions (Couch, 1938). Couch not only fully discussed the biology of Septobasidia but also gave an excellent taxonomic treatment. Only a few additional studies have been carried out since then.

All *Septobasidium* species (Fig. 2) grow internally in insect bodies, emerge from them and develop hyphal wefts covering the animals and parts of the bark of living trees and shrubs around them. During further development, corticioid to stereoid basidiomes are formed which are perennial in many species. The arid to tough fructifications are composed of efibulate hyphae with simple septal pores. Commonly the basidia are auricularioid, i.e. transversely septate. They develop from probasidia that often persist. In other cases, no structurally conspicuous probasidia are formed. Basidiospores are released forcibly at maturity, and they germinate by budding or by repetition, i.e. producing secondary ballistospores. Many spores become septate after release from the basidia.

There are several important aspects concerning tropical *Septobasidium* species. Their biodiversity is certainly much higher than documented today and, because of the fragmentary inventory, the group is in urgent need of field studies. In addition, most species are endangered, since they are associated with tropical forest ecosystems.

The taxonomic position of the Septobasidiales was discussed in detail by Couch (1938). The close relationship of the rusts, Septobasidiales, and some auricularioid taxa, especially the *Herpobasidium* group, can also be supported by ultrastructural characteristics. Dykstra (1974) and Sebald (1977) studied the ultrastructure of the septal pore apparatus in several species of the genus, which share simple septal pores.

In his monograph, Couch (1938) recognized only one genus in the Septobasidiales. Raciborski's (1909) genera *Mohortia* and *Ordonia* were listed as synonyms. However, there is much more taxonomic diversity,

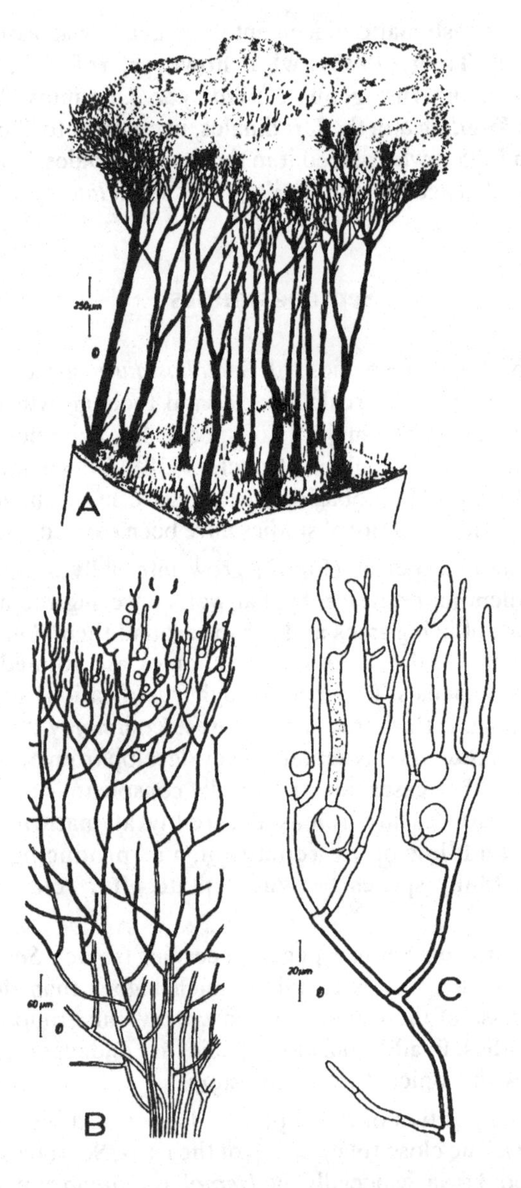

Fig. 2. *Septobasidium robustum* Boed. & Steinm. A: semi-diagrammatic drawing of the gross morphology of a portion of a basidiome showing basal hyphal stroma, pillars, and upper loose hyphal network bearing scattered basidia. B: hyphal arrangement of upper part of pillar and of top layer showing loose network and positions of basidia. C: portion of hymenium with hyphidia, probasidia, basidia of different ages, and basidiospores.

especially in tropical representatives of the Septobasidiales. *Ordonia orthobasidion* Racib. differs from *Septobasidium* species in having conspicuous setae, thick-walled probasidia and one-layered growth of basidiomes (Oberwinkler, 1989). Owing to similar structural characteristics, *Uredinella spinulosa* Couch & Petch (Couch 1941) has been transferred to *Ordonia* (Oberwinkler, 1989). Although obviously well delimited by the association with scale insects and morphological features, the Septobasidia are not isolated systematically. When proposing *Uredinella coccidiophaga*, Couch (1937) referred to its spore forms which resemble strikingly certain rusts, but in its parasitism on scale insects it resembles *Septobasidium*.

The genus *Goplana* was considered by Couch (1938) as another possible connection between *Septobasidium* and the Uredinales. The teleomorphs of the genus are similar to some teliosporeless *Septobasidium* species. However, uredospore anamorphs easily distinguish the type species, *Goplana micheliae* Racib., from *Septobasidium* species. This rust is also pathogenic on flowering plants, growing on *Michelia* (Magnoliaceae). Additional species of rthe genus are parasitic on spermatophytes. No associations with scale insects have been recorded.

Originally, *Platycarpa polypodii* (Couch) Couch was described as *Septobasidium polypodii* (Couch, 1929). However, Couch (1938) excluded this species from the genus in his monograph on *Septobasidium* because of its growth on fern sporophylls and the absence of scale insects beneath the stroma. Couch (1949) also referred to the intermediate position of *Platycarpa* between the Septobasidiales and the Auriculariales. Another unusual and inconspicuous species, *Coccidiodictyon inconspicuum* Oberwinkler, has been described from Tenerife (Oberwinkler, 1989). Morphologically it strongly resembles efibulate species of *Cystobasidium*. However, association with scale insects and septobasidioid haustoria are lacking in *Cystobasidium*. Nevertheless, it can be assumed that these taxa have evolved from a common ancestral group.

In summary, organismic diversity is high in tropical and subtropical Septobasidiales, therefore systematic and phylogenetic interpretations should be based on these taxa.

Simple-pored auricularioid taxa

The systematics of heterobasidiomycetous taxa with auricularioid basidia and simple septal pores but not belonging to rusts, smuts and Septobasidia, is not yet well understood. Again, a considerable number

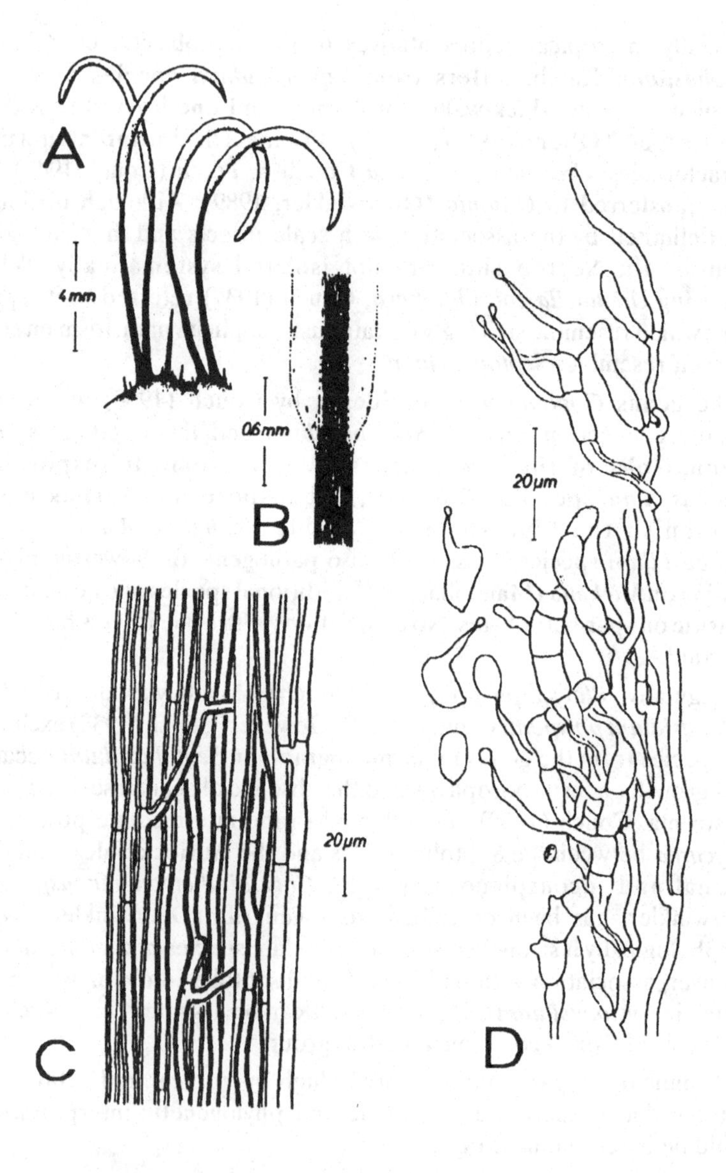

Fig. 3. *Neotyphula guianensis* Wakefield. A: habit sketches of basidiomes. B: sectional view of upper portion of stalk and lower part of hymenium, showing transgression of stalk hyphae into inner region of fertile part. C: hyphal arrangement of central part of basidiome. D: portion of hymenium, showing basidia of different ages, and basidiospores, two germinating by repetition.

of species, known only from the tropics, play an important role with regard to taxonomic questions.

Heterobasidiomycetous plant pathogens occurring on mosses and ferns may provide more information on phylogenetically primitive, auriculorioid fungi. Species of *Iola*, growing on sporophytes of mosses, are known only from the tropics. Auricularioid fern parasites are restricted to three genera, two of which, *Platycarpa and Ptechetelium* (Oberwinkler & Bandoni, 1984), occur exclusively in tropical regions.

Though rarely collected, it appears that *Eocronartium muscicola* (Pers.: Fr.) Fitzp. has a wide distribution. Two additional clavarioid, auricularioid genera, viz. *Neotyphula* (Fig. 3) and *Paraphelaria* are only known from the tropics. Both have simple pored hyphal septa, thus indicating relationship with *Eocronartium*. Species of *Paraphelaria* appear to be pathogens of bamboos, and *Neotyphula guianensis* Wakefield is considered to grow saprobically on wood, however, interactions of substratum hyphae in this species have not yet been studied.

All taxa discussed so far share important taxonomic characteristics, i.e. efibulate hyphae, simple septal pores, and predominantly parasitic growth on plants.

The tropical genus *Phyllogloea* was based on *P. singeri* Lowy (Lowy, 1971), a gelatinous, carrot-red species with lobate basidiomes and fibulate hyphae with simple septal pores, auricularioid basidia, and amphigeneous hymenia. The relationship of this taxon is unclear. A second species, *P. javanica* (Part.) Lowy, is recorded from Java. *P. tremelloidea* Lowy (Lowy, 1971), from Brazil, shares general structural features with the type species. However, the basidiospores bud in a tremelloid manner and the septa have tremelloid dolipores. The combination of such characteristics is unexpected and undoubtedly places *P. tremelloidea* in the Tremellales *sensu stricto*.

Kryptastrina inclusa Oberwinkler (Oberwinkler, 1990) was found as an intrahymenial parasite whilst examining a neotropical, corticioid fungus. This species is characterized by extremely scanty and indistinct hyphae and obscure attachment of these to the host hyphae. The basidia are apparently single and scattered in the host fruiting bodies, and the pyramid-like, tetrahedric spores are unique within heterobasidiomycetous fungi.

Such enigmatic fungi, found by chance, clearly exemplify our fragmentary knowledge of these taxa and the importance of detailed microscopic studies of collections from the tropics.

Agaricostilbales

Species of *Agaricostilbum* are easily circumscribed by the stalked-capitate, dry basidiomes and the gasteroid, auricularioid basidia capable of multiple spore production. The genus was originally described (Wright, 1970) as belonging to the Deuteromycetes, but more detailed studies (Wright *et al.*, 1981; Oberwinkler & Bandoni, 1982a) led to the conclusion that the genus is heterobasidiomycetous. *Agaricostilbum* species appear to grow only on palms. Other records are dubious and are in need of further critical examination. *Agaricostilbum* occurs in tropical and subtropical regions because of the host distribution. For phylogenetic considerations of basidiomycetes in general, *A. pulcherrimum* (Berk. & Br.) Brady, Sutton & Samson plays an important role. There are 40 to 62 nucleotide exchanges compared with other basidiomycetes (Walker, 1984; Gottschalk & Blanz, 1985), indicating an isolated position. Another unusual and unexpected characteristic is endospore production in a yeast phase. One strain producing endospores, has been studied with the transmission electron microscope. Ultrastructural characteristics of the yeast cell wall and yeast budding clearly show a basidiomycetous nature. It is rather surprising that budding cells can also produce endospores. Obviously both processes can occur in a short sequence of time but not fixed in a regular order. Budding appears to start after endospore formation. Several four-nucleate stages were found, but synaptonemal complexes could not be detected. Therefore, it is uncertain whether *Agaricostilbum* endospore formation is tied to meiosis or is the result of mitoses. It is important to answer this question in order to understand the life cycle and its nuclear phases, and phylogenetically, the origin of the basidiomycetous meiosporangium.

The ultrastructure of the septal pores of the above genus is unique and difficult to interpret. In *A. pulcherrimum* a simple pore is associated frequently with electron dense, globose vesicles which resemble ascomycetous Woronin bodies without peripheral membranes (Oberwinkler & Bauer, 1989). Additional data from meiosis and the spindle pole body cycle (Bauer *et al.* 1992) led to the conclusion that *Agaricostilbum* is derived with respect to the rusts. Its taxonomic position appears to be between the rusts, auricularioid fern parasites and associated taxa, as well as the smuts. Such an interpretation, however, contrasts with data derived from 5S rRNA sequences.

From studies of additional collections from California, Indonesia, and Australia, it appears that a morphological species concept for *Agaricostilbum* is difficult to assess. Further, culture and mating experiments should be carried out to obtain a better understanding of species delimitations.

In summary, the number of species is unknown but the genus *Agaricostilbum* is an excellent example of the importance of tropical fungal taxa for a better understanding of heterobasidiomycetous phylogeny.

Atractiellales

The Atractiellales includes auricularioid fungi with forcible spore discharge and related taxa with gasteroid basidia. Though only scattered collections are available, a world-wide distribution with a centre of diversity in the tropics can be assumed for these fungi. The order was originally introduced (Oberwinkler & Bandoni, 1982a) to accommodate those species which differ markedly from *Auricularia*, the type genus of the Auriculariales. The *Auricularia-Hirneola* group is distinguished by its peculiar basidiomes (which are unlike those of other genera currently treated in the Auriculariales) by the curved microconidia formed by germinating basidiospores, and by dolipores with parenthesomes.

Atractiella and *Agaricostilbum* species have in common basidiome gross morphology, hyphal structure, septation, and arrangement, as well as septal pore ultrastructure. They differ markedly in basidiospore ontogeny and germination, and in a wide range of various substrata. Basidiospores are budded off singly from a basidial cell in species of *Atractiella*, whilst the multiple-budding that occurs in *Agaricostilbum* species has never been observed in species of *Atractiella*.

The Atractiellales in an amended circumscription (Oberwinkler & Bauer, 1989) comprises auricularioid Heterobasidiomycetes with resupinate and stalked-capitate fructifications. The septal pore is simple but associated with vesicle-like globules, each having an electron dense peripheral layer and an electron translucent inner part. Membrane complexes, symplechosomes, are regularly present in well-developed, cytoplasmatic cells, thus characterising the Atractiellales with a unique ultrastructural feature. The interphase-prophase SPBs have discs.

There are several additional species from tropical regions, originally described in other genera, e.g. the Brazilian *Pilacrella delectans* Möller, the Javanian *Hoehnelomyces javanicus* Weese, and *Hoehnelomyces macrosporus* (Penz. & Sacc.) Boedijn (= *Sphaeronemella macrospora* Penz. & Sacc.). Both *Pilacrella and Hoehnelomyces* have to be regarded as synonyms of *Atractiella* (Oberwinkler & Bandoni, 1982a). It was surprising to find symplechosomes also in species of the genus

Helicogloea, including *Saccoblastia* (Oberwinkler & Bauer, 1989). These genera have a worldwide distribution, but *Helicogloea* appears to be considerably diverse in tropical regions (Baker, 1936, 1946).

Ustilaginales

Comparatively few smuts have been recorded from the tropics, however several of them might be of considerable importance for phylogenetic interpretations, e.g. *Melanotaenium oreophilum* H. Sydow on *Selaginella* spp. from India. The smuts are pathogenic on plants, especially angiosperms; they share probasidial teliospores (smut spores). Many smuts have a haploid and saprobic yeast phase. Meiosporangial types are very diverse, indicating that smuts may be heterogeneous. Today, 55 genera are accepted (Vánky, 1987) and there may be approximately 1000 species. There are several unusual tropical smuts with unknown basidia, for example *Angiosorus solani* Thirum. & O'Brien and *Polysaccopsis hieronymi* (Schröter) P. Hennings on neotropical Solanaceae and the monotypic *Cintractiella* on *Hypolytrum* (Cyperaceae) which is known only from New Guinea. There is also *Kuntzeomyces ustilaginoideus* (P. Hennings) P. Hennings on cyperaceous *Rhynchospora* spp. from South America, the single species of the genus, in which germination of smut spores is unknown.

Cintractia comprises predominantly tropical and subtropical species on Cyperaceae. All four known species of *Dermatosorus* occur on Cyperaceae; they are restricted to tropical and subtropical Asia and Australia. *Franzpetrakia microstegiae* Thirum. & Pavgi, the single species of the genus, is known only from the type locality, the Mussoorie Hills in northern India. Additionally, *Jamesdicksonia obesa* (H. & P. Sydow) Thirum., Pavgi & Payak on *Dichanthium annulatum* (Forsk.) Stapf, *Mundkurella heptapleuri* Thirum. on *Heptapleurum venulosum* Seem. (Araliaceae), and *Narasimhania alsimatis* Pavgi & Thirum. on *Alisma* sp., represent monotypic genera from India.

Smut systematics still suffers from insufficient information on meiosporangia. Therefore, smuts from the tropics should be collected, cultivated, and studied more extensively to elucidate phylogenetic connections.

Rather curious germination types (basidia) are reported in *Georgefischeria* (Narasimhan *et al.*, 1963), a genus with two species on Convolvulaceae in tropical south and southeast Asia. Three species of *Pericladium*, again with unusual meiosporangia, are reported from Tiliaceae and one species from Piperaceae in Africa and southern Asia.

The examples mentioned above underline the diversity of tropical smuts and their importance for a more concise understanding of natural relationships in the group.

Graphiolales

The Graphiolales represents an excellent example of a fungal group of exclusively tropical and subtropical distribution and a taxon whose biological diversity is unknown.

A study of the ultrastructural characteristics of *Graphiola phoenicis* (Moug.) Poit., typical of the genus and the most representative of the order, has revealed important basidiomycetous features, such as dikaryotic hyphae, exogenic meiospores, yeast budding with scars, and simple septal pores without Woronin bodies (Oberwinkler *et al.*, 1982). Basidia and basidiomes are unique not only in heterobasidiomycetous fungi but in all basidiomycetes.

All known species of the Graphiolales are pathogenic on palms and are therefore restricted to tropical and subtropical regions. The Graphiolales has not been monographed, therefore the taxonomy of the group is fragmentary. *Graphiola phoenicis* has mainly been recorded from species of the genus *Phoenix*. *Graphiola cylindrica* Kobayasi occurs on *Trachycarpus fortunei* (Hook.) H. Wendl. in Japan (Kobayasi, 1952). *Graphiola congesta* Berk. & Rav. is known from *Sabal palmetto* (Walt.) Lodd. in Florida and the West Indies, and *Graphiola thaxteri* E. Fischer grows on a *Sabal* sp. in Florida. In India, *Borassus flabellifer* L. harbours *Graphiola borassi* Sydow & Butler. An additional genus in the Graphiolales, *Stylina*, is also restricted to palms and its taxonomic status is uncertain.

Exobasidiales

Species of the Exobasidiales share several unique characteristics: (i) holobasidia have two to eight (occasionally more than eight) straight, stout sterigmata; (ii) basidiospores develop asymmetrically but in contrast to all other basidiomycetes, they are bent inwards; (iii) mature basidiospores are transversely septate and germinate with small and elongate microconidia; (iv) a yeast-like single cell stage can develop from microconidial propagation; (v) they are plant pathogens of angiosperms with basidia developing outside the host tissues; (vi) there is a preferential

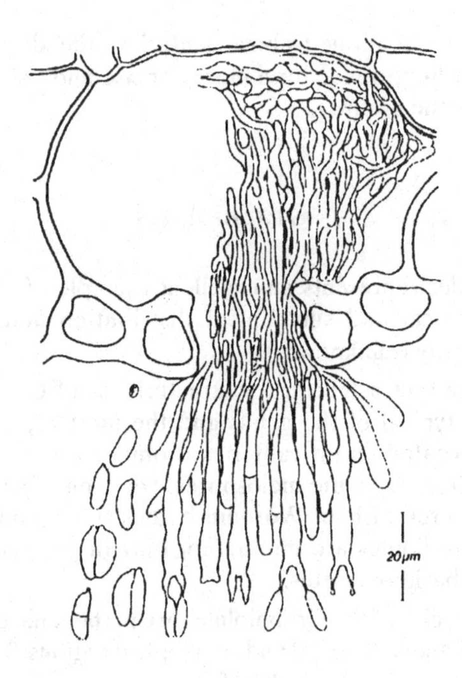

Fig. 4. *Kordyana tradescantiae* (Pat.) Racib. Section of basidiome arising through stomatal opening of *Tradescantia* sp. Note basidia at different developmental stages and basidiospores.

host range on Ericales, especially Ericaceae and Epacridaceae but also Empetraceae; additional species are known as pathogens on Lauraceae, Aquifoliaceae, Rutaceae, Symplocaceae, Theaceae, and surprisingly also on Commelinaceae, Cyperaceae, Poaceae and Palmae. Species of two genera are exclusively known from tropical regions, viz. *Kordyana* (Fig. 4) with a pantropical distribution on Commelinaceae and *Brachybasidium* from Java on *Pinanga* (Palmae). The latter has been assigned to a separate order, Brachybasidiales by Donk (1964). However, the morphology of the type species, *Kordyana tradescantiae* (Pat.) Racib. (Fig. 4) and *Brachybasidium pinangae* (Racib.) Gäumann is so similar in essential cellular details that there is no justification for such taxonomic treatment. Furthermore, the probasidial swellings of *B. pinangae* are so minute that it appears arbitrary to use this feature for taxonomic distinctions at family or even generic levels.

Descriptions of two additional, monotypic genera, *Proliferobasidium*, and *Ceraceosorus*, with tropical species have been published by

Cunningham *et al.* (1976) and assigned to the Brachybasidiaceae. *Proliferobasidium heliconiae* Cunningh. has proliferating basidia and is pathogenic on leaves of *Heliconia bihai* L. (Heliconiaceae) in Dominica. Basidial proliferation is not known in other taxa of the Exobasidiales, but it is not uncommon in certain holobasidial fungi, e.g. *Oliveonia, Repetobasidium* and *Hymenochaete. Ceraceosorus bombacis* (Bakshi) Bakshi appears to be rather similar to *Dicellomyces,* and in fact was originally placed in this genus by Bakshi (Bakshi *et al.,* 1972). The species is pathogenic on leaves of *Bombax ceiba* L. (Bombacaceae) in India. Again, it is obvious that tropical taxa play an important role in the understanding of systematic relationships. Likewise, it can be assumed that these inconspicuous fungi may be much more diverse than the few taxa that are known at present.

Cryptobasidiales

Only scattered and limited information is available for heterobasidiomycetous fungi grouped in the Cryptobasidiales. Four genera have been recognized, *Botryoconis* (= *Cryptobasidi*um), *Clinoconidium*, *Coniodyctium* and *Drepanoconis*, all recorded from tropical and subtropical regions. The single species of the genus *Coniodyctium, C. chevalieri* Har. & Pat., grows on *Ziziphus* species (Rhamnaceae) in dry areas of Africa, while species of *Botryoconis, Clinoconidium,* and *Drepanoconis* have been recorded from lauraceous hosts of the neotropics. Only *Coniodyctium chevalieri* has been studied in detail (Malençon, 1953).

There are no further taxonomic points to be added tothose made by Oberwinkler (1978) concerning comparative morphological and ultrastructural features. Gastroid holobasidia with terminal, approximately sessile spores and simple septal pores are considered to be the most important structural features. These characteristics are distinctive for placement in the Heterobasidiomycetes.

On the basis of a few records and the heterogeneous structures of cryptobasidealean fungi it is tempting to assume that there is a greater diversity than known at present. It is obvious that the tropics are the potential source for this group of fungi.

Rajendren (1969) placed three species of *Muribasidiospora* in the Muribasidiosporaceae. The type species *M. indica* Kamat & Rajendren is a leaf pathogen of *Rhus mysurensis* Heyne in India. Basidial proliferation and dimorphic basidiospores were considered of generic importance (Rajendren, 1967) and sclerotial bodies, radial fan-shaped mycelium and

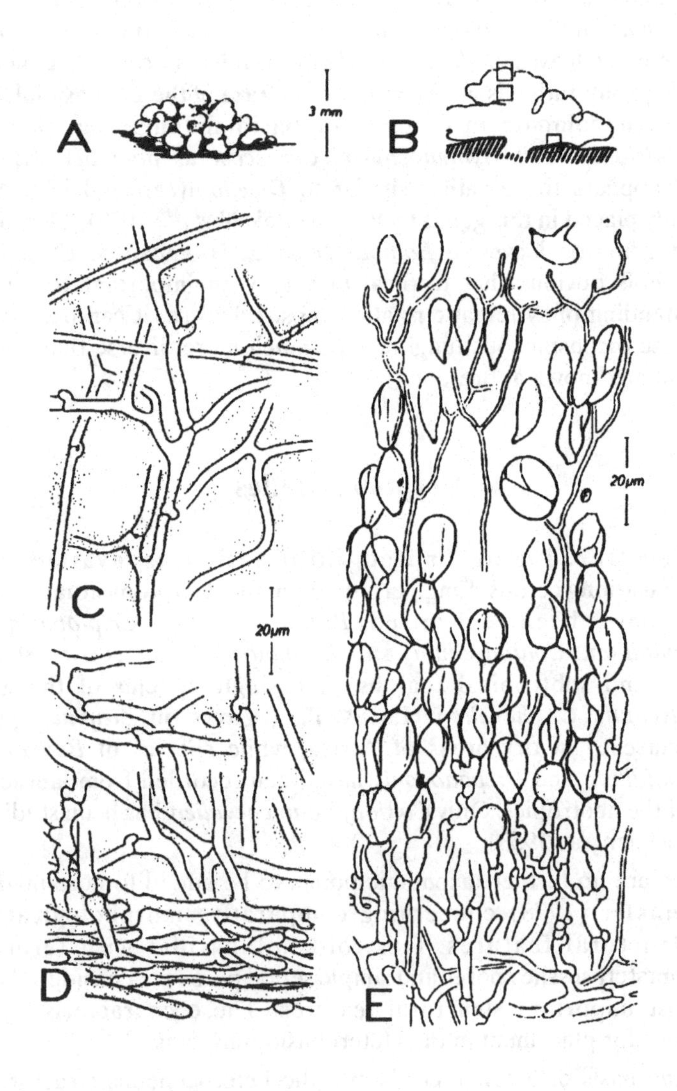

Fig. 5. *Sirobasidium sanguineum* Pat. & Lagerh. A: habit sketch of a basidiome. B: section through the basidiome with marks indicating positions of detailed drawings C-E. C: hyphal arrangement of inner part of fructification. D: detail of sterile surface of the basidiome. E: section through the hymenium with chains of basidia.

abundant blastospores in culture were used to justify a separate family in the Hymenomycetes. However, these features are not sufficient for final taxonomic conclusions.

Tremellales

The most distinctive features of species grouped in the Tremellales *sensu stricto* (Bandoni, 1984) are: (i) dolipores with parenthesomes composed of cup-shaped units; (ii) basidiospores capable of multilateral budding, and (iii) mycoparasitism with tremelloid haustoria (Oberwinkler & Bandoni, 1982b). The majority of species has tremelloid, i.e. longitudinally septate, basidia. However, meiosporangia are remarkably heterogeneous in Tremellales.

In his dissertation, Bandoni (1957a) studied some 250 taxa described as species of *Tremella*. Approximately 30 of these appear to be of tropical distribution. Considerable diversity of microscopic structures was found in a further study of some species (*T. anomala, T. auricularia, T. dysenterica, T. fibulifera* and *T. spectabilis)* described by Möller (1895) from Brazil (Bandoni & Oberwinkler, 1983).

Rick's genus *Zanchia* (Rick, 1958), also from Brazil, has not been recognized since it was described. The basidia of *Z. sanctae-mariae* Rick are sphaeropedunculate and irregularly, often transversely or obliquely, septate.

Another unique basidial type is found in species of the tropical genus *Sirobasidium* (Fig. 5): the meiosporangia are catenate, producing deciduous epibasidia (primary sessile basidiospores) which bud or germinate with sterigmata and secondary ballistospores (Bandoni, 1957b; Flegel, 1976). Since comparative studies are lacking, the taxonomy of *Sirobasidium* species is unclear.

As mentioned above, *Phyllogloea tremelloidea* Lowy from Brazil (Lowy, 1971), with auricularioid basidia, has spores which bud in the tremellaceous manner and the dolipore ultrastructure is identical with the *Tremella*-type. Thus, the scope of *Tremella* must be amended to include species with auricularioid basidia.

The genus *Holtermannia* (= *Clavariopsis*) was proposed to incorporate clavarioid species with *Tremella*-like microscopic details (Holtermann, 1898). Kobayasi (1937) monographed the genus and recognized six species, two from Japan and four from the tropics, viz. *H. damaecornis* (Möller) Kobayasi from Brazil, *H. pinguis* (Holterm.) Sacc. & Trev., the generic type from Java, *H. prolifera* (Pat.) Kobayasi from the Philippines,

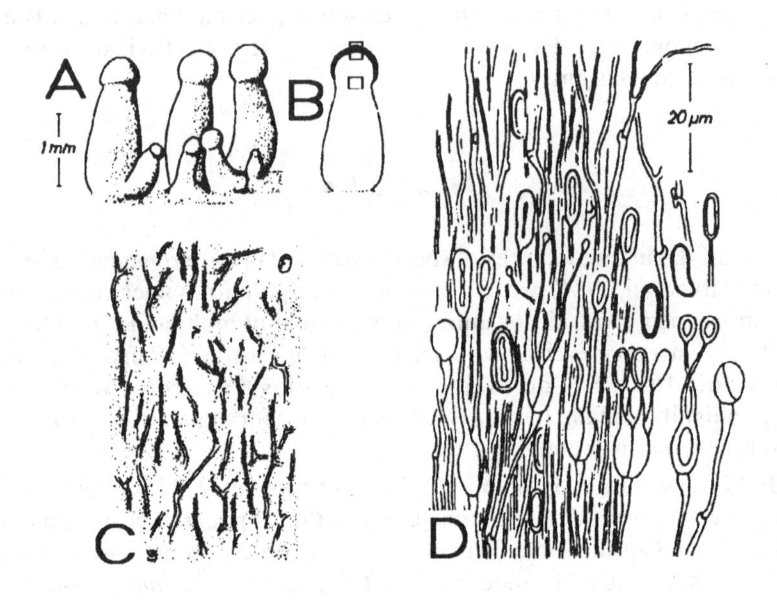

Fig. 6. *Hyaloria pilacre* Möller. A: habit sketches of basidiomes. B: section of basidiome, marks indicating positions of detail drawings. C: hyphal arrangement of inner part of fructification. D: section through the hymenium with gastroid basidia and basidiospores.

and *H. pulchella* (Pat. & Har.) Kobayasi from New Caledonia. The generic delimitation of *Holtermannia* and the taxonomy of species deserve further detailed studies.

Syzygospora alba Martin, recorded from Panama and Mexico, has been recognized as a mycoparasite with tremelloid haustoria (Oberwinkler & Lowy, 1981). The species is predominantly holobasidiate, but some basidia are apically partially septate, indicating a possible relationship with *Tremella sensu stricto*. No further tropical species are known, but monographs have been published for related taxa from temperate regions by Oberwinkler & Bandoni (1982b) and Ginns (1986).

Occurrence of species of the Tremellales *sensu stricto* appears to be very frequent in the tropics, but only a few have been studied adequately. These species show remarkably great structural diversity, investigation of which is important for better understanding of their systematic interrelationships.

Auriculariales

In an amended interpretation (Bandoni, 1984), the Auriculariales have been said to comprise phragmobasidiate species with dolipores and continuous parenthesomes; however, tremelloid haustoria, budding basidiospores and yeast stages, characteristic of Tremellales *sensu stricto*, are unknown. Thus, certain taxa with auricularioid and tremelloid basidia are united in an apparently monophyletic group. Such regrouping of more natural taxa is easier to understand when tropical species are taken into account. For example, basidia in the neotropical, resupinate *Patouillardina cinerea* Bres. are obliquely or irregularly septate. In another, yet undescribed species of *Patouillardina* from Venezuela, auricularioid and tremelloid basidia are developed in a single basidiome. Such species can be considered as missing links between structurally distinct taxa.

Another unexpected and unique meiosporangial type occurs in *Hyaloria* (Möller, 1895). The two known, neotropical species, *H. pilacre* Möller (Fig. 6) and *H. traillii* (Berk. & Cooke) Martin have gastroid basidia with long sterigmata but symmetrically attached basidiospores, embedded in a matrix of hyphidia forming the capitate part of the stalked basidiome. Collections of both species have been reported from dead or rotting palms. *Hyaloria* is the single gasteromycetous genus of the Auriculariales, thus an important taxon to exemplify convergent evolution of gasteroid fungi.

Protomerulius brasiliensis Möller was the first porioid phragmobasidiate fungus described as such. It was recorded from fallen trunks of *Jaracatia dodecaphylla* DC. (Möller, 1895). The widely distributed *Aporpium caryiae* (Schw.) Teixeira & Rogers differs from *P. brasiliensis* in lacking gloeocystidia, the absence of a thin subicular layer, and basidiospore characteristics (Bandoni *et al*., 1982). Lowy (1971) accepted and validated Rick's *Protomerulius richenii* (Rick, 1911), a species known only from Rio Grande do Sul, Brazil. Also *Protohydnum cartilagineum* Möller (Möller, 1895) appears to be restricted to the neotropics; it has been recorded from Brazil, Guyana, Panama, and Mexico (Lowy, 1971).

Several genera of the Auriculariales are much more diverse in the tropics than in temperate regions. Some of these are taxonomically rather difficult, for example species of the genus *Auricularia*. which are common in tropical forests. In a monograph, Lowy (1951) recognized nine species, seven of which are neotropical.

Heterochaete sensu lato is another important genus with mostly tropical species. In a taxonomic study, Bodman (1952) recognized 29 species, 27 of which were exclusively reported from tropical regions.

Tulasnellales

Tulasnelloid taxa share an intermediate position between hetero- and homobasidiomycetous fungi. They are characterized by holobasidia of various types, usually with thick or inflated sterigmata, spores producing secondary spores and dolipores with continuous parenthesomes. *Tulasnella* is a rather homogeneous taxon with badly defined species and only scattered records from the tropics. One of these led to the proposal of a separate genus, *Pseudotulasnella* (Lowy, 1964), because of the remarkable basidial morphology with partly septate meiosporangial apices. The genus is monotypic, the single species known only from one collection from Guatemala. It is obvious that this record is an important one for the understanding of relationships between Tulasnellales and Auriculariales. Another natural relationship can be traced between the genera *Oliveonia* and *Repetobasidium*, the latter generally placed in the Corticiaceae. Both genera are widely spread geographically, but *Oliveonia* appears to prefer tropical and warmer regions. The structural similarities in both taxa, with reference to basidial types with repetitive growths, cystidial structures and basidiome composition, were elucidated by light microscopy (Oberwinkler, 1972). Ultrastructural reinvestigations showed that there were similar dolipores with continuous parenthesomes. So far, secondary spores have not been detected in *Repetobasidium* species. These two examples point out the interesting taxonomic status of the Tulasnellales as an intermediate taxon, connecting Auriculariales and Homobasidiomycetes.

Other curious tulasnelloid taxa have been recorded from the tropics and subtropical regions, e.g. *Heteroacanthella variabile* Oberwinkler & Langer (Oberwinkler *et al.*, 1990) from Taiwan and *Monosporonella termitophila* Oberwinkler & Ryvarden (Oberwinkler & Ryvarden, 1991) from Zambia. Both species share secondary spores and dolipores with continuous parenthesomes. The basidia are either two- and one-spored (*Heteroacanthella*) or exclusively one-spored (*Monosporonella*). From my own collections in the neotropics I am aware of additional tulasnelloid fungi with aberrant meiosporangia. Unfortunately such species are commonly rather inconspicuous and therefore 'collected' only accidentally.

Dacrymycetales

All Dacrymycetales are strong wood decayers and the group has a world wide distribution. The most important characters for all representatives of the Dacrymycetales are basidial ontogeny and morphology. Young basidia first have a slightly clavate form and then expand apically to produce two thick, cylindrical and long sterigmata which taper abruptly to form spiculae on which asymmetrically-attached, large spores develop. The original apex of the young basidium remains visible between the sterigmata, a characteristic which is morphologically distinctive. In nearly all species, mature basidiospores are transversely septate. Commonly, the spores germinate by producing microconidia. Germ-tube formation is also widespread, however, secondary spores are lacking, i.e. basidiospores do not produce ballistospores. Dolipores with continuous parenthesomes (but each with a minute central opening) have been found in all species studied so far.

McNabb (1964-1973) monographed the Dacrymycetales and accepted eight genera. He recognized six species in the genus *Cerinomyces* (McNabb, 1964), two of which are tropical. In *Calocera* 12 species were treated by McNabb (1965a), three of them of tropical origin. Seven species were included (McNabb, 1965b) in *Dacryopinax*, six species occurring in tropical regions. Lowy (1971) added *D. martinii* Lowy from Columbia. *Guepiniopsis* was considered monotypic and of world wide distribution by McNabb (1965c), but few records are known from the tropics. None of the four species in *Heterotextus* have been recorded as yet from tropical areas (McNabb 1965d). One of the two species of each of the genera *Femsjonia* and *Ditiola*, accepted by McNabb (1965e, 1966), is known only from Brazil. However, Lowy (1962) recognized an additional species, *Ditiola coccinea* Lowy from Argentina, and Zang Mu (1983) described *Femsjonia rubra* Zang from high altitudes in subtropical Yunnan and in Sichuan. The genus *Dacrymyces* has a world wide distribution, and McNabb (1973) accepted 31 species in his monograph, most of them known from temperate zones. A few species with unusual spore morphology have been reported from the neotropics, e.g. *D. dictyosporus* Martin from Honduras and Mexico, and *D. falcatus* Brasf. from Panama and Jamaica. Several *Dacrymyces* species are minute and inconspicuous, one of them *D. dendrocalami* Oberwinkler (Oberwinkler & Tschen, 1989) from subtropical Taiwan is visible in the field only under very wet conditions.

In comparison with other heterobasidiomycetous fungi, the Dacrymycetales appear to be of minor diversity in tropical areas.

Acknowledgement The author wishes to thank the German Research Council (DFG) for financial support.

References

Baker, G. E. (1936). A study of the genus *Helicogloea*. *Annals of the Missouri Botanical Garden* **23**, 69-128.

Baker, G. E. (1946). Addenda to the genera *Helicogloea and Physalacria*. *Mycologia* **38**, 630-638.

Bakshi, B. R., Reddy, M. A. R., Puri, Y. N. & Singh, S. (1972). Forest disease survey final technical report 1967-72. Publications of the Forest Research Institute: Dehra Dunn, India.

Bandoni, R. J. (1957a). Taxonomic studies of the Tremellaceae. Dissertation, University of Iowa.

Bandoni, R. J. (1957b). The spores and basidia of *Sirobasidium*. *Mycologia* **49**, 250-255.

Bandoni, R. J. (1984). The Tremellales and Auriculariales, an alternative classification. *Transactions of the Mycological Society of Japan* **25**, 489-530.

Bandoni, R. J. & Oberwinkler, F. (1983). On some species of *Tremella* described by Alfred Möller. *Mycologia* **75**, 854-863.

Bandoni, R., Oberwinkler, F. & Wells, K. (1982). On the poroid genera of the Tremellaceae. *Canadian Journal of Botany* **60**, 998-1003.

Bauer, R. & Oberwinkler, F. (1991). New structures at the host-parasite interface of a mycoparasitic basidiomycete. *Botanica Acta* **104**, 53-57.

Bauer, R., Oberwinkler, F. & McLaughlin, D. J. (1992). Meiosis, spindle pole body cycle and basidium ontogeny in the heterobasidiomycete *Agaricostilbum pulcherrimum*. *Systematic and applied Microbiology* **15**, 259-274.

Bodman, M. C. (1952). A taxonomic study of the genus *Heterochaete*. *Lloydia* **15**, 193-233.

Buriticá, P. (1974). A revision of the rust genera (Uredinales) with reduced life cycles. Ph.D. thesis, Purdue University.

Buriticá, P. & Hennen, J. F. (1980). Pucciniosireae (Uredinales-Pucciniaceae). *Flora Neotropica*, Monograph No. 24: The New York Botanical Garden.

Couch, J. N. (1929). A monograph of *Septobasidium*. Part 1. Jamaican species. *Journal of Elisha Mitchell Scientific Society* **44**, 242-260.

Couch, J. N. (1937). A new fungus intermediate between the rusts and *Septobasidium*. *Mycologia* **29**, 665-673.

Couch, J. N. (1938). *The genus* Septobasidium. Waverly Press: Baltimore, MD, U.S.A.

Couch, J. N. (1941). A new *Uredinella* from Ceylon. *Mycologia* **33**, 405-410.

Couch, J. N. (1949). The taxonomy of *Septobasidium polypodii* and *S. album*. *Mycologia* **41**, 427-441.

Cummins, G. B. (1959). *Illustrated Genera of Rust Fungi*. Burgess Publishing Company: Minneapolis, Minnesota.

Cummins, G. B. & Hiratsuka, Y. (1983). *Illustrated Genera of Rust Fungi*. Revised edition. The American Phytopathological Society: St. Paul, Minnesota.

Cunningham, J. L., Bakshi, B. K., Lentz, P. L. & Gilliam, M. S. (1976). Two new genera of leaf-parasitic fungi (Basidiomycetidae, Brachybasidiaceae). *Mycologia* **68**, 640-654.

Donk, M. A. (1964). A conspectus of the families of Aphyllophorates. *Persoonia* **3**, 199-324.

Dykstra, M. J. (1974). Some ultrastructural features in the genus *Septobasidium*. *Canadian Journal of Botany* **52**, 971-972.

Flegel, T. W. (1976). Conjugation and growth of *Sirobasidium magnum* in laboratory culture. *Canadian Journal of Botany* **54**, 411-418.

Ginns, J. (1986). The genus *Syzygospora* (Heterobasidiomycetes, Syzygosporaceae). *Mycologia* **78**, 619-636.

Gottschalk, M. & Blanz, P. (1985). Untersuchungen an 5S ribosomalen Ribonukleinsauren als Beitrag zur Klarung von Systematik und Phylogenie der Basidiomyceten. *Zeitschrift für Mykologie* **51**, 205-243.

Holtermann, C. (1898). *Mykologische Untersuchungen aus den Tropen*. Berlin.

Kobayasi, Y. (1937). On the genus *Holtermannia* of Tremellaceae. *Science Reports of the Tokyo Bunrika Daigaku*, Section B, **3**, 75-82.

Kobayasi, Y. (1952). On the genus *Graphiola* found in Japan. *Nagaoa* **1**, 32-38.

Lowy, B. (1951). A morphological basis for classifying the species of *Auricularia*. *Mycologia* **44**, 656-692.

Lowy, B. (1962). Contribucion al estudio de los Tremellales de La Argentina. *Lilloa* **31**, 213-228.

Lowy, B. (1964). A new genus of the Tulasnellaceae. *Mycologia* **56**, 696-700.

Lowy, B. (1971). Tremellales. *Flora Neotropica* Monograph No. 6, 1-153.

Malençon, G. (1953). Le *Coniodyctium chevalieri* Har. et Pat. sa nature et ses affinités. *Bulletin de la Société Mycologique de France* **69**, 77-100.

McNabb, R. F. R. (1964). Taxonomic studies in the Dacrymycetaceae 1. *Cerinomyces* Martin. *New Zealand Journal of Botany* **2**, 415-424.

McNabb, R. F. R. (1965a). Taxonomic studies in the Dacrymycetaceae 11. *Calocera* (Fries) Fries. *New Zealand Journal of Botany* **3**, 31-58.

McNabb, R. F. R. (1965b). Taxonomic studies in the Dacrymycetaceae 111. *Dacryopinax* Martin. *New Zealand Journal of Botany* **3**, 59-72.

McNabb, R. F. R. (1965c). Taxonomic studies in the Dacrymycetaceae 1V. *Guepiniopsis* Patouillard. *New Zealand Journal of Botany* **3**, 159-167.

McNabb, R. F. R. (1965d). Taxonomic studies in the Dacrymycetaceae V. *Heterotextus* Lloyd. *New Zealand Journal of Botany* **3**, 215-222.

McNabb, R. F. R. (1965e). Taxonomic studies in the Dacrymycetaceae VI. *Femsjonia* Fries. *New Zealand Journal of Botany* **3**, 223-228.

McNabb, R. F. R. (1966). Taxonomic studies in the Dacrymycetaceae VII. *Ditiola* Fries. *New Zealand Journal of Botany* **4**, 546-558.

McNabb, R. F. R. (1973). Taxonomic studies in the Dacrymycetaceae VIII. *Dacrymyces* Nees ex Fries. *New Zealand Journal of Botany* **11**, 461-524.

Möller, A. (1895). Protobasidiomycetes. *Botanische Mitteilungen aus den Tropen* **Vol. 8**, (ed. A. F. W. Schimper), pp. 1-179. Gustav Fischer Verlag: Jena.

Narasimhan, H. J., Thirumalachar, M. J., Srinavasan, M. C. & Govindu, H. C. (1963). *Georgefischeria*, a new genus of the Ustilaginales. *Mycologia* **55**, 30-34.

Oberwinkler, F. (1972). The relationships between the Tremellales and the Aphyllophorales. *Persoonia* **7**, 1-16.

Oberwinkler, F. (1989). *Coccidiodictyon* gen. nov. and *Ordonia*, two genera in the Septobasidiales. *Opera Botanica* **100**, 185-191.

Oberwinkler, F. (1990). New genera of auricularioid heterobasidiomycetes. *Reports of the Tottori Mycological Institute* **28**, 113-127.

Oberwinkler, F. & Bandoni, R. (1982a). A taxonomic survey of the gasteroid, auricularioid Heterobasidiomycetes. *Canadian Journal of Botany* **60**, 1726-1750.

Oberwinkler, F. & Bandoni, R. (1982b). Carcinomycetaceae, a family in the Heterobasidiomycetes. *Nordic Journal of Botany* **2**, 501-516.

Oberwinkler, F. & Bandoni, R. (1984). *Herpobasidium* and allied genera. *Transactions of the British Mycological Society* **83**, 639-658.

Oberwinkler, F. & Bauer, R. (1989). The systematics of gasteroid, auricularioid Heterobasidiomycetes. *Sydowia* **41**, 224-256.

Oberwinkler, F. & Bauer, R. (1990). *Cryptomycocolax*, a new mycoparasitic heterobasidiomycete. *Mycologia* **82**, 671-692.

Oberwinkler, F. & Lowy, B. (1981). *Syzygospora alba*, a mycoparasitic heterobasidiomycete. *Mycologia* **73**, 1108-1115.

Oberwinkler, F. & Ryvarden, L. (1991). *Monosporonella*, a new genus in the Tulasnellaceae, Basidiomycetes. *Mycological Research* **95**, 377-379.

Oberwinkler, F. & Tschen, J. (1989). A new *Dacrymyces* species from Taiwan. *Transactions of the Mycological Society of Japan* **30**, 349-356.

Oberwinkler, F., Bauer, R. & Bandoni, R. J. (1991). *Colacogloea*, a new genus in the auricularioid Heterobasidiomycetes. *Canadian Journal of Botany* **68**, 2531-2536.

Oberwinkler, F., Langer, E., Burdsall, H. H. & Tschen, J. (1990). *Heteroacanthella*, a new genus in the Tulasnellales. *Transactions of the Mycological Society of Japan* **31**, 207-213.

Oberwinkler, F., Bandoni, R. J., Blanz, P., Deml, G. & Kisimova-Horovitz, L. (1982). Graphiolales, Basidiomycetes Parasitic on palms. *Plant Systematics and Evolution* **140**, 251-277.

Raciborski, M., (1909). Nalistne i pasorzytne grzyby Jawy. Parasitische und epiphytische Pilze Java's. *Bulletin internationale de l'Academie de Cracovie*, Cl. Sci. math. nat. 1909, 346-394.

Rajendren, R.B. (1967). Proliferating basidia in *Muribasidiospora indica*. *Experientia* **23**, 1-4.

Rajendren, R. B. (1969). Muribasidiosporaceae a new family of Hymenomycetes. *Mycologia* **61**, 1159-1160.

Rick, J. (1911). Fungi austro-americani Fasc. XI-XVIII. *Annales Mycologici* **9**, 176-184.

Rick, J. (1958). Basidiomycetes eubasidii in Rio Grande do Sul-Brasilia. *Iheringia* **2**, 1-56.

Sebald, F. (1977). Feinstrukturstudien zur Ontogenie von Arten der Uredinales und verwandter Basidiomyceten. Dissertation, Universität Tübingen.

Teng, S. C. (1940). Supplement to higher fungi of China. *Sinensia* **11**, 105-130.

Thirumalachar, M. J. & Narasimhan, M. J. (1951). Critical notes on some plant rusts. III. *Sydowia* **5**, 476-483.

Vánky, K. (1987). *Illustrated Genera of Smut Fungi*. Gustav Fischer Verlag: Stuttgart & New York.

Walker, W. F. (1984). 5S rRNA sequences from Atractiellales, and basidiomycetous yeasts and fungi imperfecti. *Systematic and Applied Microbiology* **5**, 352-359.

Wright, J. E. (1970). *Agaricostilbum*, a new genus of Deuteromycetes on palm spathes from Argentina. *Mycologia* **62**, 679-682.

Wright, J. E., Bandoni, R. & Oberwinkler, F. (1981). *Agaricostilbum*, an auricularioid basidiomycete. *Mycologia* **73**, 880-886.

Zang Mu (1983). Notes on the genus *Femsjonia* in China. *Mycologia* **75**, 468-471.

Chapter 9

Tropical polypores

Leif Ryvarden

Botany Department, University of Oslo, P.O. Box 1045 Blindern, N-0316 Oslo, Norway

Introduction

Although the tropical mycobiota is still only fragmentarily known, the polypores have been better studied than other groups. Most polypores are easy to collect and dry without being degraded. Thus, they were among the first fungi to be sent home from the tropics. Swartz (1791) and Afzelius (Fries, 1860) started the stream from Jamaica and Nigeria respectively, around 1790, and it has continued unabated to this day. American and European mycologists in the last century and the early part of the present had the notion that polypores were as diversified as higher plants, and this led them to describe new species continuously as collections from the tropics were received in Europe and North America. For example, when a mycologist received a collection from New Guinea, he never troubled to check what had been described from South America. As most polypores have very wide distributions, the same species were repeatedly described as new. *Trametes elegans* (Fr.) Fr. and *Phellinus gilvus*, (Schw.:Fr.) Pat. being very common and widespread in the tropics have, for example, both been named or described as new more than 20 times. Type studies over the last 20 years have solved the problem of the multiplicity of synonyms. The use of names for the more common and widespread species has now stabilized, but the taxonomy of the tropical polypores is still in a state of flux quite unlike the situation in the Northern Hemisphere where a generally accepted taxonomic system is used at generic level. The reason for the unstable system in the tropics is that new species are still being described - often blurring the previously accepted lines between the genera. Many of the currently accepted polypore genera were for historic reasons based on species from the temperate zone, further contributing to the unsuitability of the generic delimitations.

Manuals and monographs

Before intensive and detailed studies can be started on any groups of organism, manuals or monographs are necessary, because these are the gateway to the literature. The lack of pertinent manuals and monographs has been the most important barrier to studies of tropical fungi. This situation is changing though it is far from satisfactory. A preliminary polypore flora of East Africa by Ryvarden & Johansen (1980) is still the most comprehensive and popular manual of tropical polypores. Corner's (1983, 1984, 1987, 1989 a, b and 1991) publications cover many polypores reported from tropical East Asia and will play a very important role in further studies in that and adjacent areas. In addition, the tropical polypores of northern Australia are well covered by Cunningham (1965) in his flora of the polypores of New Zealand and Australia.

However, the polypores of the neotropics are the least known. The only larger manual available covers Venezuela (Dennis 1970) and the polypore part is a compilation of smaller publications. It includes many names that have fallen into synonymy, making it very difficult to use. Few systematic expeditions have been undertaken to the rain forests of South America so only scattered contributions are available, e.g. Venezuela (Fidalgo & Fidalgo, 1968), Colombia (Setliff & Ryvarden, 1983) and Costa Rica (Carranza-Morse, 1991 & 1992). From Brazil considerable collections especially those of O. Fidalgo await determination in the Sao Paulo Herbarium, but the largest gap is still the Amazonian area where very few Aphyllophorales have been collected. The most recent paper on polypores from this area (Ryvarden, 1987) was based on collections made mainly by mycologists interested in ascomycetes. A complete list of mycological studies on polypores in the neotropics can be found in Nishida (1989).

Status and knowledge of tropical polypores

Even if we have a certain knowledge of the more common polypores in the tropics, much has still to be done both in terms of taxonomy and ecology. As for undescribed species, exact figures of course are not available but may be estimated from Corner's latest publications (Corner, 1983, 1984, 1987, 1989 a,b and 1991). He described 288 new species from East Asia, based on collections over some 30-40 years. Even then, many resupinate white polypores (*'Poria'* in the old sense) were omitted. A similar number of new species from Africa and probably even more from tropical America may therefore be expected should accurate and thorough collecting be undertaken. This is partly confirmed by some 55 tropical African species awaiting description in Oslo alone.

Though new species will ultimately be described and critical studies increase the number of polypores from the tropics, some general preliminary trends can be discerned. This is the subject of the following sections.

Diversity

In the following it is necessary to stress that the term 'polypores' refers to all fungi with a poroid hymenophore except members of Boletales and a few fleshy members of Agaricales such as *Favolaschia, Poromycena* and similar genera. Thus, the definition is the same as in Ryvarden (1991) where a survey of the families is presented. Those interested in the taxonomy at family and generic levels are referred to this book.

The latest checklist of polypores of Europe (Ryvarden & Gilbertson, 1992) includes some 330 species whilst that of Africa (Ryvarden, 1975, 1978; Ryvarden & Johansen, 1980) includes some 350 species.

It may come as a surprise that the difference is so small, but it should be remembered that the polypores of Europe have been intensively studied, especially in the last 40 years, while no one to my knowledge has collected extensively in East Africa since 1972 when I spent 2 months in the area. The ratio for the vascular plants in the two areas is approximately 1:4 (I. Nordal pers. comm.), but the reason for the far less diversity of the polypores lies in the fact that the greatest part of the life cycle is spent inside a host or deep in the soil. The fungi are protected against the environment in a different way from that of higher plants and are less susceptible to short term changes. Thus, fungi tend to have much wider distributions than most higher plants.

Spore morphology

Within some other families of basidiomycetes cylindric to allantoid spores have been considered to be advanced compared with globose spores (Corner, 1972; Pegler & Young, 1981; Høiland, 1987), based on the argument that proportionally less globose spores can be produced per unit area of the hymenium.

The distribution of three different spore types in species selected areas where fairly comprehensive floras are available is shown in Table 1. The

Table 1. Percentage of species of different spore morphology in different regions

Region	Cylindric	Ellipsoid	Globose
Europe	50	30	20
N. America	46	32	22
East Africa	37	37	26

Table 2. The ten polypore species with the largest spore sizes in Europe and Africa (Ryvarden & Gilbertson, 1992; Ryvarden & Johansen, 1980).

Species	spore length (μm)
Europe	
Fomes fomentarius (L.:Fr.) Kickx.	15–20
Perenniporia ochroleuca (Berk.) Ryv.	12–17
Polyporus tuberaster Jacq.: Fr.	12–16
Polyporus squamosus Huds.: Fr.	12–16
Hexagonia nitida Dur. & Mont.	10·5–15
Pachykytospora tuberculosa (Fr.) Kotl. & Pouz.	10–13
Dichomitus campestris (Quél.) Dom.	9–14
Coltricia montagnei (Fr.) Murr.	9–14
Antrodia albida (Fr.) Donk	9·5–14
Coriolopsis gallica (Bull.:Fr.) Ryv.	8–14
Africa	
Humphreya eminii (Henn.) Stey.	25–33
Polyporus miquelii Mont.	15–20
Grammothelopsis macrospora (Ryv.) Jül.	15-20
Hexagonia tenuis (Hook.) Fr.	14–20
Humphreya lloydii (Pat.) Stey.	14–18
Polyporus retirugis (Bres.) Ryv.	14–20
Hexagonia speciosa Fr.	13·5–16
Polyporus cucullatus Mont.	13–16
Amauroderma argentofulvum (van Byl.) Reid	13–18
Hexagonia pobequinii Har.	12·5–15·5

information has been compiled from Gilbertson & Ryvarden (1986, 1987; North America), Ryvarden & Gilbertson (1992; Europe) and Ryvarden & Johansen (1980; East Africa). The screening was, through necessity, rather rough and the spore types were designated as follows:

- globose to subglobose = globose
- oblong to broadly ellipsoid = ellipsoid
- cylindric to allantoid = cylindric

Ellipsoid and globose spores are slightly more common in tropical than in temperate and boreal zones where cylindric to allantoid spores predominate. This is because of the proliferation of species in the Hymenochaetaceae in tropical countries where cylindric spores are rare in the tropical representatives.

Spore size

The results of a comparison of the ten species with the largest spore sizes in Africa and Europe are shown in Table 2. Tropical species clearly have larger spores than temperate ones. The reason for this trend is unknown. An examination of spore size variation conducted by Steyaert (1975) on *Ganoderma tornatum* (Pers.) Pat. (*G. australe* (Fr.) Pat. *sensu* Ryvarden & Johansen, 1980) showed that the spores increased in size with increasing latitude and altitude. He suggested ...'one might perhaps consider bigger spores as a bigger reserve for germination, thus ensuring an accrued life potential in less favourable environments'.

Other significant facts relating toTable 2 are that out of the 20 species listed only two,*Antrodia albida* and *Perenniporia ochroleuca*, are common to both regions; all except one (*Antrodia albida*) cause white-rot; 19 species have clamped generative hyphae; 17 species have tetrapolar sexuality; all except *Coltricia montagnei* have complex hyphal systems, being either di- or tri-mitic, and all species in the Polyporaceae (*sensu stricto*) have cylindric spores.

Correlation between spore size and other features

It is also noteworthy that many of the listed species (Table 2)have pileate basidiomes, often developed high above the ground. The correlation between exposed basidiomes and large spores is also seen in the corticoid genera *Aleurodiscus, Laeticorticium*, *Peniophora* and *Vuilleminia.*

The reason for this tendency is unknown, though it may be speculated that these species need much more energy in the spores enabling them to grow rapidly into the wood, since if they remained on the surface for a long time they would risk desiccation and death. With high levels of energy

Table 3. Genera of Polyporaceae *sensu stricto* with exclusively simple septate generative hyphae in the basidiomes

Bondarzewia	*Castanoporus*
Ceriporia	*Flavodon**
*Henningsia**	*Heterobasidion*
*Hydnopolyporus**	*Laetiporus*
Leptoporus	*Leucophellinus*
*Melanoporiella**	*Meripilus*
*Nigrofomes**	*Oxyporus*
Phaeolus	*Physisporinus*
Pycnoporellus	*Rigidoporopsis**
Rigidoporus	*Wolfiporia*

* = tropical genus. Genera with both simple septate and clamped species are excluded.

Table 4. Genera with cystidiate species

Abortiporus (1)	G	*Amylocystis* (1)	M
Auriporia (all)	G	*Australoporus* (1)	S*
Climacocystis (1)	M	*Cyclomyces* (all)	Se*
Daedalea (all)	S	*Echinochaete* (all)	P*
Echinodontium (all)	M	*Flaviporus* (all)	S*
Flavodon (1)	S*	*Fomitopsis* (1)	H
Gloeophyllum (all)	H	*Gloiothele* (all)	G*
Hexagonia (1)	S*	*Hydnochaete* (all)	Se
Inonotus (many)	Se	*Irpex* (all)	S
Junghuhnia (all)	S	*Lentinus* (many)	S,M
Leucophellinus (1)	H	*Microporellus* (3)	M*
Nigrofomes (1)	M*	*Oligoporus* (2)	G,M
Oxyporus (all)	M,H	*Phaeolus* (1)	G
Phellinus (many)	Se	*Pycnoporellus* (all)	H
Rigidoporus (many)	M,H	*Schizopora* (all)	H
Steccherinum (all)	S	*Trechispora* (1)	H
Trichaptum (all)	M	*Wrightoporia* (1)	G*

Numbers in parenthesis indicate the number of species with cystidia. G = gloeocystidia, H = hyphoid cystidia, M = metuloids, P = pseudosetae, Se = setae, and S = skeletocystidia.

* = tropical genus

the hyphae will be able to grow deep into the host before they begin energy production by degrading the wood of the host. The deeper the germinating hyphae can grow the better they are protected against desiccation. Through this ability these species will be able to establish themselves in ecological niches where other species with smaller spores may be excluded.

Septation

The septation of the generative hyphae is accepted as a basic character for generic delimitation among the polypores. Table 3 shows all genera where the type species have simple septa. The number of tropical genera where the type species have simple septate generative hyphae is about 30%, while the total number of tropical genera is about 35%. Thus, there is no significant difference or overweight for the tropical genera with regard to this character.

Cystidia

Genera of polypores with cystidiate species are shown in Table 4. Of the 34 genera listed, 10 (33%) are tropical, the same percentage as for tropical genera compared with the total number of accepted genera. Thus there is no significant difference with regard to this character between the tropical genera and the other groups.

Category of wood-rot

Wood consists mainly of three components, lignin, cellulose and hemicellulose. Conifer wood has a higher lignin content (27-35%) than that of hardwoods (19-35%). It is well known that the polypores include both parasitic and saprotrophic species. Some species are quite aggressive and attack the wood of living trees or newly dead ones, while others are secondary invaders establishing themselves only on partially or well decayed wood. During this attack they secrete enzymes to degrade the constituents of the tree.

Wood-rotting basidiomycetes can be grouped into two categories, white-rot fungi and brown-rot fungi, according to the wood decay enzymes they produce. White-rot fungi produce cellulase and lignase that enable

Table 5. A condensed survey of all poroid and non-poroid families where brown-rot species have been recognised

Agaricales (No tropical representatives)

Hypsizygus | *Neolentinus*
Helocybe | *Ossicaulis*

Auriculariales

Helicobasidium corticoides Bandoni

Coniophoraceae (All representatives of this family produce a brown rot)

Genus	Total species	Tropical species
Coniophora	15	3
Corneromyces	1	1
Gyrodontium	5	2
Leucogyrophana	9	0
Meruliporia	1	0
Pseudomerulius	1	0
Serpula	3	1

Paxillaceae (very few tropical representatives)

Hygrophoropsis | *Paxillus*
Tapinella

Corticiaceae (very few tropical species)

Amylocorticium | *Chaetoderma*
Crustoderma | *Columnocystis*
Dacryobolus | *Veluticeps*

Dacrymycetales (very few tropical representatives)

Calocera | *Dacrymyces*
Dacryopinax | *Cerinomyces*

Polyporaceae

Amylocystis† | *Amylosporus**
Anomoporia | *Antrodia*
Auriporia† | *Daedalea*
Fistulina† | *Fomitopsis*†
Gloeophyllum | *Laetiporus*
Leptoporus† | *Melanoporia*†
Oligoporus† | *Parmastomyces*†
Phaeolus | *Piptoporus*†
Pycnoporellus† | *Wolfiporia*
*Wrightoporia**

Sparassidaceae

Sparassis

† = Boreal genus, * = tropical genus. Further details can be found in: Gilbertson (1980, 1981, Corticiaceae, Sparassidaceae), Ginns (1982, Coniophoraceae), Seifert (1983, Dacrymycetales), Redhead & Ginns (1985, Agaricales).

them to degrade all components of wood cell walls. This is not the place to relate how fast and under which temperature and humidity conditions these fungi remove the components. The reader is again referred to Ryvarden (1991) for further details and Rayner & Boddy (1988) gave an excellent account of these conditions. White-rot fungi eventually decay wood completely and the residues are not stable components of forest soils.

Brown-rot fungi selectively remove cellulose and hemicellulose from wood. Wood decayed by brown-rot fungi loses its strength rapidly and undergoes drastic shrinkage and cracking across the grain. In the advanced stages the wood is reduced to a residue of amorphous, crumbly, brown cubical pieces composed largely of slightly modified lignin.

The number of brown-rot fungi is remarkably small compared with white-rot fungi. Gilbertson (1981) estimated that approximately 6% of the wood-rotting North American basidiomycetes cause a brown rot. The percentage is probably of a similar order in Europe. The number of species in any group of fungi is of course a reflection of the taxonomic philosophy or principles of the monographers. Thus, the numbers in the following tables must not be taken as absolute, they may well vary from one author to another. Nevertheless, the numbers are according to a generally accepted classification, and for the polypores a large variation which could disturb the general trends is not expected. The brown-rot fungi are found in a small number of orders and families of the basidiomycetes. The survey in Table 5 is based largely on figures for North America taken from Gilbertson (1980, 1981), with additions for Europe (Ryvarden & Gilbertson, 1992); 19 out of 132 accepted polypore genera are brown-rot fungi, i.e. about 15%. Only two of them (10%) are tropical, i.e. far less than the total number of genera in the tropics, which represents 33% of all the polypore genera known. It is evident that although brown-rot species are scattered over many systematic groups, there are very few generathat are restricted to the tropics.

Host relationships of brown-rot fungi

As pointed out by Nobles (1958, 1971) and Gilbertson (1980, 1981) most brown-rot fungi are associated with conifers (Table 6) It is evident that approximately 80% of all brown-rot fungi are associated with conifers. Since the tropical zone has few large stands of coniferous trees, the number of such species in the zone is low. Table 6 also shows that 75% of the brown-rot polypores are primarily on conifers while 23 brown-rot polypores exclusively or prefe:ably occur on angiosperms. It is

Table 6. Host relationships of brown-rot fungi

	Brown-rot species	Species living exclusively or primarily on conifer
(a) across Orders and Families		
Agaricales	18	15
Auriculariales	1	–
Coniophoraceae	17	17
Corticiaceae	8	8
Dacrymycetales	8	7
Polyporaceae	92	69
Sparassidaceae	2	2
Totals	146	118
(b) Across genera and species of polypores		
Amylocystis	1	1
Amylosporus	1	–
Anomoporia	2	2
Antrodia	24	21
Auriporia	2	2
Daedalea	1	–
Fistulina	2	–
Fomitopsis	10	4
Gloeophyllum	8	7
Laetiporus	2	–
Leptoporus	1	1
Melanoporia	1	–
Meruliporia	1	1
Oligoporus	25	24
Parmastomyces	1	1
Phaeolus	1	1
Piptoporus	4	–
Pycnoporellus	2	2
Wolfiporia	2	1
Wrightoporia	1	1
Totals	92	69

remarkable that most brown-rot polypores not associated with conifers occur on *Quercus* or other genera of Fagales. This suggests that the brown-rot enzyme system was developed fairly early in the development of fungi. When the tropical vegetation forced these boreal host genera towards the

Table 7. Brown-rot polypores in regional mycobiotas

Region	Total*	Brf†	%Brf‡	Reference
West U.S. & Canada	247	60	24	Lowe & Gilbertson (1961a)
Europe	188	41	22	Ryvarden & Gilbertson (1992)
North America	400	79	20	Gilbertson & Ryvarden (1986)
former U.S.S.R.	254	42	17	Bondarzev (1953)
Southeast U.S.A.	276	45	16	Lowe & Gilbertson (1961b)
Japan	246	32	13	Ito (1955)
Iran	93	7	8	Hallenberg (1981)
New Zealand	242	15	6	Cunningham (1965)
East Africa	332	17	5	Ryvarden & Johansen (1980)
Zimbabwe	150	5	2·5	A. W. Mswaka (pers. comm.)

* = Total number of species recorded in the area.
† = brown rot fungi in the region.
‡ = % of brown rot fungi.

north, the polypores already associated with them had to coevolve or perish. Reverse mutations are not expected in such a complex feature as a change in type of rot would indicate.

Geographical distribution of brown-rot fungi

On a geographic basis, brown-rot fungi are primarily distributed north of the Tropic of Cancer (23·5° N latitude). The greatest concentration is in fact north of the 35° N latitude. Brown-rot fungi thus in general may be said to have a boreal distribution. South of the 35° N latitude, concentrations of brown-rot fungi occur at high elevations in coniferous forest regions like the southern Rocky Mountains (U.S.A.) or Himalayas, and pine forest regions such as the Gulf Coast of the United States and Vietnam and Thailand. In the tropics and in the southern temperate zone brown rot fungi are few. This is clearly shown by a compilation of brown-rot polypores in regional mycobiotas as shown in Table 7. Ryvarden & Gilbertson (1992 and unpublished data) have recorded approximately 330 polypores from Europe of which 74 (23%) cause a brown rot. This is by far the highest percentage of brown-rot fungi in all families where both types of rot occur. Hymenochaetaceae in the two boreal areas includes 85 poroid white-rot species. If these are excluded, the percentage of brown-rot fungi of the polypores will increase to 26%.

Geographical distribution of polypores

Like any cosmopolitan group of organisms, there exists within the Polyporaceae many types of distribution patterns. In the following sections examples of the main tropical generic patterns are shown. As expected there are transitions from one group to another. This is especially true for groups seemingly restricted by climate. A climate is never sharply defined and this is reflected in the distribution of many genera (and of course species).

The following mycogeographical classification is not meant to be exhaustive, that would extend this Chapter far beyond reason. However, for the sake of completeness, all accepted genera are put into geographical groups although the main emphasis is placed on the tropical genera. Many polypores have still poorly known distributions, being reported from only one to a handful of localities. Thus, the patterns reported in the following have to be viewed cautiously and with a critical mind.

The genera of polypores will be classified geographically as cosmopolitan or climate dependent.

Cosmopolitan genera

This group contains genera which are known from all continents and from all major climatic zones, including the following 30 genera (about 22% of the total): *Abortiporus*, *Antrodiella*, *Bondarzewia*, *Bjerkandera*, *Ceriporia*, *Daedalea*, *Datronia*, *Dichomitus*, *Ganoderma*, *Gloeophyllum*, *Gloeoporus*, *Inonotus*, *Irpex*, *Junghuhnia*, *Lentinus*, *Laetiporus*, *Phellinus*, *Phylloporia*, *Perenniporia*, *Polyporus*, *Protomerulius*, *Pycnoporus*, *Pyrofomes*, *Rigidoporus*, *Schizopora*, *Trechispora*, *Trichaptum*, *Trametes*, *Tyromyces* and *Wrightoporia*.

Climate dependent genera

Boreal/temperate genera

Circumpolar genera This group includes genera whose main distribution falls inside the northern boreal temperate zone and those which are entirely circumboreal. The group includes the following 34 (25% of the total) genera: *Albatrellus*, *Amylocystis*, *Anomoporia*, *Antrodia*, *Auriporia*, *Byssoporia*, *Boletopsis*, *Ceriporiopsis*, *Cerrena*, *Chaetoporellus*, *Climacocystis*, *Daedaleopsis*, *Diplomitoporus*, *Donkipora*, *Fistulina*, *Fomes*, *Fomitopsis*, *Grifola*, *Hapalopilus*, *Haploporus*, *Heterobasidion*, *Irpicodon*, *Ischnoderma*, *Jahnoporus*, *Leptoporus*, *Meripilus*, *Oligoporus*, *Piptoporus*, *Pachykytospora*, *Parmastomyces*, *Phaeolus*, *Pycnoporellus*, *Physisporinus* and *Spongipellis*.

Gondwanaland genera The name Gondwanaland describes the coherent continent existing before the current continents split up and moved to their present positions. In terms of present distributions the region encompasses Australia–southern South America–southern Africa. The only genus in this group is *Phaeotrametes*, which is subtropical in distribution.

American boreal-temperate genera *Globifomes*, *Melanoporella*, *Meruliporia*, *Polyporoletus* and *Wolfiporia*.

European genera Only one genus, *Podofomes*.

Asian boreal-temperate genera *Porodontia*, *Castanoporus* and *Protodaedalea*.

Eurasian boreal-temperate genera There is only one genus with this distribution, *Piloporia*.

American-Asian boreal-temperate genera *Cryptoporus*, *Echinodontium*, *Hydnochaete* and *Pyrrhoderma*.

Tropical genera

There are 33 genera which are generally restricted to the tropical zones and they are divided into three subgroups, i.e. pantropical, paleotropical and neotropical genera.

Pantropical genera The genera are distributed throughout the forested parts of the tropical zones and include the following 21 genera which constitute about 64% of all tropical genera: *Amauroderma*, *Amylosporus*, *Aurificaria*, *Coltriciella*, *Cyclomyces*, *Diacanthodes*, *Earliella*, *Echinochaete*, *Flavodon*, *Flaviporus*, *Grammothele*, *Haddowia*, *Hexagonia*, *Humphreya*, *Megasporoporia*, *Microporellus*, *Navisporus*, *Nigrofomes*, *Nigroporus*, *Porodisculus* and *Porogramme*.

Paleotropical genera These genera are restricted to the tropics of the old world (Africa–Asia–N. Australia): *Amylonotus*, *Leucophellinus*, *Lignosus*, *Macrohyporia* and *Microporus*.

African genera Genera which are endemic to tropical Africa: *Gloiothele*, *Grammothelopsis*, *Pseudopiptoporus*, *Rigidoporiopsis*, *Theleporus* and *Xerotus*.

Asian genera Genera endemic to tropical Asia: *Elmerina*, *Flabellophora*, *Hymenogramme*, *Paratrichaptum* and *Sparsitubus*.

Neotropical genera These genera are endemic to tropical America: *Echinopora*, *Fomitella*, *Fuscocerrena*, *Henningsia*, *Hydnopolyporus*, *Lamelloporus*, *Melanoporia*, *Nigrohydnum* and *Stiptophyllum*.

Australian genus Just one genus, *Australoporus*.

There are considerable difficulties in explaining all the patterns shown above. The paucity of fossils bars any insight into the evolutionary history of the genera involved. It is however striking to note the wide distribution of the majority of genera. Even the endemic genera have a very wide distribution in the continent to which they are restricted.

The wide distribution of many polypore genera is intriguing as it seems to indicate:

- that the group was already very diverse when the former continent of Gondwana broke up about 150 million years ago and that evolutionary processes have been slow thereafter;
- that many genera have more effective spore dispersal mechanisms than commonly assumed.

If we assume that most of the cosmopolitan and pantropical genera were developed before Gondwanaland broke up and have only followed the continents to their current positions, we are then confronted with a difficult paradox. The vascular plants developed with the Coniferopsida, as possible hosts for polypores in late Permian, approximately 230 million years ago. This gives the group some 80–100 million years to develop before continental drift split the old genetic stock. If this concept is right, it indicates a very rapid evolution in the first 100 million years and virtually no evolution in the last 100 million years. We are then left with numerous cosmopolitan genera and species, the latter still being compatible on a world scale.

If we reject this hypothesis we then have to accept that the spore dispersal mechanisms are very effective, at least in some groups. The question arises why this is so in some groups and not in others with exactly the same type of spores? Stated differently, if spore dispersal mechanisms were so effective that they eventually overcame intercontinental distances and barriers, why then are not all tropical genera pantropical and all boreal-temperate genera circumpolar?

There are no satisfactory answers to these questions and, clearly, there are mechanisms at work which we do not currently understand.

An examination of the widely distributed genera does not give many clues to an understanding of the evolutionary history of the polypores. These genera include most of *Trametes* and *Polyporus* groups which, based on other arguments, are assumed to be advanced. Do these have very effective spore dispersal or were they already evolved when Gondwanaland broke up?

As shown already (Table 6), most of the brown-rot genera are circumpolar in the boreal zone. For example *Gloeophyllum*, a brown-rot genus with a cosmopolitan distribution, has only one species in the tropics and 7–8 species in the boreal zone, again showing that the diversity among the brown-rot genera is restricted to the conifer ecosystem.

Some genera or species with a restricted distribution, either historically, host or climatically related, may shed some light on the questions raised above. Scattered examples are given to illustrate this.

The distribution of the genus *Phaeotrametes* is shown in Fig. 1. It is restricted to parts of the old Gondwanaland and is characterized by its thick-walled pigmented truncate spores. This is a very unusual feature among the polypores and it indicates an ancient origin (see Ryvarden, 1991 for further details). I regard *Phaeotrametes* as a living fossil.

Pyrofomes demidoffii, a parasitic species with a wide distribution, is restricted to living tree-forming *Juniperus* spp., following the hosts almost wherever they occur. Its distribution is shown in Fig. 1. This species clearly represents a case of the principle that the antiquity of the host reflects that of the parasite (principle number 14 of Savile, 1955). This would indicate that its thick-walled truncate spores are primitive, as suggested also for *Phaeotrametes*, and that white rot and complex hyphal systems were early achievements in the history of polypores. Its distribution again

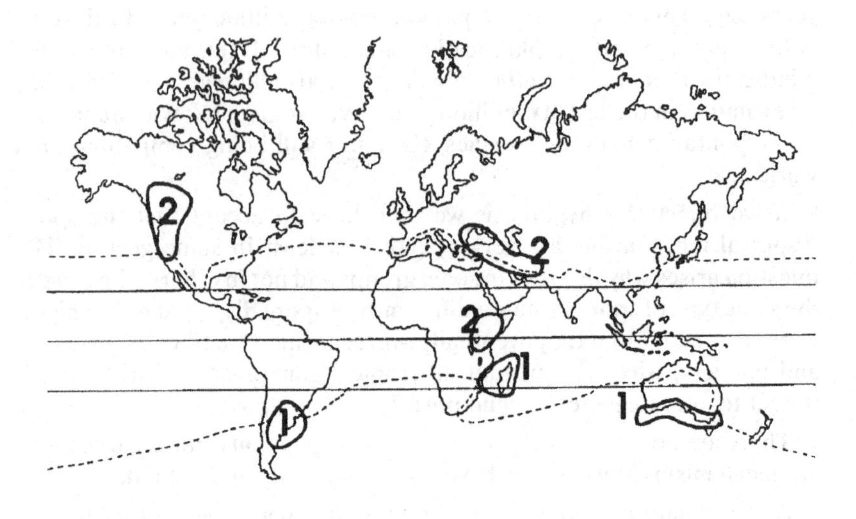

Fig. 1. Distribution of *Phaeotrametes decipiens* (1) and *Pyrofomes demidoffii* (2), the latter restricted to arboriform *Juniperus* species. Dashed line = palm line.

illustrates the paradox outlined above. If the species has followed *Juniperus* for over 125 million years, why then has there not been any specific diversity in the very widespread areas of distribution? It may be of course that an infrageneric incompatibility mating system in this species and a high outbreeding level resulted in a very restricted speciation when these systems were first established. Cultural work on specimens from different parts of the world has not been conducted to test such a theory. If the species has spread to its present hosts at a later stage, it indicates that the spores are very effective dispersal units, being eventually able to cross the Atlantic ocean rather easily.

Microporus is a highly characteristic tropical genus, both macro- and microscopically and is widespread in the paleotropics where it has a wide ecological amplitude. The distribution is shown in Fig. 2. The genus is ecologically versatile, so why has it not spread to the neotropical areas where favourable niches should be abundantly available? This suggests that the genus evolved in the paleotropical area after America split from Africa and its spores have not been able to cross the Atlantic Ocean, for

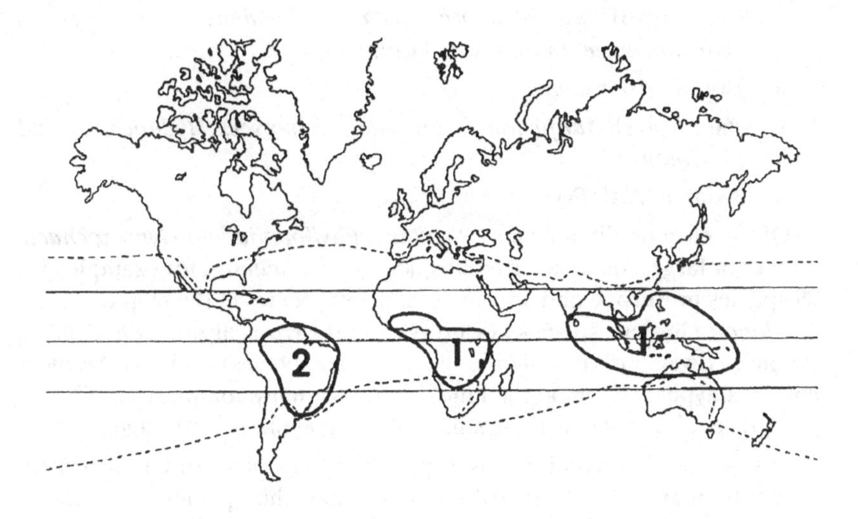

Fig. 2. Distribution of *Microporus* and *Lignosus* (1) and *Stiptophyllum* (2). Dashed line = palm line.

one reason or another. This implies that the stipitate basidiome is a rather recent development.

Lignosus is a similar case to *Microporus*, see Fig. 2. It is a trimitic genus with a stipitate basidiome and a sclerotium. Its sclerotium seems well adapted to survive periods without rain and there is no other logical explanation except a historical reason for its absence in the neotropics. This again points to the stipitate basidiome as an advanced character, at least in this group.

Stiptophyllum, see Fig. 2. for distribution, is restricted to the neotropics and has a trimitic hyphal system, pigmented hyphae and cylindric thin-walled spores. It is well adapted to resist drought and occupies much the same niche as *Lentinus* spp. Thus, its endemic distribution points to a recent origin and again underscores the stipitate basidiome and cylindric thin-walled spores as advanced characters.

The *Hymenochaetaceae* is a highly characteristic family with all indications of a monophyletic origin. Mycogeographically, the family (here including all genera) can be broken down as follows:

- Cosmopolitan: *Asterostromella*, *Phellinus*, *Phylloporia*, *Hymenochaete*, *Inonotus* and *Coltricia*.
- Boreal: *Asterodon*.
- Pantropical: *Aurificaria*, *Coltriciella*, *Cyclomyces*, *Hydnochaete* and *Phaeolopsis*.
- Asian boreal: *Pyrrhoderma*.

Of the cosmopolitan genera *Phellinus*, *Phylloporia* and *Hymenochaete* have their largest diversity in the tropical zone. *Phellinus*, for example, has 29 species in Europe and approximately 130 species in the tropical zone. *Phylloporia* has one species in Europe, 4 in the tropical zone. *Phellinus* is the most diverse of all polypore genera (200 species) and has the most complex hyphal system in the family. With its many complexes of closely related species, it seems to be still in a strong evolutionary stage.

Inonotus is a smaller genus (approx. 85 species) and is distinctly boreal-temperate in its distribution. Usually the species are easy to determine and there are no difficult species complexes like those seen in *Phellinus* (for example *P. igniarius* and *P. rimosus* complexes). Possibly, *Inonotus* species have had a longer time for species differentiation than in *Phellinus*. It is tempting to guess that the monomitic hyphal system which makes basidiomes rather soft and vulnerable is the main reason why they were 'forced' away from the tropics where predation by insects apparently is an important factor for their survival. *Phellinus*, with its dimitic hyphal system making the basidiomes woody to bone hard, was well suited to develop in the tropics, thus we see the greatest diversity of the genus just in this zone. If this argument is correct, it is an indication that the complex hyphal system, at least in this family, is an advanced character.

The stipitate *Coltricia perennis* is of interest since it is the only polypore demonstrated to be facultatively ectomycorrhizal (Danielson 1984). This complex metabolism which was established only 50 million years ago in the early Tertiary period, underscores again that the stipitate basidiome is an advanced character.

Tropical polypores in medicine

In some African countries, polypores are used for medical purposes. Morris (1987) and Sikombwa & Piearce (1985) reported the use of *Perenniporia mundula* (Wakef.) Ryv. in the treatment of pleurisy and impotence. As far as I know, there have been no chemical or medical investigations of any active compounds from this species. The stipitate basidiomes of *Lignosus sacer* (Fr.) Ryv. arise from a buried sclerotium in the ground and I have been told that this sclerotium was used for stomach problems in Kenya and Tanzania. Again there are no reports on the active compounds in the sclerotium. The most comprehensive treatment of polypores for medical use is that of Ying *et al.* (1987) whose beautiful book *Icones of Medical Fungi from China* has coloured drawings and descriptions of 77 polypores used for medical purpose. Most of the species are pantropical and are used for treatment of a very long list of diseases and conditions. Interested readers are referred to this publication for further details. In tropical Asia many polypores are used in medicine by local people, but no systematic studies have been published (Schumacher, pers. comm.).

Tropical polypores and insects

It is well known that termites have a close relationship with wood-rotting fungi, especially polypores, (see Sands (1969) and Becker (1975) for reviews of the literature). In many cases the termites were strongly attracted to wood attacked by polypores, especially brown rot fungi (Gilbertson, 1980). On the other hand, wood rotted by *Ganoderma applanatum* (Pers.: Wallr.) Pat. was toxic to termites and unattractive to them (Kovoor, 1964). Whether this is also the case with the many strictly tropical *Ganoderma* species is unknown.

For the temperate zone there are numerous studies of polypores as a habitat for insects, especially beetles (see Hanski, 1989, for a review of the literature), but I have seen no similar reports based on tropical polypores.

Pathological importance of tropical polypores

The tropical polypores as in other zones play a very important role in recycling the nutrients locked up in all types of wood. During this process it is unavoidable that they also attack economically important timber and wooden constructions. Many polypores growing on the ground, such as *Amauroderma*, *Ganoderma* and *Phellinus* species cause butt- and heart-rot in living trees (Corner, 1983; Masuka, 1991; Mswaka, 1991). There have been a large number of pathological studies involving tropical polypores, but it is outside the scope of this chapter to review the literature. A typical paper showing the importance of this aspect is Piearce (1986) treating only the tree diseases of Zambesi Teak (*Baikiaea plurijuga* Harms). A serious pathogen, apparently on the spread in tropical Africa

(Ofosu-Asiedu, 1975), is *Laetiporus baudonii* (Pat.) Ryv. which is a facultative brown-rot parasite on many indigenous and exotic trees. It seems to kill the host rather rapidly based on the authors own observations in Zimbabwe.

Tropical polypores as food

In the temperate zone several polypores, especially in the genus *Albatrellus*, are eaten as food. The same is true for *Laetiporus sulphureus* (Bull.: Fr.) Murs. in North America (Gilbertson, 1980). In Japan, many polypores are cultivated such as *Grifola frondosa* (Dicks.: Fr.) S. F. Gray and *Meripilus giganteus* (Pers.: Fr.) Pil. There are no comprehensive studies of edible wild fungi from any tropical area, but *Polyporus tenuiculus* (Beauv.) Fr. (*Favolus brasiliensis* (Fr.) Fr.) and *P. moluccensis* (Mont.) Ryv. are both eaten in Zambia and Malawi (Piearce, 1981; Morris, 1987). Representatives of *Lentinus* (which is a polypore genus despite its lamellate hymenophore) are collected in many tropical areas for consumption and sale (Chang & Quimio, 1982; Morris, 1987).

References

Becker, G. (1975). Termites and fungi. *Material und Organismem* **3**, 465-478.

Bondarzev, A. (1953). *Polyporaceae of the European part of the USSR and the Caucasus*. Izdatel'stvo Akademii Nauk SSSR: Russia.

Carranza-Morse, J. (1991). Pore fungi of Costa Rica. *Mycotaxon* **41**, 345-370.

Carranza-Morse, J. (1992). Pore fungi of Costa Rica II. *Mycotaxon* **43**, 351-369.

Chang, S. T. & Quimio, T. H. (eds.) (1982). *Tropical Mushrooms, Biological Nature and Cultivation Methods*. The Chinese University Press: Hong Kong.

Corner, E. J. H. (1972). Studies in the basidium, spore spacing and the *Boletus* spore. *Gardens' Bulletin of the Straits Settlement* **26**, 195-228.

Corner, E. J. H. (1983). Ad Polyporaceas I. *Beiheft Nova Hedwigia* **75**, 1-182.

Corner, E. J. H. (1984). Ad Polyporaceas II & III. *Beiheft Nova Hedwigia* **78**, 1-222.

Corner, E. J. H. (1987). Ad Polyporaceas IV. *Beiheft Nova Hedwigia* **86**, 1-265.

Corner, E. J. H. (1989a). Ad Polyporaceas V. *Beiheft Nova Hedwigia* **96**, 1-218.

Corner, E. J. H. (1989b). Ad Polyporaceas VI. *Beiheft Nova Hedwigia* **97**, 1-197.

Corner, E. J. H. (1991). Ad Polyporaceas VII. *Beiheft Nova Hedwigia* **101**, 1-175.

Cunningham, G. H. (1965). Polyporaceae of New Zealand. *New Zealand Department of Science and Industrial Research Bulletin* **164**, 1-304.

Danielson, R. M. (1984). Ectomycorrhizal associations in Jack pine stands in northern Alberta. *Canadian Journal of Botany* **62**, 932-939.

Dennis, R. W. G. (1970). Fungus flora of Venezuela and adjacent countries. *Kew Bulletin, Additional Series III*, 1-531.

Fidalgo, O. & Fidalgo, M. E. P. K. (1968). Polyporaceae from Venezuela I. *Memoirs of the New York Botanical Garden* **17** (2), 1-34.

Fries, E. (1860). *Reliquiae Afzelinae sistentes icones fungorum quos in Guinea collegit et in aere incias*. Uppsala: Sweden.

Gilbertson, R. L. (1980). Wood-rotting fungi of North America. *Mycologia* **72**, 1-49.

Gilbertson, R. L. (1981). North American wood-rotting fungi that cause brown rots. *Mycotaxon* **12**, 372-412.

Gilbertson, R. L. & Ryvarden, L. (1986). *North American polypores Abortiporus–Lindtneria*. Fungiflora: Oslo, Norway.

Gilbertson, R. L. & Ryvarden, L. (1987). *North American polypores, Megasporoporia–Wrightoporia*. Fungiflora: Oslo, Norway.

Ginns, J. H. (1982). A monograph of the genus *Coniophora*. *Opera Botanica* **61**, 1-61.

Hallenberg, N. (1981). *Wood-inhabiting Aphyllophorales (Basidiomycetes) and Heterobasidiomycetes in N. Iran*. Thesis, Göteborg.

Hanski, I. (1989). Fungivory, fungi, insects and ecology. In *Insect–Fungus Interactions*, (ed N. Wilding, N. M. Collins, P. M. Hammond & J. F. Webber), pp. 25-68. Academic Press: New York.

Høiland, K. (1987). A new approach to the phylogeny of the order Boletales. *Nordic Journal of Botany* **7**, 705-718.

Ito, S. (1955). *Mycological flora of Japan. Vol. II. Basidiomycetes no 4, Auriculariales, Tremellales, Dacrymycetales, Aphyllophorales*. Yohendo: Japan.

Koovoor, J. (1964). Modifications chimiques provoques par une termitide dans du bois de peuplier sain ou partiellement degrade par des champignons. *Bulletin Biologique France Belgique* **98**, 491-510.

Lowe, J. L. & Gilbertson, R. L. (1961a). Synopsis of the Polyporaceae of the western United States and Canada. *Mycologia* **53**, 474-511.

Lowe, J. L. & Gilbertson, R. L. (1961b). Synopsis of the Polyporaceae of the southeastern United States. *Journal of the Elisha Mitchell Scientific Society* **77**, 43-61.

Masuka, A. J. (1991). *Aphyllophorales in pine and eucalypt plantations in Zimbabwe, their taxonomy and ecology*. Thesis, University of Zimbabwe.

Morris, B. (1987). *Common Mushrooms of Malawi*. Fungiflora: Oslo, Norway.

Mswaka, A. Y. (1991). *A mycofloristic survey of the Polyporaceae s. lato of indigenous forests habitats in Zimbabwe*. Thesis, University of Zimbabwe.

Nishida, F. H. (1989). Review of mycological studies in the Neotropics. In *Floristic Inventory of Tropical Countries*, (ed. D. Campbell & D. H. Hammond), pp. 495-522. New York Botanical Garden: U.S.A.

Nobles, M. K. (1958). Cultural characters as a guide to the taxonomy and phylogeny of the Polyporaceae. *Canadian Journal of Botany* **36**, 883-926.

Nobles, M. K. (1971). Cultural characters as a guide to the taxonomy of the Polyporaceae. In *Evolution in the Higher Basidiomycetes*, (ed. R. Petersen), pp. 169-196. Tennesee University Press: U.S.A.

Ofosu-Asiedu, A. (1975). A new disease of *Eucalyptus* in Ghana. *Transactions of the British Mycological Society* **65**, 285-289.

Pegler, D. N. & Young, T. W. K. (1981). A natural arrangement of the Boletales with reference to spore morphology. *Transactions of the British Mycological Society* **76**, 103-146.

Piearce, G. D. (1981). *An introduction to Zambia's wild mushrooms and how to use them*. Forest Department: Zambia.

Piearce, G. D. (1986). Tree diseases and disorders in the Zambezi Teak Forests. in *The Zambezi Teak Forests*, (ed. G. D. Piearce), pp. 239-257. Forest Department: Zambia.

Rayner, A. D. M. & Boddy, L. (1988). *Fungal Decomposition of Wood, its Biology and Ecology*. J. Wiley & Sons: Chichester, U.K.

Redhead, S. A. & Ginns, J. H. (1985). A reappraisal of agaric genera associated with brown rot. *Transactions of the Mycological Society of Japan* **26**, 349-381.

Ryvarden, L. (1975). Studies in the Aphyllophorales of Africa 3. Three new polypores from Zaire. *Bulletin Jardin Botanique Belgique* **45**, 197-203.

Ryvarden, L. (1978). Studies in the Aphyllophorales of Africa 6. Some species from eastern Central Africa. *Bulletin Jardin Botanique Belgique* **48**, 79-119.

Ryvarden, L. (1987). New and noteworthy polypores from tropical America. *Mycotaxon* **28**, 525-541.

Ryvarden, L. (1991). Genera of polypores. Nomenclature and taxonomy. *Synopsis Fungorum* **5**, 1-373.

Ryvarden, L. & Gilbertson, R. L. (1992). European polypores part 1. *Synopsis Fungorum* **6**, 1-344.

Ryvarden, L. & Johansen, I. (1980). *A Preliminary Polypore Flora of East Africa*. Fungiflora: Norway.

Sands, W. A. (1969). The association of termites and fungi. In *Biology of Termites*, (ed. K. Krishna & F. M. Weesner), pp. 495-524. Academic Press: New York.

Savile, D. B. O. (1955). A phylogeny of the basidiomycetes. *Canadian Journal of Botany* **33**, 60-104.

Saville, D. B. O. (1968). Possible relationship between fungal groups. In *The Fungi, An Advanced Treatise*, vol. **III**, (ed. G. C. Ainsworth, F. K. Sparrow & A. S. Sussman), pp. 649-675. Academic Press: New York.

Seifert, K. A. (1983). Decay of wood by the Dacrymycetales. *Mycologia* **75**, 1101-1118.

Setliff, E. C. & Ryvarden, L. (1983). Los hongos de Colombia 7. Some Aphyllophorales wood-inhabiting fungi. *Mycotaxon* **18**, 509-526.

Sikombwa, N. & Piearce, G. D. (1985). *Vanderbylia ungulata* and its medicinal use in Zambia. *Bulletin of the British Mycological Society* **19**, 124-125.

Steyaert, R. L. (1975). The concept and circumscription of *Ganoderma tornatum*. *Transactions of the British Mycological Society* **65**, 451-467.

Swartz, O. (1791). *Flora India Occidentalis*. B. Williams & Sons: London.

Ying, J., Mao, X., Ma, Q., Zong, Y. & Wen, H. (1987). *Icones of Medical Fungi from China*. Science Press: Beijing, China.

Chapter 10

Comparison of the macromycete biotas in selected tropical areas of Africa and Australia

Roy Watling

Royal Botanic Garden, Inverleith Row, Edinburgh EH3 5LR

Introduction

During the expansive period of the European exploitation of the world at the end of the nineteenth century and beginning of the present, many fungal collections were made. The majority were accompanied by little ecological information, and although the collections were made rather piece-meal they indicated the diversity of larger fungi to be found outside the western hemisphere. Fortunately with better communications and travel this gap in our knowledge is now being addressed, but unfortunately mycologists are very thin on the ground and, as plants and animals call heavily on the available funding, the accumulation of data on fungi is extremely slow. A few detailed observations have been made over several years by expatriates, and some short term studies which can at least give pointers to further studies. Thus, one such individual, Professor E. J. H. Corner, to whom this article is dedicated, has amassed a huge well-documented collection of larger fungi from Malaysia, Singapore and other tropical areas. The many publications which these specimens have spawned are extraordinarily important in assisting mycologists in the developing world to document their fungal biotas.

In contrast, the present study is an example of intensive but short studies. The data presented here are based on short collecting trips to two sites, one near Kitwe, Zambia and another near Gympie, Queensland, Australia (both supported by longer term collections found in the Herbaria at Riverside Forest Research Station, Kitwe and Department of Primary Industry, Indooropilly (BRIP) respectively) and a third site in Cameroon. In the last country three intensive and two minor collecting trips have been made to the Korup Forest Reserve over the period 1984-91, in collaboration with Dr I. Alexander.

This account therefore draws together the observations made in two continents and acts as a base-line for further and more extensive discussion, and some possible speculation. The origin of the two areas, either as parts of a fragmented and long dispersed archaic continent, or the edges of such a continent (Hoffman, 1991), has led to their very different position today in relation to the equator and the poles. Connected with this history is the development of arrays of quite different cormophytes exhibited by the two continents.

The fungal biota of the Gympie site is compared with both the miombo woodland which covers large areas of E. Central Africa and the lowland rainforest of Korup. This is instructive because of the unique nature of the Queensland site (Cooloola Sand mass) which consists of a series of high dunes stretching 10 km inland from the Pacific Coast, with rainforest development in the depressions between the dunes.

The dune crests and upper plateaus are dominated by *Eucalyptus* spp. (*Myrtaceae*) which parallels the legume dominated miombo woodland of Zambia. In contrast the rainforest developed within the sand-mass can be compared with the Korup site in the rich intermixture of lianas, nanophylls, buttressed rooted trees, etc. However, at Cooloola the dominants include *Agathis* and *Araucaria* (Araucariaceae), *Tristania* and *Syncarpia* (Myrtaceae) and *Tieghemopanax* (= *Polyscias*: Araliaceae) whereas at Korup are emergents amongst a very rich community of caesalpinoid legumes in the old *Macrolobium* consortium, e.g. *Berlinia*, *Microberlinia*, *Tetraberlinia*.

Site data

Cooloola sand-mass

The study area lay near the northern tip of a whole series of overlapping dunes in the Noosa River drainage system and have accumulated over many thousands of years; it was close to Rainbow Beach 175 km north of Brisbane. The oldest dunes were probably formed as much as 400,000 years ago. The area has been studied during a 7-year interdisciplinary research project instigated by CSIRO, Division of Soils and started in 1974. Many technical reports have been published, e.g. Thompson & Moore (1984), but the reader is urged to read the popular accounts by Seymour (1982) or Thompson (1981).

A summary only is offered here to set the scene. The Cooloola sand mass stretches 60 km along the Pacific coast (Table 1). A six-part dune system is exposed along the coast near Rainbow beach and has not been covered by younger deposits. A relative age sequence is exhibited on

Table 1. Climatic and topographic features of study sites.

	Cooloola	Korup	Kitwe
Area (ha)	24,000	6,000	5,000
Elevation (m)	50 - 250	50 - 100	1,350
Temperature range (°C)	28 - 12	30 - 24	27 - 5
Annual rainfall (mm)	1,200 - 1,600	5,500	1,250
Highest rainfall	March	July to September	November to March
Lowest rainfall	July to November	December to February	May to October

which podzols varying from 1 to 20 m deep have formed, and which probably extends at least to the last interglacial period.

The six-part system consists of young mobile dunes interspersed with dunes colonized by low, dense shrubby woodland on embryonic podzols (Mutyi) characterised by *Banksia integrifolia* L., *Tristania*, *Acacia* and *Pultenaea*, and five other elements, Chalambar, Burwilla, Warrawonga, Mundu and Kabali. The Chalambar dunes rise about 100 m above sea level and support a low grassy forest characterised by *Eucalyptus signata* F. Muell. The Burwilla dunes, a little higher, however, are dominated by *E. intermedia* R. T. Baker; these dunes have suffered greatly from erosion. *E. intermedia*, but this time subordinate to *E. pilularis* Smith, is again important on the Warrawonga dunes which rise to nearly 250 m. *E. signata* enters again in System 5, the Mundu dunes, characterised by tall but shrubby woodland far less impressive than the Warrawonga. The final system, Kabali, is composed of the most ancient dunes, whale-back sand hills rising only 40-80 m above sea level and characterised by *E. signata* and *Banksia aemula* R. Br. (Proteaceae), forming a rather shrubby woodland often on extremely deep podzols; the plants may only grow 2 m tall indicating the extreme lack of nutrients. Indeed unlike System 1 nutrients are poorly supplied from atmospheric accession. Warrawonga is the climax of the system with increasing numbers of tree, shrub and understorey species, and good supplies of nutrients which encourage rich growth. System 6 abutts the Como scarp of late Triassic to early Jurassic age.

Litter of leaves, twigs and bark gathers and, on decomposition, releases additional nutrient which, with rain water, moves between the dunes. In the valleys of 4 systems (2, 4, 5 & 6) this has encouraged the development of rainforest especially in the Warrawonga system with *Agathis robusta* C. Moore and *Ficus watkinsiana* Bail. as prominent members of a nothophyll vine scrub or microphyll vine forest.

Korup rain forest

This study area is a natural lowland rainforest (Table 1) lying between the Akva Korup and Ndian rivers in southwest Cameroon and within the Guinea-Congian refugia area into which the forest was reduced during the Pleistocene (Gartlan, 1974). It is characterised by both an assemblage of legumes in subfamily Caesalpinoideae, tribes Amherstieae and Detarieae, on freely draining acidic, sandy soils. The area was selected for intensive study, by an interdisciplinary team, the results of which have been published and should be referred to by the reader (Gartlan *et al.*, 1986; Newbery *et al.*, 1988).

The rainforest is at a low elevation (*ca* 100 m) and consists of an estimated 700 different arborescent taxa; 411 of the flowering plants within defined subplots have been recognised amongst all living stems >30 cm girth at breast height (1·3 m), although only 66% were identified to given taxa. The subplots were 40 x 40 m with the main plots (80 m^2) divided along two transects 5 km long at 150 m intervals. Only one transect (P) was intensively monitored.

The transect covered a gravelly plateau with low litter accumulation except in depressions where twigs, bark, leaves and flower-parts often gather in permanently damp conditions, or along running water. In the rainy season many of these areas are inundated. The transect is characterised by groves of *Oubanguia alata* Bak. f. locally common in the forest, *Lecomtedoxa klaineana* (Pierre ex Engl.) Dubard but more importantly by *Tetraberlinia bifoliata* (Harms.) Haum., *T. moreliana* Aubr., *Microberlinia bisulcata* A. Chev., *Didelotia africana* Baill., *Berlinia bracteosa* Benth. and *Anthonotha fragrans* (Bak. f.) Excll. & Hillcoat, all ectotrophic Leguminosae. Small populations of *Afzelia* spp. and *Gilbertodendron* spp. are also found and in the wet areas *Uapaca staudtii* Pax. (Uapacaceae: Euphorbiaceae).

Kitwe & Misaka Forest Reserves

The Copperbelt in which the two study areas are situated forms the north eastern region of the international boundary between Zambia and

Zaire (Table 1). It is a gently sloping peneplain broken by low granite or quartzite hills and infrequent 'Kopjes', and is drained by the Kafue river which runs in a wide curve served by several large tributary streams. The soils are slightly acidic, structureless, fine-textured, pale or pallid, intermixed or overlain with quartz rubble and concretionary ironstone. They are both poor in nutrients and humus and much of the organic cycling is governed by termite activity; indeed, termite mounds are a marked feature of the Copperbelt. The Zambian Forest Department has issued an account of this area (Anon., 1975).

The Misaka Forest Reserve is typical miombo woodland consisting of *Brachystegia* spp. especially *B. longifolia* Benth. and *B. utilis* Burtt Davy & Hutchinson and *Isoberlinia angolensis* (Benth.) Hoyle & Brenan and *Julbernardia paniculata* (Benth.) Troupin, all ectomycorrhizal Leguminosae. *Marquesia macroura* Gilg. (Dipterocarpaceae) is in evidence in this vegetation, but unlike at Korup, where *Uapaca* spp. are found, they are part of the dry forest and include *U. kirkiana* Müll. Arg. and *U. nitida* Müll. Arg. Misaka is an example of moderate quality woodland and the constituents may reach 18 m high. Both communities are open grassy woodlands maintained by regular burning and therefore they reflect the history of this type of African vegetation, which is being monitored in study plots at Ndola.

Methods

Members of the Agaricales *s. lato* especially in the family Boletaceae *s. lato* viz. the poroid fleshy fungi, were monitored in the field for all three study sites, supported in the first and last by additional dried material. In Korup the monitoring was carried out over five field expeditions. Several taxonomic studies have ensued and the results are presented here to ascertain whether or not differences and similarities exist between the areas. Dr J. Aberdeen was instrumental in gathering together collections from Cooloola (Aberdeen, Ross & Thompson, 1989), as were Drs M. Ivory and G. Piearce from the Copperbelt (Piearce, 1981).

Results

In Cooloola the greatest numbers of basidiomes appear during the months of highest mean rainfall in late summer and early autumn. Amanitaceae and boletes tend to dominate the warmer months whilst in the rainy, cooler months members of the genus *Cortinarius* are common and some

Table 2. Number of genera and species in the Cooloola soil-landscapes considered by Watling & Gregory (1986, 1989, 1991).

	Direction towards ocean ⇒ Soil-Landscapes							No. of	No. of
	Noosa	K	Mu	W	B	C	M	species	collections
Boletaceae									
Austroboletus	-	2	1	1	3	-	-	5	10
Boletellus	2	1(R)	1	7(2R)	3	3(2R)	-	12	18
Boletochaete	-	-	-	1	-	-	-	1	1
Boletus inc. *Xerocomus*	1	-	3	3(1R)	1	1	-	9	9
Fistulinella	-	-	1	-	-	-	-	1	1
Gyroporus	-	-	1	-	-	-	-	1	1
Heimiella	-	-	2	-	-	1	-	2	3
Phlebopus	-	-	-	-	-	1(R)	-	1	1
Pulveroboletus	-	-	-	1		-	-	1	1
Rubinoboletus	-	-	-	2	-	-	-	2	2
Strobilomyces	-	-	-	-	1	1	-	2	2
Tylopilus	5(2R)	-	1	4(1R)	3	-	-	10	14
Gomphidiaceae									
Chroogomphus	-	-	-	-	-	-	-	-	-
Gomphidius	-	-	-	-	-	-	-	-	-
Paxillaceae									
Paxillus	-	-	-	-	-	-	-	-	-
Phylloporus	-	-	-	3	1	2	1	6	7
Totals	10	3	9	27	12	11	1	53	70

Abbreviations: B = Burwilla; C = Chalambar; K = Kabali; M = Mutyi; Mu = Mundu; R = Rainforest; W = Warrawonga

Amanita spp. also appear but boletes are less frequent (Aberdeen *et al.*, 1989). In Korup, all members of the ectomycorrhizal species were apparent in the early part of the rainy period particularly some of the more fleshy Russulaceae. Basidiomes, however, may be found from the beginning of the year until May. In the Copperbelt the best fruiting would appear to be in the months either side of Christmas and New Year although basidiomes are still found under favourable conditions into April.

Table 3. Number of species in eucalpyt and rainforest communities in Cooloola.

Genera	Number of species			Notes
	Total	Eucalypt	Rainforest	
Boletaceae				
Austroboletus	5	5	-	
Boletellus	12	9	5	one species common to both communities
Boletochaete	1	1	-	
Boletus inc. *Xerocomus*	9	8	1	
Fistulinella	1	1	1	conspecific?
Gyroporus	1	1	-	
Heimiella	2	1	1	including *H. fruticola*, formerly in *Austroboletus*
Phlebopus	1	-	1	
Pulveroboletus	1	1	-	
Rubinoboletus	2	2	-	
Strobilomyces	2	2	-	
Tylopilus	10	8	3	*T. olivaceoporus* wet sclerophyll or margin of rainforest
Paxillaceae				
Phylloporus	6	6	-	

A total of 308 collections have been examined from Korup, 114 boletes from the Copperbelt and 102 boletes from Cooloola. In addition other suspected mycorrhizal species have been examined from Zambia and Australia. The main results based both on published or unpublished data are summarized in Tables 2 to 4. In the Cooloola sand-mass system members of the genera *Tylopilus* and *Boletellus* are dominant with the former associated with the Noosa river plains as much as with the tall forest on dunes (Warrawonga) where *Boletellus* becomes prominent. Indeed the climax eucalypt vegetation associated with the Warrawonga hosts twenty-three boletes but the developing climax found on Burwilla only twelve taxa. Boletes are poorly represented in the low grassy forest of the Chalambar and in the post-climax vegetation of the Mundu and Kabali. Thus the boletes are concentrated with the greatest development of the eucalypts (Table 2).

Rainforest developing between the dunes also supports boletes with *Boletellus* being the important genus represented in the Chalambar, Warrawonga and Kabali systems, (5 taxa); *Tylopilus* is represented by three taxa in the Warrawonga rainforest and in the established vegetation of the Noosa river system (Table 3). *Boletellus* is accompanied by *Phlebopus marginatus* (Berk.) Watl. & Gregory in the Chalambar rainforest which is significant since *Phlebopus silvaticus* Heinem. is found under similar conditions in the Korup rainforest. Other small bolete genera are represented by single or small numbers of taxa but as might be expected *Boletus*, when xerocomoid elements are included, is prominent. *Heimiella* is a very significant member because along with *Pulveroboletus ravenelii* (Berk. & Curt.) Murr. it shows some links with the rainforests of New Guinea and Malaysia (Corner, 1972; Watling & Hollands, 1990). Although not uniquely Australian, the high numbers of species of *Austroboletus* noted there highlights the significant role this genus has in the biota. There are very few species of bolete which are found throughout the sand-dune system or in eucalypt and rainforest communities.

When the Cooloola collections are compared with two areas south of the Sahara (a) the W Central African rainforest (e.g. Korup) and (b) the miombo woodland of Zambia (e.g. Kitwe/Misaka systems) the importance of the boletes in Australia becomes clear. On the basis of published records this generalization still holds. The wide planting of members of the Pinaceae in both continents has resulted in records of *Suillus* spp., which are specifically limited to the family. These are introduced boletes, but since only a few and then often the same taxa are involved the overall picture remains unchanged.

Tylopilus appears to be as important in the *Brachystegia* forests of the Copperbelt (Table 4) as it is in the eucalypt forests of Queensland, but *Tuboseta* is uniquely African. *Rubinoboletus* represented by only a few species is a very prominent member of the Korup biota. *Strobilomyces* which is a common genus in Australia would appear to be replaced in Zambia by *Afroboletus* although the former genus is in fact represented by several species in Zaire (Heinemann, 1954, 1966). *Boletellus* on the other hand is less well represented in Africa. There are apparently more species of bolete in Kitwe, viz. the *Brachystegia* forest, than in Korup, and Zaire is also rich in Boletaceae with some 82 recorded; (see Heinemann, 1954 & 1966, Heinemann & Rammeloo, 1980, 1983 & 1989). This contrasts markedly with only 13 taxa in East Africa (Pegler, 1977), 3 in Sri Lanka (Pegler, 1986) and 18 in the Lesser Antilles (Pegler, 1983)(see Table 5).

Comparison of the African and Australian native fungal biotas emphasises their difference from those of north temperate regions. Thus

Table 4. Generic profiles showing the number of species at Cooloola, Korup and Kitwe/Misaka woodlands.

Genera	Number of species		
	Cooloola (rainforest in brackets)	Korup rainforest	Kitwe/Misaka(M) Miombo
Afroboletus	-	-	1 (only at M)
Aureoboletus	-	-	1
Austroboletus	5	-	-
Boletellus	12(5)	3	1
Boletochaete	1	-	-
Boletus incl. *Xerocomus*	9(1)	6	13 (1 only at M) & 1 at Chati
Chalciporus	-	2	-
Fistulinella	(1)	2	-
Gyroporus	1	1	1
Heimiella	2(1)	-	-
Phlebopus	(1)	1	1
Pulveroboletus	1	1	5 (all but 1 at M also at Chati)
Rubinoboletus	2	4	1
Strobilomyces	2	-	-
Tuboseta	-	1	1 (also at Chati)
Tylopilus	10(3)	5	11
Veloporphyrellus	-	-	1
Paxillus	-	-	-
Phylloporus	6	1	1

See text for the third Miombo locality, Chati.

Leccinum is a dominant member of woods containing Betulaceae and Salicaceae and in North America Pinaceae; a few taxa are associated with Fagaceae. Corner (1972, 1974) recorded 4 *Leccinum* spp. from Malaysia and Heinemann (1966) *L. foetidum* Heinemann and 5 others from Zaire originally placed in *Krombholzia* (Heinemann, 1954) and showing aberrant features. Pegler (1977) recorded *L. foetidum* from East Africa. Very recently Bougher & Thiers (1991) have reported an indigenous taxon from Queensland associated with *Eucalyptus grandis* Hill ex. Maid. *Suillus* is also basically a North temperate genus and when found in the Southern Hemisphere is associated with planted conifers; members of the

Table 5. Comparison of boletoid fungi from tropical areas.

	Sri Lanka (Pegler, 1986)	East Africa (Pegler, 1977)	Central Africa*	Cameroon (this survey)	Lesser Antilles (Pegler, 1983)
Phylloporus	-	-	8	2	-
Boletoid/Xerocomoid	3	13(-3)	56	19	15
Tylopiloid	-	-	10	7	-
Boletellus	-	-	8	3	3

* *Flore Illustrée des Champignons d'Afrique Centrale.*
Figure in brackets shows questionable records.

Gomphidiaceae are also closely tied to conifers but unlike *Suillus* the family has not as yet been recorded from Central Africa or Australia. In New Zealand McNabb (1970) reported *G. maculatus* (Scop.: Fr.) Fr. with introduced larch. Amongst the lamellate boletes *Phylloporus* is strongly represented in Australia (Watling & Gregory, 1991), but very few species have been found in either the Korup or Kitwe sites (Heinemann & Rammeloo, 1987). *Paxillus* is very common in northern forests and again has been introduced extensively in Australia but was not found during the present study of the African communities outlined in this Chapter (Heinemann & Rammeloo, 1986).

Finally, there appears to be a distinct parallel between the boletes of northern fagaceous forests and southern hemisphere areas, although if not in species certainly in genera, e.g. *Aureoboletus, Strobilomyces* .

Non-bolete studies

Unfortunately, comparative data to those on boletes have not been similarly accumulated for other groups of larger fungi in the three areas studied. The members of Russulaceae in the eucalypt forests of Cooloola were so overwhelming that no data were collected (J. E. C. Aberdeen, pers. comm.). Personal observations also mirror the bolete picture with large numbers of taxa in the main vegetational types, viz. Warrawonga and Burwilla, falling off to the Kabali and to the Mutyi. In the Cooloola rainforests 9 taxa were recorded (Watling, unpublished data). However, for the Korup rainforest at least 41 different taxa were recorded until 1990, and there is every likelihood that another 60 will be added to this. This is similar to records for Zaire (103; Buyck, 1991), and in Zambia with *Brachystegia* etc. there are parallels judging from my own collections and those of Buyck (pers. comm.) made in 1991.

In Korup there are many annulate members of both *Russula* and *Lactarius*, (e.g. *R. annulata* Heim and *L. zenkeri* (Henn.) Sing.). Whereas some of these *Russula* species are confined to Central Africa, especially many of those placed in the Pelliculariae, others such as *L. pandani* have a wide range extending into East Central Africa and Madagascar. Pegler (1977) recorded a single species of *Lactarius* from East Africa and none from Sri Lanka (Pegler, 1986), and Heim (1955) recorded eighteen from Zaire. Pegler (1983) noted ten from the Lesser Antilles, including some taxa very closely related to those in Korup. Comparative figures for *Russula* are five in East Africa, three in Sri Lanka and twelve in the Lesser Antilles. Buyck (1990) in his thesis described 103, 70 new to science, from Zaire and adjacent areas.

Amanitaceae are also important members of the three communities. Aberdeen, Ross & Thompson (1989) recorded 39 taxa from Cooloola and 30 were recorded from Korup (Watling unpublished data); Beeli (1935) reported 26 from Zaire.

One genus of agarics absent from Australia but important both in the Copperbelt of Zambia and in Korup is *Termitomyces*. Although termites are found in Cooloola (20 are recorded; 2 new to science, Seymour, 1982) they are not in general associated with *Termitomyces*. The Copperbelt is very rich in taxa (Piearce, 1987). Four species have been recorded from Korup, viz. *T. clypeatus* Heim, *T. globulus* Heim & Goos, *T. microcarpus* (Berk. & Br.) Heim and *T. striatus* (Beeli) Heim. In addition *T. medius* Heim has been found in the adjacent area of Kumba in Cameroon. Heim (1958) recorded 10 species from the Congo.

In Cooloola, 47 potentially distinct taxa of *Cortinarius* have been collected, several of them new species, and undoubtedly many more are to be found (Høiland & Watling, 1990); *Inocybe* is also well represented in Australia. Significantly in both Korup and Kitwe no *Cortinarius* spp. have been found and only a handful of *Inocybe* spp.

Cantharellaceae is very strongly represented throughout Central Africa from the Bight of Biafra through Zaire to Zambia (Heinemann, 1959); over a dozen taxa are known, many occurring in huge troops which form an important source of food for native peoples. In Cooloola only 3 taxa were found during the survey. In contrast Sri Lanka has only two species (Pegler, 1986) and the Lesser Antilles one (Pegler, 1983).

Three species of potentially ectomycorrhizal hypogeous fungi have been collected in Korup, one being the unusual and little known

Corditubera staudtii Henn. In Zambia subterranean species are found with *Brachystegia* spp. including *Dendrogaster congolense* Dissing & Lange originally described from Zaire and seen at Chati Forest Reserve. In contrast Cooloola only sports *Hymenogaster aureus* Rodw.

Amongst the Ramariaceae, *Ramaria* is very well represented in Cooloola (Petersen & Watling, 1990); they are not known from Central Africa although they form an important part of the biota in Malaysia (Corner, 1950). Corner & Heinemann (1967) record 25 clavarioid fungi from Zaire covering 11 genera; *Clavulina cavipes* var. *ramosior* Corner has been found in both Korup and Zambia. Of the other branched clavarioid fungi *Lachnocladium zenkeri* P. Henn. was originally described from Bipinde in the Cameroon (mat. in E); *L. schweinfurthianum* P. Henn. is recorded from Zaire (Corner & Heinemann, 1967). *Coltricia cinnamomea* (Jacq.: Gray) Murr. and *C. spathulata* (Hook.) Murr., found with *Uapaca* are both found in Korup, whilst *C. oblectens* (Berk.) Cunn. occurs with *Tristania* and *Pultenaea* at Cooloola. Thoen & Ba (1989) listed the first in their account of ectomycorrhizal species in southern Senegal. The *Ganoderma lucidum* group, *Amauroderma* spp., *Pycnoporus* spp. and *Microporus* spp. are prominent in all three areas, indeed taxa in the last along with *Podoscypha* spp. are the most widespread genera in the Korup rainforest. *Microporus* is commonly found in the rainforest communities between the dunes in Cooloola. The occurrence of species of *Trametes* in the three areas is uncannily similar. Twenty-eight potential species of *Phellinus* are known from Cooloola, although this may be an exaggeration (Aberdeen *et al.*, 1989). *P. alliardii* (Bres.) Ryv. and *P. durissimus* (Lloyd) Ryv. are the commonest in Korup, where there are eleven authenticated taxa. Poroid and resupinate members of the Hymenochaetaceae are also very common.

Finally, *Lentinus* should be mentioned as this is an important component of the Korup forest where *L. tuber-regium* (Fr.) Fr. provides two sources of food. The large sclerotia are ground up to make a flour for thickening sauces and the basidiomes are sliced and added to stew. *L. squarrulosus* Mont., probably the most common species, is also eaten throughout Central Africa. Other species so far recorded include *L. velutinus* Fr., *L. brunneofloccosus* Pegler, and *L. sajor-caju* (Fr.) Fr. Pegler (1977) recorded twelve species of *Lentinus* from East Africa, nine from Zaire (Pegler, 1972), six from the Lesser Antilles (Pegler, 1983) and eleven from Sri Lanka (Pegler, 1986).

Australia is also rich in *Lentinus* spp. but none was found at Cooloola; only two species of *Pleurotus* were recorded there, poor in comparison with the rainforests of Korup, and a single undetermined species of *Armillaria*. *A. fuscipes* Petch has been found in Korup (Watling, 1992).

The saprotrophic agarics include, at all three sites, species of *Marasmiellus*, *Marasmius* and *Mycena* especially the second, but only in Korup are aerial hammocks composed of marasmioid mycelium found (Hedger, 1990). In both Cooloola and Kitwe the grassy forests are too open for such structures to be formed.

South American studies

During the same period as the author's study of Cameroon fungi, André de Meijer has studied the fungi of the Atlantic coast of Brazil intensively. The sites studied included dense ombrophilous forests; mixed ombrophilous forests with *Araucaria angustifolia* (Bert.) O. Kuntze; seasonal deciduous alluvial irregularly flooded forests in the valley of the Iguaçu river; arboreal savanna with gallery forest and 'restinga' on marine sands.

At least 9 species of bolete were recorded viz *Austroboletus festivus* (Sing.) Wolfe from the restinga, *Gyrodon exiguus* Singer & Digillio from both kinds of ombrophilous forest, *G. rompelii* (Pat. & Rick) Singer from both the mixed and seasonal deciduous forests and *Phaeogyroporus beniensis* Singer & Digillio from the pluvial forest; the last species is equivalent to the *Phlebopus marginatus* and *P. silvaticus* from wet sclerophyll and rainforest of Cooloola and Korup respectively.

Neopaxillus echinospermus (Speg.) Singer is the only member of the Paxillaceae associated with the *Araucaria* forest; *Phylloporus* and *Paxillus* are known from other more temperate areas of South America (Horak, 1979). *Neopaxillus* is interesting as the genus is recorded along with *Paxillus russuloides* Petch from Sri Lanka by Pegler (1986; as *N. reticulatus* (Petch) Pegler). Paxillaceae is neither recorded from East Africa (Pegler, 1977) nor the Lesser Antilles (Pegler, 1983), except the white-spored genus *Hygrophoropsis*.

In common with the observations in Australia and Africa discussed above, three species of *Suillus* growing with exotic conifers have been recognized, *S. cothurnatus* Singer, *S. granulatus* (L.: Fr.) Kuntze and *S. luteus* (L.: Fr.) S.F. Gray. In addition *Chalciporus piperatus* (Bull.: Fr.) Bat., which is usually found either in birch woods or mixed pine-birch stands in the British Isles, was also recorded from pine plantations in Australia and South America. All these four taxa have been introduced into South America. The author has also been sent for determination from de Meijer a bolete very close to *Xerocomus coccolobae* Pegler, described originally from the Lesser Antilles (Pegler, 1983); it comes close to *X. brasiliensis* Singer and *X. hypoxanthus* Singer both from the same geographic area. At least two other members of the *Boletus-Xerocomus* complex occur in the vegetation types under observation by de Meijer (unpublished data). In addition to *Coltricia spathulata* also found on several occasions in Korup, often with *Uapaca*, de Meijer recorded two species of *Cantharellus*, including *C. guyanensis* Mont., and *Gomphus subclavaeformis* Corner (Gomphaceae; see *Ramaria* above) not seen in either Korup or Kitwe. The last genus is known from Zaire (Heinemann, 1959 as *Neurophyllum*).

Some of the stipitate hydnaceous fungi found by de Meijer are worth further investigation as some are known to be ectomycorrhizal; several were described by Geesteranus (1979) from Zaire and *Hydnum repandum* Fr. has been collected at Cooloola. The saprotrophic *Donkia sanguinea* Geesteranus is a feature of Korup and probably found there more often than in all other localities put together.

Thus the bolete biota of these Brazilian communities compared with the African and Australian sites studied is exceedingly poor. A similar pattern is seen in the Russulaceae where eight *Lactarius* spp. including *L. panuoides* Singer and *L. venezuelensis* Dennis and five *Russula* spp. including *R. puiggari* (Speg.) Singer and material close to *R. theissenii* Rick are known from de Meijer's sites, and by the Cortinariaceae three each in *Cortinarius* and *Inocybe*. It is significant that *R. puiggari* is in the same group (PellicularIae) as many of the Korup taxa. Examination of the potential ectomycorrhizal hosts of these communities shows the reasons for the absence of large numbers of ectomycorrhizal genera. The presence of ectomycorrhizal legumes is the important factor in Korup and Kitwe and myrtaceous plants in Cooloola; *Pultenaea* (Leguminosae) may also be ectomycorrhizal in Cooloola (Watling & Gregory, 1989). Amanitaceae is only represented in de Meijer's studies by the introduced *A. muscaria* (L. Fr.) Hook. *Araucaria angustifolia* is endomycorrhizal.

Similar observations to those of de Meijer for Russulaceae, Boletaceae, etc have been made by Courtecuisse in former French Guinea (Courtecuisse, 1990; pers. comm.), again emphasising the importance of the constituents of the aborescent flora. Nowhere was it demonstrated more clearly to the author than when passing from rainforest on the Lamington Plateau, Queensland into groves of Antarctic Beech (*Nothofagus moorei* (F. Muell.) Krasser) on rocky promontories and high ridges. On so doing Russulaceae were seen immediately. *Nothofagus* is ectomycorrhizal!

The results for ectomycorrhizal agarics contrast markedly with those for families such as the Bolbitiacaceae where 42 taxa have been recognised amongst de Meijer's collections (Watling, in press) whereas only 21 adventitious species are recorded from Zaire (Watling, 1974) and little more than a handful in Cameroon, although this includes *C. ochraceodisca* Watl. described from Central Africa.

Discussion

As regards diversity, the eucalypt woodlands of Cooloola can be compared favourably with the miombo woodland of the Copperbelt of W. Central Africa, the major components of the vegetation viz. Myrtaceae and Leguminosae (Caesalpinoideae), are ectomycorrhizal and support a full range of Boletaceae *s. lato*, Russulaceae, Amanitaceae, etc. In the former, the Cortinariaceae play an important role but are insignificant in miombo woodland. Experimental results suggest that the dominants of these same woodlands may support endomycorrhizal fungi under certain conditions something which is probably very rare or absent in boreal woodlands of Europe and North America dominated as they are by Fagaceae and Coniferae, and in many parts of temperate Australia where *Nothofagus* (Fagaceae) is the climax tree.

However, the rainforest of Cooloola is totally different from that of the Korup lowland rainforest. This may be a direct result of the constituent cormophytes. Not only are the numbers of basidiomes of larger fleshy fungi low in the rainforest of the Cooloola sand-dune system but the families found include only a few members of species which one would suspect as ectomycorrhizal. In contrast, the Korup rainforest supports Amanitaceae, boletes, at least 10 species of Cantharellaceae and innumerable Russulaceae; many species are undoubtedly new to science.

In Korup the Cortinariaceae is dramatically reduced, only six members were recorded in the present survey, and one (*Inocybe* sp.) growing with caesalpinoid legumes in a nursery 10 km from the rainforest. Whereas *Cortinarius* is widespread at the Australian site, with many suspected primitive elements, in the Cameroon and the two Zambian areas the genus is unknown or extremely rare. *Cortinarius* alone contrasts the Australian biota with the two African sites. This difference is very dramatic and is repeated also in the biotas of East Africa (Pegler, 1977) and Sri Lanka (Pegler, 1986) where there are no *Cortinarius* spp. recorded and in *Inocybe* only one introduced species in the former and three in Sri Lanka. In the Lesser Antilles only 10 species of *Inocybe* are recorded (Pegler, 1983) but no species of *Cortinarius*. One member of the genus *Inocybe* (subgenus *Mallocybe*; Kuyper, 1986) found at Korup is very common occurring in many populations and has been found from 1984 onwards.

In the three study areas *Amanita* spp. covered the main sections of the genus from those with amyloid basidiospores to those with inamyloid basidiospores and those lacking a ring to those with either a membranous or friable powdery universal veil.

The African and Australian sites were similar in the high numbers of tylopiloid boletes found in contrast to Europe's two or three and North America's small number, the majority in the *T. felleus* complex, whilst *Fistulinella* and the two large sections of *Boletellus* represented, viz. those with faintly ornamental basidiospores and those with strongly developed costae, have distinct members in Korup and Cooloola areas. *Austroboletus* is confined to the Australian site. *Afroboletus* is found in Zambia whereas *Strobilomyces* occupies a similar niche in Cooloola; neither was found in Korup. *Phlebopus* is a feature of all three vegetation types but *Rubinoboletus* is a particular feature of the Korup rainforest. *Pulveroboletus* occurs in the sclerophyll forests and not the rainforest thus resembling the drier oak forests of Central North America.

Finally, for the Russulaceae one outstanding difference between the African sites and Cooloola is the prominence in Korup of ephemeral, delicate, annulate, often strongly pectinate members of the genus *Russula*. Annulate members of the genus *Lactarius* characteristic of both Korup and the Copperbelt are lacking in Cooloola. In all areas species with a basidiome facies which would be considered more normal by European mycologists also occur. The annulate, pectinate *Russula* spp. are considered rather primitive (Heim., 1937; Singer 1951) and are also found in Central and South American localities suggesting some earlier link between the areas in parallel to some of the angiosperm genera. There is some suggestion that the *Brachystegia* forest is a derivative of the rainforest by attenuation and modification of the forest margin communities to savannah and park-land (Sowunmi in Lawson, 1986). *Lactarius pandani* and related species may have been associated with this supposed movement and speciation of *Brachystegia* in the waves of drying out and expansion of wetter regimes.

From personal observations the rainforest of Korup is more like the temperate wet sclerophyll of Tasmania than the rainforest of Cooloola. In Tasmania the dominant trees are *Eucalyptus* and *Nothofagus* with their accompanying boletes and Russulaceae, many of the latter family being gasteroid in nature. Furthermore the rainforest of Korup resembles that of Malaysia in that the dominant trees in many communities are also ectomycorrhizal but belong, however, to a different family, the Dipterocarpaceae; see above for the discussion on the miombo where *Marquesia* (Dipterocarpaceae) is a component (Singer & Singh, 1971; W. Smits pers. comm.) These two communities, however, differ markedly. In Korup the caesalpinoid legumes are generally in groves scattered in a non-ectomycorrhizal area, whereas in Peninsula Malaysia the dipterocarps are scattered through the entire vegetation. This explains the patch-work basidiome production in the former and the blanket-like coverage found in the latter (Watling unpubl. data).

In northern communities arborescent willows often form corridors through and between larger forest types otherwise lacking members of the Salicaceae. They follow water courses or damper areas and often form islands within communities such as grasslands or dune-systems otherwise lacking arborescent species. It has been suggested that in Britain the creeping willow (*Salix repens* L.), more akin to the arborescent forms than the dwarf willows of European mountains and arctic-alpine tundra, associates with the fungi one might expect of the vegetation which would have naturally existed prior to any extensive clearance and grazing (Watling, 1981). Arborescent willows, although they may be either ecto- or endomycorrhizal or both in any one locality, support a rich macromycete biota associated with their roots. Many of the fungi involved are recorded from other plant-communities, indeed mycorrhizas have been synthesised in the laboratory using non-willow hosts. It might be hypothesised, therefore, that the willows act as a bank of potential ectomycorrhizal species which can, given the opportunity, colonize other suitable hosts (Watling, 1992).

In the Korup rainforest, field observations suggest that *Uapaca staudtii* plays the same role as the willows in the North temperate zone. *U. staudtii*, parallel with the willows in Europe, colonizes stream-sides, soaks and wet patches which in Korup may even be completely inundated at certain periods of the year. This tree is found in the lower reaches of the Mana/Ndian river system in an interlocking network with leguminous, potentially ectomycorrhizal hosts which form groves on generally slightly higher ground. In the dry season the basidiomes of members of the Russulaceae, Boletaceae and Amanitaceae are found consistently associated with *Uapaca* in ones and twos but where a fairly wide spectrum of taxa is found; with the legumes in the same areas basidiomes were only found with trees on the rather damper soils. In the early periods of the rainy season, before the soil becomes saturated and in places flooded, basidiomes are found extensively both with *Uapaca* and with the leguminous hosts but this time not in small numbers but as hundreds of basidiomes per taxon, and many populations of the same species in a given area. Perhaps *Uapaca* is playing the same role as the willows in the temperate zone and acting as a bank for ectomycorrhizal fungi.

Just as *Laccaria laccata* (Scop.: Fr.) Cooke and *Cortinarius uliginosus* Berk., etc. fruit on mossy branches of fallen or curved willows in carrs under deep shade above the standing summer water level, so *Boletellus* spp. and *Boletus* spp. fruit on the stilt roots of *Uapaca*, perhaps finding a suitable water capacity or other favourable factor there. Fruiting on wood can similarly be seen in Europe in many ectomycorrhizal fungi, eg. *Paxillus involutus* (Batsch: Fr.) Fr., *Boletus badius* Fr. The author has seen *Boletellus emodensis* (Berk.) Sing. fruiting at shoulder-level and above on the trunk of a living stringy bark in the Lamington Rainforest National Park, Queensland perhaps finding a suitable site above the large accumulation of woody debris, where otherwise discharged spores would be hindered in their dispersal, and a soft putrescent basidiome attacked by hyperparasites in such a poorly ventilated place. Christy, Sollins & Trappe (1982) have even speculated that fallen wood permeated by potential mycorrhizal roots and hyphae of ectomycorrhizal fungi is an important nursery bed for developing seedlings in the conifer forests of North Western America.

Conclusion

Early collections of larger fungi from tropical sites lacked adequate field data, but from the information outlined above for ectomycorrhizal fungi it can be seen that specific relationships exist. The same undoubtedly also applies to saprotrophic larger fungi. In future it will be essential to carry out careful analysis of the cormophyte biota as a prerequisite to forming any opinion on the total ecosystem. To place the fieldwork on a more scientific basis there is now a need to expand root and inoculum studies and ascertain whether rainforest/fungal relationships are facultative or obligate in nature (Read, 1991). The litter decomposing abilities of these ectomycorrhizas also need to be critically assessed in the light of work in South America (Singer & Araujo 1979).

Acknowledgements I am very grateful for the help given to me by Drs J. A. Aberdeen and C. Thompson in Queensland, to Hudson Muthali in Zambia, A. Allo and Drs S. Gartlan and J. Rother in Cameroon, to André de Meijer for sharing his notes and sending me important, instructive material from Brazil, and especially to Dr Ian Alexander for help, encouragement and interesting discussions especially whilst journeying through Cameroon and Zambia. Finally I would like to thank Evelyn Turnbull for assistance in the examination of the material. This work would not have been possible but for travel grants from the British Council and the Royal Society of Edinburgh.

References

Aberdeen, J. E. C., Ross, D. J. & Thompson, C. H. (1989). *Studies in Landscape Dynamics in the Cooloola-Noosa River Area, Queensland. 7 Larger Fungi.* CSIRO Division of Soils Divisional Report No. 100, 1-93. CSIRO: St. Lucia.

Anon. (1975). *Vegetation of the Copperbelt.* Research Bulletin No. 20. Forest Department: Kitwe.

Beeli, M. (1935). *Flore Iconographique des Champignons du Congo.* Fasc. 1. *Amanita & Volvaria*, pp. 11-27. Le Jardin Botanique de l'État: Brussels.

Bougher, N. L. & Thiers, H. (1991). An indigenous species of *Leccinum* (Boletaceae) from Australia. *Mycotaxon* **42**, 255-262.

Buyck, B. (1990). *Revision du genre* Russula *Persoon en Afrique Centrale* Ph.D. Thesis, University of Ghent.

Buyck, B. (1991). Russulales at International Mycological Congress-4. *Russulales Newsletter* **1**, 4-6.

Christy, E. J., Sollins, P. & Trappe, J. M. (1982). First-year survival of *Tsuga heterophylla* without mycorrhizal and subsequent ectomycorrhizal development on decaying logs and mineral soil. *Canadian Journal of Botany* **60**, 1601-1605.

Corner, E. J. H. (1950). A monograph of *Clavaria* and allied genera. *Annals of Botany Memoir* **1**, 1-740.

Corner, E. J. H. (1972). Boletus *in Malaysia.* Government Printing Office: Singapore.

Corner, E. J. H. (1974). *Boletus* and *Phylloporus* in Malaysia: further notes and descriptions. *Gardens' Bulletin, Singapore, Supplement* **27** , 1-16.

Corner, E. J. H. & Heinemann, P. (1967). *Flore Iconographique des Champignons du Congo.* Fasc. 16. *Clavaires et* Thelephora, pp. 309-324 . Le Jardin Botanique de l'État: Brussels.

Courtecuisse, R. (1990). Mycological Inventory in French Guiana. First Results & Plans. In *Abstracts, Fourth International Mycological Congress*, (ed. A. Reisinger & A. Bresinsky), 1A-15/1 . University of Regensberg: Regensberg, Germany.

Gartlan, J. S. (1974). The African forests and problems in conservation. *Symposium of the 5th Congress of the International Primate Society*, (ed. Y. Kendo), pp. 509-524. Japanese Science Press: Nagoya.

Gartlan, J. S., Newbery, D. M., Thomas, D. W. & Waterman, P. G. (1986). The influence of topography and soil phosphorus on the vegetation of Korup Forest Reserve, Cameroon. *Vegetatio* **65**, 131-148.

Geesteranus, R. Maas (1979). *Flore Iconographique des Champignons du Congo.* Fasc. 17. *Hydnum*, pp. 325-338 . Le Jardin Botanique de l'État: Brussels.

Hedger, J. N. (1990). Fungi in the tropical forest canopy. *The Mycologist* **4**, 200-202.

Heim, R. (1937). Les Lactario Russules du domaine oriental de Madagascar. Essai sur la classification et la phytogéine des Asterosporales. *Prodrome à une Flore Mycologique de Madagascar* **1**, 21-196.

Heim, R. (1955). *Flore Iconographique des Champignons du Congo. Fasc. 4. Lactarius*, pp. 83-97. Le Jardin Botanique de l'État: Brussels.

Heim, R. (1958). *Flore Iconographique des Champignons du Congo. Fasc. 7. Termitomyces*, pp. 139-151. Le Jardin Botanique de l'État: Brussels.

Heinemann, P. (1954). *Flore Iconographique des Champignons du Congo. Fasc. 3. Boletineae*, pp. 51-78. Le Jardin Botanique de l'État: Brussels.

Heinemann, P. (1959). *Flore Iconographique des Champignons du Congo. Fasc. 8.* Cantharellineae, pp. 153-165. Le Jardin Botanique de l'État: Brussels.

Heinemann, P. (1966). *Flore Iconographique des Champignons du Congo. Fasc. 15.* Hygrophoraceae, *Laccaria* & Boletineae II (complément), pp. 279-308. Le Jardin Botanique de l'État: Brussels.

Heinemann, P. & Rammeloo, J. (1980). *Flore Illustrée des Champignons d'Afrique Centrale.* Fasc. 7. Boletineae, pp. 123-131. Le Jardin Botanique de l'État: Brussels.

Heinemann, P. & Rammeloo, J. (1983). *Flore Illustrée des Champignons d'Afrique Centrale.* Fasc. 10. Gyrodontaceae, pp. 173-198. Le Jardin Botanique de l'État: Brussels.

Heinemann, P. & Rammeloo, J. (1986). *Flore Illustrée des Champignons d'Afrique Centrale.* Ibid. Fasc. 12. Agariceae, pp. 249-271. Le Jardin Botanique de l'État: Brussels.

Heinemann, P. & Rammeloo, J. (1987). *Flore Illustrée des Champignons d'Afrique Centrale.* Fasc. 13. *Phylloporus*, pp. 277-309. Le Jardin Botanique de l'État: Brussels.

Heinemann, P. & Rammeloo, J. (1989). *Flore Illustrée des Champignons d'Afrique Centrale.* Fasc. 14. *Suillus* & *Tubosaeta*, pp. 313-335. Le Jardin Botanique de l'État: Brussels.

Hoffmann, P. F. (1991). Did the Break of Laurentia turn Gondwanaland inside-out? *Science* **252**, 1409-1411.

Høiland, K. & Watling, R. (1990). Some *Cortinarius* spp. (Agaricales) of the Cooloola Sand-Mass, Queensland, Australia. *Plant Systematics and Evolution* **171**, 135-146.

Horak, E. (1979). Paxilloid Agaricales in Australasia. *Sydowia* **32**, 154-166.

Kuyper, T. W. (1986). A revision of the genus *Inocybe* in Europe. *Persoonia* **3** (supplement), 1-247.

McNabb, R. F. R. (1970). A Record of *Gomphidius maculatus* (Agaricales) in New Zealand. *New Zealand Journal of Botany* **8**, 320-325.

Morris, B. (1987). *Common Mushrooms of Malawi*. Fungiflora: Oslo.

Newbery, D. M., Alexander, I. J., Thomas, D. W. & Gartlan, J. S. (1988). Ectomycorrhizal rainforest legumes and soil phosphorus in Korup National Park, Cameroon. *New Phytologist* **109**, 433-450.

Pegler, D. N. (1972). *Flore Illustrée des Champignons d'Afrique Centrale*. Fasc. 1. 1-26. Lentineae (Polyporaceae), Schizophyllaceae et espèces lentinoides et pleurotoides des Tricholomataceae. Le Jardin Botanique de l'État: Brussels.

Pegler, D. N. (1977). A Preliminary Agaric Flora of East Africa. *Kew Bulletin, Additional Series* **6**, 1-615.

Pegler, D. N. (1983). Agaric Flora of the Lesser Antilles. *Kew Bulletin, Additional Series* **9**, 1-668.

Pegler, D. N. (1986). Agaric Flora of Sri Lanka. *Kew Bulletin Additional Series* **12**, 1-519.

Petersen, R. H. & Watling, R. (1990). New or Interesting *Ramaria* Taxa from Australia. *Notes from the Royal Botanic Garden, Edinburgh* **46**, 141-159.

Piearce, G. (1981). *An Introduction to Zambia's Wild Mushrooms.* Forest Department Production: Zambia.

Piearce, G. (1987). The Genus *Termitomyces* in Zambia. *The Mycologist* **1**, 111-116.

Read, D. J. (1991). Mycorrhizas in ecosystems - Nature's response to the 'Law of the Minimum'. In *Frontiers in Mycology*, (ed. D. L. Hawksworth) pp. 101-130. CAB International: Wallingford.

Seymour, J. (1982). The Dunes of Cooloola. *Ecos* **30**, 3-11.

Singer, R. (1951). Agaricales in Modern Taxonomy. *Lilloa* (1949) **22**, 1-831.

Singer, R. & Araujo, I. J. S. (1979). Litter decomposition and ectomycorrhiza in amazonian forests. *Acta Amazonica* **9**, 25-41.

Singer, R. & Singh, B. (1971). Two new ectotroph-forming boletes from India. *Mycopathologia et Mycologia Applicata* **43**, 25-33.

Sowunmi, M. A. (1986). Change of vegetation with time. In *Plant Ecology in West Africa*, (ed. G. W. Lawson), pp. 273-307. J. Wiley & Sons: Chichester.

Thoen, D. & Ba, A. M. (1989). Ectomycorrhizas and putative ectomycorrhizal fungi of *Afzelia africana* Sm. and *Uapaca guineensis* Müll. Arg. in southern Senegal. *New Phytolologist* **113**, 549-559.

Thompson, C. H. (1981). Podzol chronosequences on coastal dunes of eastern Australia. *Nature* **291**, 59-61.

Thompson, C. H. & Moore, A. W. (1984). Studies in Landscape Dynamics in the Cooloola-Noosa River Area, Queensland. *CSIRO, Division of Soils Divisional Report*, **73**, 1-93. CSIRO: St. Lucia.

Watling, R. (1974). *Flore Illustrée des Champignons d'Afrique Centrale* Fasc. 3. Bolbitiaceae, pp. 55-71. Le Jardin Botanique de l'État: Brussels.

Watling, R. (1981). Relationships between macromycetes and the development of higher plant communities. In *The Fungal Community*, (ed. D. T. Wicklow & G. C. Carroll), pp. 427-458. Marcel Dekker: New York.

Watling, R. (1992). Macrofungi associated with British Willows. *Proceedings of the Royal Society of Edinburgh* **98B**, 135-147.

Watling, R. (1992). *Armillaria staude* in the Cameroon Republic. *Persoonia* **14**, 483-491.

Watling, R. & Gregory, N. M. (1986). Observations on the Boletes of the Cooloola Sandmass, Queensland & Notes on their Distribution in Australia Part 1.Introduction and keys. *Proceedings of the Royal Society of Queensland* **97**, 97-128.

Watling, R. & Gregory, N. M. (1989). Observations on the Boletes of the Cooloola Sandmass, Queensland & Notes on their Distribution in Australia Part 2D. *Proceedings of the Royal Society of Queensland* **100**, 31-47.

Watling, R. & Gregory, N. M. (1991) Observations on the Boletes of the Cooloola Sandmass, Queensland & Notes on their Distribution in Australia Part 3. Lamellate taxa. *Edinburgh Journal of Botany*, **48**, 353-391.

Watling, R. & Hollands, R. (1990). Boletes from Sarawak. *Notes from the Royal Botanic Garden, Edinburgh* **46**, 405-422.

Chapter 11

Looking for ectomycorrhizal trees and ectomycorrhizal fungi in tropical Africa

Daniel Thoen

Fondation Universitaire Luxembourgeoise, Avenue de Longwy 185, B-6700 Arlon, Belgium

Introduction

The first tropical ectomycorrhizas (ECM) were discovered by Peyronel & Fassi (1957) on *Gilbertiodendron dewevrei* (De Wild.) J. Léonard, a dominant legume tree of the Congolian lowland rainforest, and on a liana, *Gnetum africanum* Welw., found in evergreen and gallery forests (Fassi, 1957). Later, the same workers discovered four new ectotrophic genera, namely *Anthonotha*, *Julbernardia*, *Monopetalanthus* and *Paramacrolobium* (Peyronel & Fassi, 1960), whereas Redhead (1960) discovered ECM on *Afzelia*, all belonging to the Caesalpinioids. In addition, Högberg & Nylund (1981) discovered ECM on woodland trees, bringing the number of ectomycorrhizal species to 14. Since then, our knowledge has significantly increased. Over the last ten years, with contributions by Alexander & Högberg (1986), Newbery *et al.* (1988), Thoen & Ba (1989), Thoen & Ducousso (1989), the number of ectomycorrhizal genera has risen to 19 and the number of species to 51. However, the mycorrhizal status of most tropical trees is still unknown and needs to be examined. A good historical review on ECM in tropical Africa has been published recently (Fassi & Moser, 1991).

The ectomycorrhizal trees in tropical Africa

Most of the known ectomycorrhizal trees belong to the Leguminosae (Caesalpinioideae), followed by the Euphorbiaceae, the Dipterocarpaceae, Proteaceae and the Leguminosae (Papilionoideae) (Table 1). If an extrapolation of all the species of known African ectomycorrhizal genera is made then about 300 ectotrophs can be listed. Some other genera might be ectotrophic in Africa: *Baphia*, *Swartzia*

Table 1. Native ectomycorrhizal trees of tropical Africa

Taxonomic groups	Genus	Number of species	Number of ECM species*	% ECM species
Caesalpinioideae				
Detarieae	*Afzelia*	7	6	86
Amherstieae	*Anthonotha*	30	6	20
	Aphanocalyx	3	1	33
	Berlinia	18	3	17
	Brachystegia	36	10	28
	Didelotia	12	1	8
	Gilbertiodendron	28	1	3.5
	Isoberlinia	5	1	20
	Julbernardia	11	3	27
	Microberlinia	2	1	50
	Monopetalanthus	20	2	10
	Paramacrolobium	1	1	100
	Tetraberlinia	3	2	66
Papilionoideae				
Sophoreae	*Pericopsis*	3	1	33
Subtotals		*179*†	*39*	22
Dipterocarpaceae				
	Marquesia	4	1	25
	Monotes	36	2	5.5
Euphorbiaceae				
	Uapaca	50	7	14
Proteaceae				
	Faurea	18	1	5.5
Subtotals		*108*‡	*11*	*10*
Grand totals		**287**	**50**	**17.5**

* Compiled from: Peyronel & Fassi, 1957, 1960; Fassi & Fontana, 1962; Jenik & Mensah, 1967; Redhead, 1968a, 1974; Högberg, 1982, 1986; Högberg & Piearce, 1986; Newbery *et al.*, 1988; Thoen & Ba, 1989; Thoen & Ducousso, 1989.

† Lock (1989).

‡ Mabberley (1987).

(Papilionoideae), *Eugenia* (Myrtaceae), *Salix* (Salicaceae) and *Allophyllus* (Sapindaceae).

Regarding the distribution of ectomycorrhizal legume trees (Lock, 1989) within the phytochoria defined by White (1983), the Guineo-Congolian region has the highest number of species, followed by

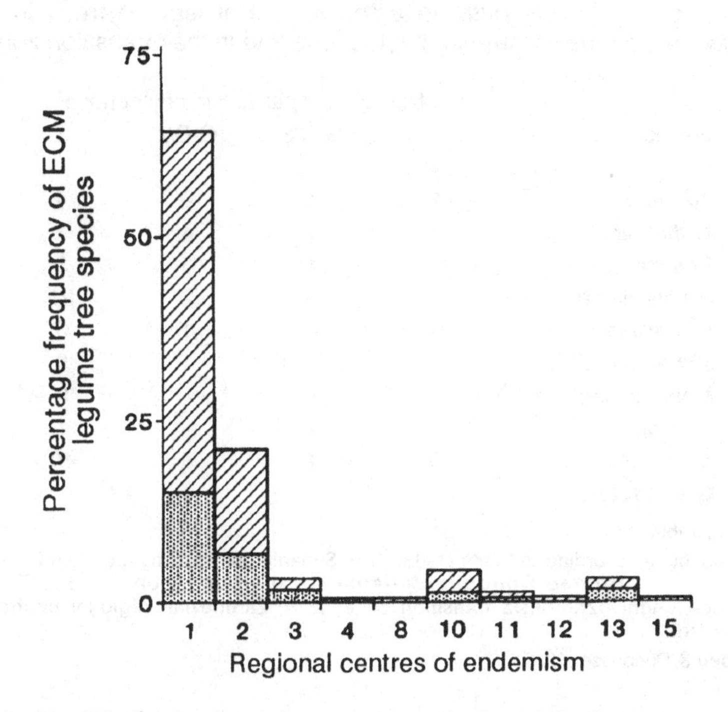

Fig. 1. Frequency of putative (oblique lines) and known (dotted lines) ectomycorrhizal legume trees in tropical Africa according to the phytochoria defined by White (1983) and the data of Lock (1989). Regional centres of endemism: 1, Guineo-Gongolian; 2, Zambezian; 3, Sudanian; 4, Somalia-Masai; 8, Afromontane. Regional transition zones: 10, Guineo-Congolia/Zambezia; 11, Guineo-Congolia/Sudania. Regional mosaics: 12, Lake Victoria; 13, Zanzibar-Inhambane; 15, Tongaland-Pondoland.

the Zambezian region (Fig. 1). Compared with the latter the Sudanian region is poor in ectotrophs. The same trends occur in the transition zones (4 versus 8 legume species; Table 2). Only three ectotrophic legumes are common to both Sudanian and Zambezian regions (White, 1965). Two important ectotrophic genera of the Zambezian region, *Brachystegia* and *Julbernardia*, are absent in the Sudanian phytochoria and in the corresponding transition zone. Fig. 2 shows that ectomycorrhizal legumes occur in several vegetational types (Lock, 1989) with the forest type having the highest number of species, followed by woodlands, the main

Table 2. Number of putative ectomycorrhizal legume trees in the Sudanian and the Zambezian phytochoria and in their transition zones

	Number of species* in phytochoria†			
Genus	S	G-C/S	G-C/Z	Z
Afzelia	1	2‡	–	4
Anthonotha	1	1	1	1
Berlinia	–	1	3	3
Brachystegia	–	–	2	21
Isoberlinia	3	–	1	3
Julbernardia	–	–	1	5
Monopetalanthus	–	–	–	2
Pericopsis	1	–	–	1
Subtotals	*6*	*4*	*8*	*40*
Grand totals	**10**		**48**	

* Lock, 1989.

† Phytochoria according to White (1983): S = Sudanian regional centre of endemism, G-C/S = Guineo-Congolia/Sudania transition zone, G-C/Z = Guineo-Congolia/Zambesia transition zone, Z = Zambezian regional centre of endemism.

‡ Thoen & Ducousso, 1989.

---

vegetational type of both Zambezian and Sudanian regions (Högberg & Nylund, 1981).

In the lowland rain forest, some ectotrophs occur over a wide area and dominate the canopy, as for instance *Gilbertiodendron dewevrei* (see Fig. 47 in *Distributiones Plantarum Africanum* **2**, Anon.197). Figure 3 shows the main regions, in the author's opinion, where ectotrophs occur in tropical Africa with the number of putative ectomycorrhizal legumes given within brackets (Lock, 1989). Compared with the map published by Read (1991), the extent of tropical ectotrophs is much greater. Ectotrophs represent only a small percentage of all tree species, but in terms of number of trees and surface cover they often dominate in both rain forest (Newbery *et al.*, 1988) and woodland (Högberg & Piearce, 1986).

The author's experience is limited to the Shaba district in Zaire (Thoen, 1974 and unpublished data), and in western Africa to Senegal (Thoen & Ba, 1989), to the Fouta Djalon plateau in French Guinea (Thoen & Ducousso, 1989) and to Niger (Ibrahim, Thoen & Piérart, unpublished). In the Shaba district ectomycorrhizal fungi were common in the *Brachystegia* woodland (Miombo), the *Marquesia macroura* Gilg.

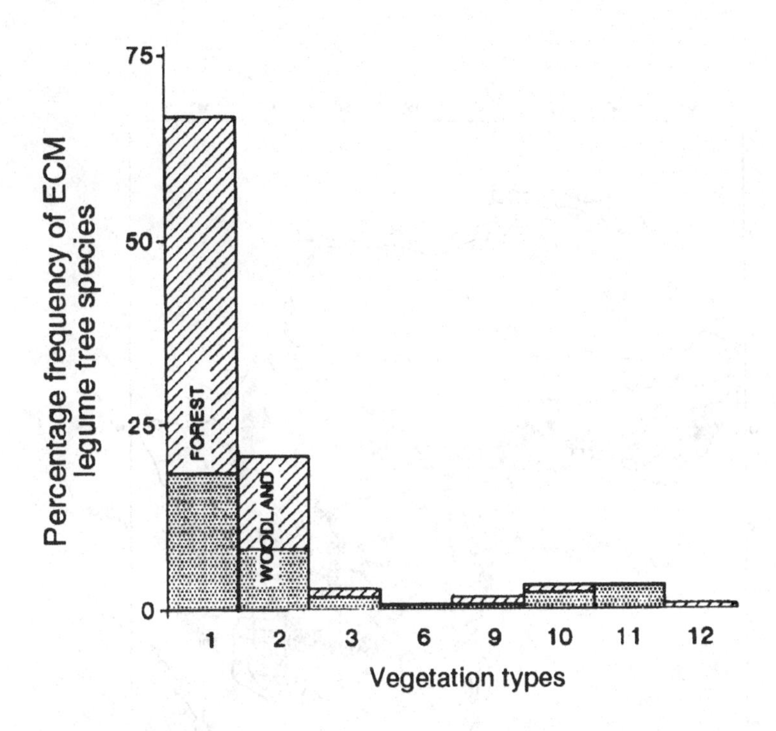

Fig. 2. Frequency of putative (oblique lines) and known (dotted lines) ectomycorrhizal legume trees in tropical Africa according to the vegetation types defined by White (1983) and the data of Lock (1989). Vegetation types: 1, forest; 2, woodland; 3, bushland and thicket: 6, wooded grassland; 9, scrub forest: 10, transition woodland; 11, scrub woodland; 12, strand vegetation.

---

woodland and in the dry evergreen forest (Muhulu). Planted stands of eucalypts and species of pine were shown to be ectomycorrhizal (Thoen, 1974).

In western Africa the roots of about one hundred tree species were examined to establish their mycorrhizal status. Among the exotics, ECM were found on *Pinus*, *Eucalyptus*, *Melaleuca*, *Casuarina*, Australian phyllodous *Acacia* and *Cinnamomum*.

Species of tree genera reported as ectomycorrhizal outside Africa, such as *Cassia*, *Dalbergia*, *Dialium* and *Erythrophleum* (Chalermpongse, 1984), *Allophyllus* (Alexander, 1981), *Acacia* and *Parkia* (Warcup, 1980; Reddel

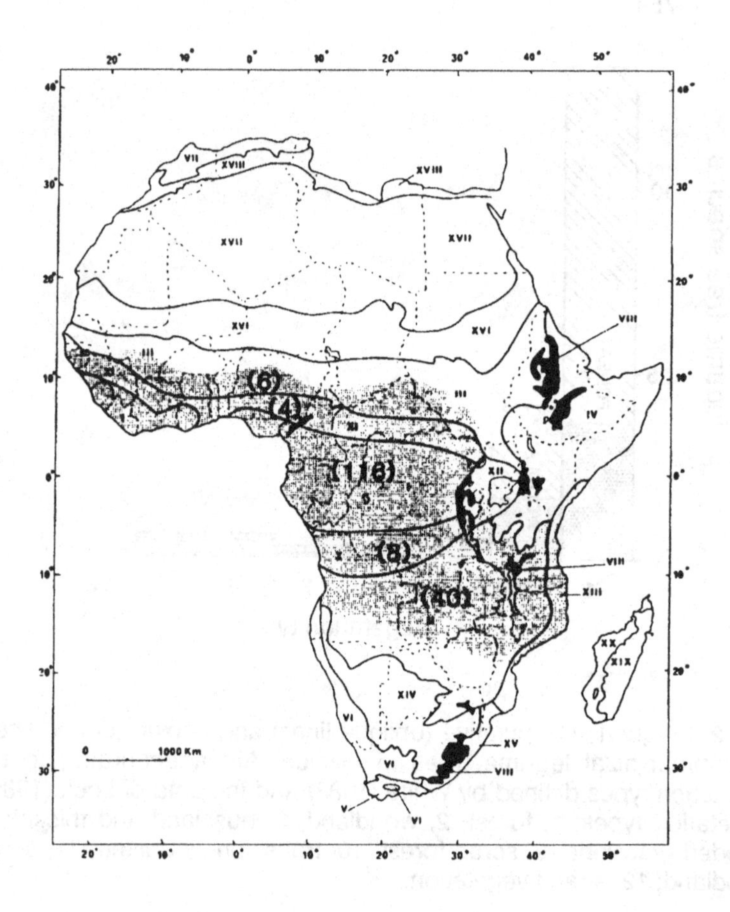

Fig. 3. Phytochoria of Africa (White, 1983) and indicative area (grey zone) where ectotrophs occur in tropical Africa. Estimated numbers of ectotrophic legume trees (Lock, 1989) in the main phytochoria (I, II, X, XI, XII) are given in brackets. I. Guineo-Congolian region. II. Zambesian region. III. Sudanian region. IV. Somalia-Masai region. V. Cape region. VI. Karoo-Namib region. VII Mediterranean region. VIII Afromontane archipelago-like region, including IX, Afroalpine archipelago-like region of extreme floristic impoverishment (not shown separately). X. Guinea-Congolian/Zambesia transition zone. XI. Guinea-Congolian/Sudania transition zone. XII. Lake Victoria mosaic. XIII. Zanzibar-Inhambane mosaic. XIV. Kalahari-Highveld transition zone. XV. Tongaland-Pondoland mosaic. XVI. Sahel transition zone. XVII. Sahara transition zone. XVIII. Mediterranean/Sahara transition zone. XIX. East Malagasy region. XX. West Malagasy region.

& Warren, 1987), and *Terminalia* (Pegler, 1983) were also examined, but they showed only endomycorrhizal species in West Africa. The mycorrhizal status of species belonging to the same genus might therefore have evolved differently depending on the continent where they developed. This seems especially evident within the genus *Acacia*. African species with true leaves are endomycorrhizal whereas Australian acacias, with phyllodes, have ECM, endomycorrhizas (VAM) or both types of mycorrhizas (Reddel & Warren, 1987). The mycorrhizal status may also change with soil conditions or with the age of the tree (Warcup, 1980) and this may explain some conflicting observations in published reports. Some species of the genera *Afzelia*, *Gilbertiodendron* and *Uapaca* may bear both VAM and ECM (Newbery *et al.*, 1988).

Six of the seven African *Afzelia* species are known to be ectomycorrhizal in addition to *Afzelia xylocarpa* Craib. which occurs in Thailand (Chalermpongse, 1987). The mycorrhizal status of a few other *Afzelia* species reported from Malaysia (*A. palembica* Baker), Bangladesh (*A. retusa* Kurz), Pacific Islands (*A. bijuga* (Colebr.) Gray) (Uphof, 1968) and Australia (Warcup, 1980) is still unknown.

In the Sahel wooded grassland and in the Sudanian undifferentiated woodland of Senegal (Thoen & Ducousso, unpublished) and Niger (Ibrahim, unpublished) all sampled trees were endomycorrhizal. A similar situation was described in austral Africa for the Kalahari Sand vegetation, where VAM are the dominant type of symbiosis (Högberg & Piearce, 1986).

In western Africa, native ectotrophs were found only in the Guineo-Congolia/Sudanian transition zone; these are *Afzelia africana* Sm., *A. bracteata* Benth., *Anthonotha crassifolia* (Baillon) J. Léonard, *Uapaca guineensis* Müll. Arg. and *U. chevalieri* Baille where *U. guineensis* and *A. bracteata* were confined to riverside or swamp forests.

Which tropical fungi are ectomycorrhizal?

It is generally not easy to state whether a fungal species is ectomycorrhizal or not. Nevertheless, hundreds of putative ectomycorrhizal tropical fungi are present in Africa. Until now, only two African species of *Amanita* and two of *Scleroderma* were shown to form ECM on *Afzelia africana* in axenic trials (Ba & Thoen, 1990). In some cases, field observations showed evident links between basidiomes and mycorrhizal roots, as for instance between *Austrogautieria* sp. and *Uapaca guineensis*. Careful study of morphological and anatomical features of ECM, rhizomorphs and can provide strong evidence regarding the fungal associates. The ECM formed by *Scleroderma* have a strong smell when handled which is reminiscent of freshly cut basidiomes of *Scleroderma*, whereas those of *Lactarius* have lactifers. The ECM of *Coltricia cinnamomea* (Pers.) Murr. have the same rust-brown colour and the same bristles as the basidiomes. Indirect evidence of the symbiotic nature of fungi is also given by the patterns of distribution of the basidiomes. In Senegal, the ectomycorrhizal fungi of *Afzelia africana* and of *Uapaca guineensis* disappeared when the cormophytes were absent. Whereas an isolated *Afzelia*, surrounded by endomycorrhizal trees, showed some putative ectomycorrhizal fungi at its foot (Thoen & Ba, 1989). It is also presumed that fungal genera, known to be ectomycorrhizal in the temperate zone, are also symbiotic in tropical areas. Some small and slender *Lactarius* and *Russula* species of the rain forest seem to be saprotrophic on rotten wood, but it must be remembered that rotten wood is a favourable niche for many ectomycorrhizal roots (e.g. Maser & Trappe, 1984). Fructifications on wood, above the ground level, may also be an adaptation to seasonal overflooded sites (Singer & Araujo, 1986). However, it is not excluded that some small *Russula* or *Lactarius* species could be parasitic, e.g. *Russula carmesina* Heim var. *parasitica* (Heim) Buyck, known from Gabon (Heim, 1970; Buyck, 1990), and *Lactarius igapoensis* Singer, known from Brazil (Singer, 1984). Nevertheless, the symbiotic habit of many genera such as representatives of *Phylloporus*, *Boletellus*, and *Strobilomyces* still remains questionable.

The main putative ectomycorrhizal groups, based on species number, are the Amanitaceae, the Boletales, the Russulales and the Cantharellales. These groups occur in several vegetation types distributed from western to central and east Africa. The author has seen them in seven vegetation types including riverside forests and, from the literature, they would also appear to be very common in the lowland rain forest of Zaire. It is noteworthy that the same warmth-adapted ectomycorrhizal groups of fungi are reported from tropical America (Dennis, 1970; Singer & Araujo, 1979; Pegler, 1983) and Asia (Pegler, 1986; Chalermpongse, 1987) where they seem less well represented than in Africa. A few mycorrhizal fungi belong to the Gasteromycetes (e.g. *Scleroderma* and *Astraeus*), the Cortinariaceae (*Inocybe*) (Redhead, 1968b) and to the Aphyllophorales (e.g. *Coltricia* and *Sarcodon*).

The subterranean niche has not yet received enough attention in Africa. Nevertheless, hypogeous fungi have been found in forest, woodland and in exotic stands. *Elasmomyces* sp. is a symbiont of *Uapaca guineensis* (Thoen & Ba, 1989) and a *Sclerogaster* sp. is found under both *Afzelia africana* and *Anthonotha crassifolia* (Thoen & Ducousso, 1989). *Corditubera* sp. is associated with *Pinus kesiya* Royle in Guinea (Thoen & Ducousso, 1989) and *Scleroderma capense* Lloyd forms white mycorrhizas with *Eucalyptus robusta* Sm. (Thoen & Ducousso, unpublished).

Hydnangium nigricans Kalchbr. var. *longispinosum* de Vries is reported from a rain forest in Madagascar (de Vries, 1981) and *Elasmomyces densus* Heim, from a *Dipterocarpus* forest in Thailand (Heim, 1959). Hypogeous fungi, including secotioid forms, are thus components of tropical forests, although they have not been reported yet from Amazonian forests (Singer & Araujo, 1979).

Distribution of ectomycorrhizal fungi in tropical Africa

The distribution of ectomycorrhizal fungi is still imperfectly known. Nevertheless, it is clear that at least some ectomycorrhizal fungi can occur over a very large area and in a wide range of habitats and share different host trees. A good example of this is *Tubosaeta brunneosetosa* (Singer) Horak which occurs in the Guineo-Congolian forest and in Zambezian woodland (see Fig. 202 in Heinemann & Rammeloo, 1989) under several host trees: *Gilbertiodendron dewevrei*, *Paramacrolobium* sp., *Uapaca guineensis*, *Brachystegia* sp., *Julbernardia* sp., *Isoberlinia* sp., *Marquesia macroura* (Heinemann & Rammeloo, 1989; Thoen & Ba, 1989; Thoen & Ducousso, 1989). Other species seem to be linked only to the rain forest, or only to the woodland (Buyck, pers. comm.). Yet other ectomycorrhizal fungi with a wide distribution include, for instance, *Amanita crassiconus* Bas, *Amanita hemibapha* (Berk. & Br.) Sacc., *Lactarius gymnocarpus* Heim, *Russula roseoalba* Buyck and *Cantharellus rufopunctatus* (Beeli) Heinem.

A few fungi (e.g. *Scleroderma verrucosum* Pers., *Coltricia cinnamomea*) are cosmopolitan or pantropical and are probably very old species which appeared before continental drift. *Pulveroboletus ravenelii* (Berk. & Curt.) Murr. presents a curious disjunction. It is known from eastern North America and eastern Asia (Singer, 1947) and from Guinea (Thoen & Ducousso, 1989).

Most warmth-adapted ectomycorrhizal fungi of Africa seem to be endemic to this continent. Following Buyck (1990), all the tropical species of *Russula* of Africa are endemic and because of the very large number of species, Africa must be the centre of origin of the genus. Other genera with high species diversity in Africa, such as *Cantharellus* and *Amanita*, probably also contain endemic species. Some ectomycorrhizal fungi might be endemic to a given phytochoria, such as the Guineo-Congolian one, but available information is too scarce to be fully affirmative.

In exotic stands, the fungal diversity is very low. Native fungi from the surrounding forest or woodland disappear or are very scarce. One of these is *Scleroderma verrucosum* which is found in both native and exotic stands.

In *Pinus* stands, *Suillus* spp. and *Amanita* spp. can also be found; most of these fungi were introduced fortuitously. This is certainly the case for *Pisolithus* sp., a native from the Australian region which is now commonly found in Senegal under eucalypts, *Casuarina equisetifolia* Forst., *Melaleuca leucodendron* L. and *Acacia holosericea* A. Cunn. ex G. Donn. This fungus has expanded throughout Africa in stands of Australian exotics (Thoen, 1985).

Conclusions

Ectotrophic trees and putative ectomycorrhizal fungi are widely distributed in forests and woodlands of tropical Africa. The fungal diversity of ectotrophic trees is important, perhaps as important as for temperate ectotrophs. *Afzelia africana* and *Uapaca guineensis*, for instance, have about twenty or more putative symbionts (Thoen & Ba, 1989). On the other hand, fungal diversity of exotic stands remains very low.

Important vegetation types (e.g. the Sudanian woodland of *Isoberlinia doka* Craib. & Stapf) and large countries (e.g. Ivory Coast, Madagascar) need to be urgently studied before human activities destroy the native forests.

Significant ectomycorrhizal groups such as the genus *Amanita* need to be studied in the tropics and hypogeous symbionts have been and are still overlooked. There is a striking need for long-term field work allowing the patterns of fungal successions in regenerating forests to be studied. Today it is not known what happens to ectomycorrhizal fungi during the forest cycle, from the gap phase through to the development of mature stages (Whitmore, 1982).

Do ectomycorrhizal species change with ageing of their host tree? Are ectomycorrhizal fungi important for the survival of seedlings? Does host specificity occur within the tropical ectomycorrhizal fungi? What is the influence of fires on ectomycorrhizal fungi in woodlands? How do ectomycorrhizal propagules spread? Are they able to survive in disturbed areas and for how long? What are the light requirements of ectotrophic trees? Do ectotrophs need more light than anectotrophs? Are ectomycorrhizal fungi of the rain forest really able to translocate nutrients from the litter layer to the host trees? These are many of thequestions that need further investigations in order to understand and help tropical forests to survive in the near future.

References

Alexander, I. J. (1981). Mycorrhizas in the Sapindaceae. *Abstracts, Fifth North American Conference on Mycorrhizae*. Université Laval: Québec.

Alexander, I. J. & Högberg, P. (1986). Ectomycorrhizas of tropical angiospermous trees. *New Phytologist* **102**, 541-549.

Anon. (1970). *Distributiones Plantarum Africanum* **2**. Jardin Botanique National de Belgique.

Ba, A. M. & Thoen, D. (1990). First syntheses of ectomycorrhizas between *Afzelia africana* Sm. (Caesalpinioideae) and native fungi from West Africa. *New Phytologist* **114**, 99-103.

Buyck, B. (1990). *Revision du genre Russula Persoon en Afrique Centrale. I. Partie générale*. Ph.D. thesis, University of Ghent.

Chalermpongse, A. (1987). Mycorrhizal survey of dry-deciduous and semi-evergreen Dipterocarp forest ecosystems in Thailand. In *The Third Round Table Conference on Dipterocarps*, (ed. A. J. G. H. Kostermans), pp. 81-103. Bogor: Indonesia.

Dennis, R. W. G. (1970). *Fungus Flora of Venezuela and Adjacent Countries*. Royal Botanic Gardens: Kew.

de Vries, G. A. (1981). *Hydnangium nigricans* var. *longispinosum*, a new hypogeous fungus from Madagascar. *Persoonia* **11**, 381-384.

Fassi, B. (1957). Ectomycorrhizie chez le *Gnetum africanum* Welw. due à *Scleroderma* sp. Bulletin de la Société Mycologique de France 73, 280-286.

Fassi, B. & Fontana. A. (1962). Micorrize ectotrofiche di *Brachystegia laurentii* e di alcune altre Cesalpiniacee minori del Congo. *Allionia* **8**, 121-131.

Heim, R. (1959). Une espèce nouvelle de Gastrolactarié en Thailande. *Revue de Mycologie* **24**, 93-102.

Fassi, B. & Moser, M. (1991). Mycorrhizae in the natural forest of tropical Africa and the neotropics. In *Funghi, Piante e Suolo*, (ed. A. Fontana), pp. 183-202. Centro di Studio sulla Micologia del Terreno del Consiglio Nazionalle delle Ricerche: Torino, Italy.

Heim, R. (1970). Particularités remarquables des Russules tropicales *Pellicularíae* lilliputiennes: les complexes *annulata* et *radicans*. *Bulletin de la Société Mycologique de France* **86**, 59-77.

Heinemann, P. & Rammeloo. J. (1989). *Tubosaeta* (Xerocomaceae, Boletineae). *Flore Illustrée des Champignons d'Afrique Centrale*, fasc. **14**. Ministère de l'Agriculture-Jardin Botanique National de Belgique: Meise, Belgium.

Högberg, P. (1982). Mycorrhizal associations in some woodland and forest trees and shrubs in Tanzania. *New Phytologist* **92**, 407-415.

Högberg, P. (1986). Nitrogen-fixation and nutrient relations in savanna woodland trees (Tanzania). *Journal of Applied Ecology* **23**, 675-688.

Högberg. P. & Nylund. J.-E. (1981). Ectomycorrhizae in coastal miombo woodland of Tanzania. *Plant and Soil* **63**, 283-289.

Högberg, P. & Piearce, G. D. (1986). Mycorrhizas in Zambian trees in relation to host taxonomy, vegetation type and successional patterns. *Journal of Ecology* **74**, 775-785.

Jenik, J. & Mensah, K. O. A. (1967). Root system of tropical trees. I. Ectotrophic mycorrhizae of *Afzelia africana* Sm. *Preslia* **39**, 59-65.

Lock, J. M. (1989). *Legumes of Africa: a check-list*. Royal Botanic Gardens: Kew.

Mabberley. D. J. (1987). *The Plant Book. A Portable Dictionary of the Higher Plants*. Cambridge University Press: Cambridge, U.K.

Maser, C. & Trappe, J.M. (1984). The fallen tree. A source of diversity. In *New Forests for a Changing World*, Proceedings of the Society of American Foresters, National Conference, Portland, Oregon, 16–20 October 1983, pp. 335-339. Society of American Foresters: Bethesda, MD, U.S.A.

Newbery, D. M., Alexander, I. J., Thomas, D. W. & Gartlan, J. S. (1988). Ectomycorrhizal rain-forest legumes and soil phosphorus in Korup National Park, Cameroon. *New Phytologist* **109**. 433-450.

Pegler, D. (1983). *Agaric Flora of the Lesser Antilles*. Royal Botanic Gardens: Kew.

Pegler, D. (1986). *Agaric Flora of Sri Lanka*. Royal Botanic Gardens: Kew.

Peyronel, B. & Fassi, B. (1957). Micorrize ectotrofiche in una Cesalpiniacea del Congo Belga. *Atti della Accademia delle Scienze di Torino* **91**, 1-8.

Peyronel, B. & Fassi, B. (1960). Nuovi casi di simbiosi ectomicorrizica in Leguminose della famiglia delle Cesalpiniacee. *Atti della Accademia delle Scienze di Torino* **94**, 1-3.

Read, D. J. (1991). Mycorrhizas in ecosystems. *Experientia* **47**, 376-391.

Reddel, P. & Warren, R. (1987). Inoculation of Acacias with mycorrhizal fungi: potential benefits. In *Australian Acacias in Developing Countries*, Australian Centre for International Agricultural Reserach Proceedings 16, (ed. J. W. Turnbull), pp. 50-53.ACIAR: Canberra, Australia.

Redhead, J. F. (1960). *A study of mycorrhizal associations in some trees of Western Nigeria*. Diploma in Forestry Thesis, University of Oxford.

Redhead. J. F. (1968a). Mycorrhizal associations in some Nigerian forest trees. *Transactions of the British Mycological Society* **51**, 377-387.

Redhead, J. F. (1968b). *Inocybe* sp. associated with ectotrophic mycorrhiza on *Afzelia bella* in Nigeria. *Commonwealth Forestry Review* **47**, 63-65.

Redhead, J. F. (1974). *Aspects of the biology of mycorrhizal associations occurring on tree species in Nigeria*. Ph.D. thesis, University of Ibadan.

Singer, R. (1947). The Boletineae of Florida with notes on extralimital species III. *The American Midland Naturalist* **37**, 129-263.

Singer, R. (1984). Tropical Russulaceae II. *Lactarius* section *Panuoidei*. *Nova Hedwigia* **40**, 435-447.

Singer, R. & Araujo, I. A. (1979). Litter decomposition and ectomycorrhiza in Amazonian forests. I. A comparison of litter-decomposing and ectomycorrhizal Basidiomycetes in latosol-terra-firme rain forest and white sand podzol Campinarana. *Acta Amazonica* **9**, 25-41.

Singer, R. & Araujo, I. A. (1986). Litter decomposition and ectomycorrhized Basidiomycetes in an igapo forest. *Plant Systematics and Evolution* **153**, 107-117.

Thoen, D. (1974). Premières indications sur les mycorrhizes et les champignons mycorrhiziques des plantations d'exotiques du Haut-Shaba (République du Zaïre). *Bulletin des Recherches Agronomiques de Gembloux* **9**, 215-227.

Thoen, D. (1985). Is *Pisolithus* found under *Eucalyptus camaldulensis* in Senegal conspecific with the mycorrhizal *Pisolithus tinctorius*? Abstracts, IX Congress of

European Mycologists, Oslo, 15–21 August 1985, p. 68, University of Oslo: Oslo, Norway.

Thoen, D. & Ba, A. M. (1989). Ectomycorrhizas and putative ectomycorrhizal fungi of *Afzelia africana* Sm. and *Uapaca guineensis* Müll. Arg. in southern Senegal. *New Phytologist* **113**, 549-559.

Thoen, D. & Ducousso, M. (1989). Champignons et ectomycorrhizes du Fouta Djalon. *Bois et Forêts des Tropiques* **221**, 45-63.

Uphof, J. C. Th. (1968). *Dictionary of Economic Plants*. J. Cramer: Lehre.

Warcup, J. H. (1980). Ectomycorrhizal associations of Australian indigenous plants. *New Phytologist* **85**, 531-535.

White. F. (1965). The savanna woodlands of the Zambezian and the Sudanian domains – An ecological and phytogeographical comparison. *Webbia* **19**, 657-681.

White, F. (1983). *The Vegetation of Africa*. UNESCO: Paris.

Whitmore, T. C. (1982). On pattern and process in forests. In *The Plant Community as a Working Mechanism*, (ed. R. I. Newman), pp. 45-59. Blackwell Scientific Publications: Oxford & London, U.K.

Chapter 12

Armillaria in tropical Africa

C. Mohammed & J. J. Guillaumin*

*Oxford Forestry Institute, South Parks Road, Oxford, OX1 3RB, U.K. and *Centre INRA de Clermont-Ferrand, Unité de Mycologie, 12 Avenue du Brézet, 63039, Clermont-Ferrand, France*

Introduction

Fungal species of the genus *Armillaria* (Fr.:Fr.) Staude can provoke a root rot disease resulting in extensive economic loss to a woody crop. The disease is found worldwide in a range of different hosts and habitats; natural and planted softwood and hardwood forest, orchards, vineyards, tropical cash crop plantations, parks and gardens (Kile, McDonald & Byler, 1991; Hood, Redfern & Kile, 1991). Since the beginning of 1990 work on African *Armillaria* has been undertaken by a group of researchers working both in Europe and Africa (The Congo, Gabon, Kenya, Zimbabwe and La Réunion). This collaborative effort has wide ranging research objectives from the more fundamental work on taxonomy and sexuality to the establishment of effective control methods.

Historical perspective

For over a century, the confusion surrounding the nomenclature and taxonomy of the genus *Armillaria* greatly hindered the study of this fungal pathogen. Until the 1980s most pathologists attributed the disease syndrome to one particular *Armillaria* species, *A. mellea* (Vahl: Fr.) Kummer. Only when the riddle of sexuality was solved in *Armillaria* was it possible to unravel the different species of this fungal genus.

Hintikka (1973) was the first to show that *Armillaria* is heterothallic with a bifactorial (tetrapolar) sexual system; he established that single-spore isolates are cottony in appearance whereas cultures from basidiome tissue or the mycelium found under the bark are flat and crustose. In sib-matings of compatible single-spore mycelia the cottony isolates fuse to form a crustose thallus. Although cells of both unmated monospore isolates and basidiome tissue (or mycelium from wood) are monokaryotic, nuclei in crustose mycelia have been proved to be diploid (Korhonen & Hintikka, 1974; Peabody & Peabody, 1984).

Hintikka's research opened the way to further investigation of sexual compatibility and mating-type alleles (Ullrich & Anderson, 1978; Kile, 1983; Guillaumin, Berthelay & Savin, 1983). Subsequently, pairing tests based on both sexual behaviour and somatic incompatibility have been developed which can establish either inter- or intra- specific incompatibility (Korhonen, 1978; Guillaumin & Berthelay, 1990). Inter-specific incompatibility is indicated by blackline formation in the medium whatever the pairing combination *i.e.* haploid/haploid, haploid/diploid, or diploid/diploid. Intra-specific incompatibility between two diploids is shown by the formation of a sparse barrage zone of mycelium between the paired thalli.

Research during the 1980s exploited pairing tests, especially for identifying species. These tests led to the clear distinction of six 'biological species' of European *Armillaria*, some of which corresponded to morphological species described by classical taxonomists (Korhonen, 1978; Guillaumin *et al.*, 1985). It became apparent that these species differed in distribution, host range, pathogenicity and the type and frequence of rhizomorphs found in the soil. Cultural characteristics such as vegetative mat morphology, ability to fruit in culture and optimum temperature for growth also proved to be species related criteria (Guillaumin, Mohammed & Berthelay, 1989). *Armillaria* species in Australia, North America and Japan were also shown to be heterothallic and tetrapolar. The application of the 'biological species' concept has aided investigation of *Armillaria* species present in these continents; as in Europe, the morphological (taxonomic) and biological species concepts were reconciled (Kile & Watling, 1983, 1988; Anderson & Ullrich, 1979; Guillaumin *et al.*, 1989).

Armillaria root rot has been widely reported from the wetter regions of Africa, principally in exotic softwood and cash crop plantations (reviewed by Hood, Redfern & Kile, 1991). Many of these reports are rather dated sporadic records of the disease and are vague as to its economic impact. In fact few up-to-date appraisals of the incidence of *Armillaria* have been made. Discussions with researchers familiar with the situation in Africa indicated that damage to a wide range of crops, although considerable, was written off. Indeed from the recent literature available it would seem as if *Armillaria* is a serious obstacle to increasing the productivity of crops of both local and export importance, including cassava (Makambila & Bakala, 1986), coffee (Blaha, 1978), cocoa (Saccas, 1975; Rishbeth, 1980), rubber (Wastie, 1986; Petit-Renaud, 1991) in West and Central Africa; citrus (Seif & Whittle, 1984), tea (Onsando, 1988), and introduced trees such as teak and pine (Ivory, 1987) in East and South Africa.

The behaviour of African *Armillaria* was most comprehensively described in the 1960s and 1970s (Pichel, 1956; Dadant, 1963; Gibson, 1960; Goodchild, 1960; Gibson & Goodchild, 1960; Gibson & Corbett, 1964; Swift, 1968, 1970, 1972; Olembo, 1972; Blaha, 1978). The literature spans West, Central and East Africa and the attack of various hosts of economic importance. One generality is evident: indigenous forest tree roots harbour superficial or latent *Armillaria* lesions. When the indigenous forest is cleared, stumps with superficial lesions are invaded by *Armillaria* and serve as sources of inoculum for the subsequent plantation. It is, however, easier to detect differences than similarities in the literature of this period. *Armillaria* tends to infect plantations established above 1000 m in East Africa where the climate is cooler and wetter. In Central and West Africa the disease occurs in crops at both high and low altitudes. Rhizomorphs in the soil may be totally absent or abundant (usually at higher altitudes). Reports also vary concerning the degree of pathogenicity, the cultural characteristics of isolates and the effectiveness of trenching, ringbarking or other control measures.

Experience with temperate *Armillaria* suggests that much of this contradictory information might be explained by the presence of several different species. However (as hitherto with temperate *Armillaria*) most cases of *Armillaria* have been arbitrarily attributed to *A.mellea*, described in Uganda and Kenya by Pegler (1977). Notwithstanding the size of Africa and the economic damage caused by *Armillaria*, only one other species has been described in any detail. This situation may perhaps be partly explained by the low frequency with which wild basidiomes are found in a number of East and South African countries (Gibson, 1960; Ivory, 1987). More frequent basidiome production has been reported from Madagascar (Dadant, 1963), the Cameroon (Blaha, 1978) and the Congo (C. Makambila, pers. comm.). In fact Heim described *Clitocybe elegans* Heim (subgenus *Armillariella*) from basidiomes in Madagascar (Heim, 1963; Kile & Watling, 1988). Heim also recorded this species from the Cameroon, Ivory Coast, Central African Republic and New Guinea (Heim, 1963; 1967). French mycologists did not transfer the taxon to *Armillaria* although the close affinity of *Clitocybe elegans* to the type of the genus *Armillaria* was generally recognised (Watling, Kile & Gregory, 1982; Watling, 1992). Pegler described basidiomes similar to *Clitocybe elegans* from Tanzania, Uganda and Zambia (Pegler, 1977; M.H. Ivory, pers. comm.). Pegler realised that *Clitocybe elegans* would be pre-dated if transferred to *Armillaria* as Beeli (1927) had already described *Armillaria elegans* (now known as belonging to the genus *Cystoderma*) from the Congo. Pegler adopted the epithet *heimii* in recognition of Heim's contribution to agaricology. More recently both Pegler (1986) and Watling (1992) have reduced *A. heimii* to synonomy under *A. fuscipes* Petch, first described by Petch (1909) from Sri Lanka and supposedly introduced to this country from West Africa (Pegler, 1986; Kile & Watling, 1988; Watling, 1992). The authors however rest prudent about this synonomy and continue to use the nomenclature *A. heimii*.

In 1988, 47 *Armillaria* isolates originating from 16 different African countries (mainly East African) were investigated in a preliminary taxonomic study (Mohammed, Guillaumin & Berthelay, 1989). Methods used to attempt species delimitation were the same as those used for European *Armillaria* species. All groups or 'species' were incompatible with all temperate species with the exception of the African *mellea* which appears partially compatible with its counterpart species in Europe and the U.S.A. Although the majority of isolates fell into two groups which could be equated with species already described from Africa - *A. mellea* and *A. heimii*, it was clear that certain isolates could not be classified with either of these two species.

The present study

Although much of the research concerning African *Armillaria* is still in progress, significant advances have already been made in the area of taxonomy and sexuality. The methods used to differentiate European *Armillaria* did enable some classification of African *Armillaria* in the preliminary taxonomic studies of Mohammed *et al.* (1989), but they were less discriminative than expected. The pairing test method, so necessary in the determination of the biological species of temperate *Armillaria*, is not entirely transferrable to the study of African isolates because many do not have a haploid form. Recent studies include biochemical and molecular analytical techniques which complement the more traditional investigatory methods, facilitating the task of interpretation. The methods employed in investigating the taxonomy and sexuality of African *Armillaria* isolates are summarised in Table 1.

Sixty West, Central, East and South African *Armillaria* isolates available at the beginning of 1991 have been screened for differences in protein, enzymes or DNA. Other types of taxonomic investigation and species characterisation are more labour-, materials- and space-intensive and therefore were concentrated on 25 isolates carefully chosen to be as representative as possible of the isolates available. Four main intersterile biological groups were defined. These four intersterile groups or 'species'

Table 1. Techniques used to group different *Armillaria* isolates from tropical Africa

- Fruiting in culture (Guillaumin *et al.*, 1989)
- Observation of: morphology of basidiomes, conditions inducive to fruiting, frequency of fruiting, morphology of single-spore isolates
- Pairing tests (Guillaumin, Anderson & Korhonen, 1991) to investigate: sexuality (Hintikka, 1973), inter-specificity (Korhonen, 1978), intra-specific incompatibility (Guillaumin & Berthelay, 1990)
- Comparison of culture mat morphology (Mohammed, 1987; Guillaumin *et al.*, 1989)
- Comparison of growth in culture at different temperatures (Rishbeth, 1986; Mohammed, 1987)
- Comparison of subterranean rhizomorphs obtained in a mist case (Mohammed, 1985; 1987)
- Biochemical and molecular analysis of: protein (SDS PAGE)(Laemmli, 1970), esterase (Market & Hunter, 1959; Mwangi, Lin & Hubbes, 1989), proteinase (Lockwood *et al.*, 1987), whole cell DNA by (i) hybridisation of restriction digests with mtDNA European *Armillaria* and *Agaricus bisporus* probes (Anderson, Petsche & Smith, 1987; Hintz, Anderson & Horgen, 1988), and (ii) Random Amplified Polymorphic DNA analysis (Williams *et al.*, 1990).

---

were also incompatible with all temperate species with the exception of group 2 which appeared to be partially compatible with European and North American *A. mellea*.

1. *Armillaria heimii sensu lato* - a large group of 46 isolates from West, Central, East and South Africa. This group could be further divided into 3 subgroups:

    1a 7 isolates from West Africa (Gabon, Cameroon, Ivory Coast and Liberia)

    1b 5 isolates from the Congo

    1c 34 East and South African isolates (from Kenya, Tanzania, Uganda, Malawi, Zambia, Zimbabwe, South Africa, Madagascar and La Réunion)

2. A small but distinct group of 8 isolates - the African form of *A. mellea* (from São Tomé, Tanzania and Kenya)

3. 5 isolates from high altitude (2000 m) Kenya

4. 1 isolate also from high altitude Kenya

Isolates in group 1 did not show any incompatibility in diploid/diploid pairing tests. Protein profiles of group 1 isolates were similar and different from those of isolates in other groups. Group 1 isolates fruited easily in culture. Although the morphology of basidiomes obtained in culture might be atypical, in general, isolates of group 1 (whether from West or East Africa) gave rise to small pale brownbasidiomes similar to *A. heimii* as described by Heim (1963) and Pegler (1977). The pilei were distinguished by small, erect, rufous brown squamules, crowded towards the centre. The ring, if present, was floccose.

Unlike protein analysis, that of certain enzymes and DNA showed variability among the isolates of group 1, only grouping isolates of a similar regional origin. Other regional differences existed between group 1 isolates. Group 1a and 1b isolates (especially the Congo isolates) grew better at higher temperatures than group 1c isolates. Group 1a single-spore isolates of West African origin were cottony. Group 1b and 1c single-spore isolates were crustose like the parent diploid isolate. The crustose single-spore isolates themselves gave rise to viable basidiomes. Single-spore, sib-matings indicated the existence of heterothallism for group 1a and a possible bipolarity. Cytological studies showed that the homothallism indicated by the morphology and behaviour of single spores isolated from group 1c was a true homothallism (not apomixis) with a meiosis in the basidium. Single-spore isolates representative of group 1a were diploidised by diploids of groups 1b and 1c in a compatible confrontation analogous in *Armillaria* to the Buller phenomenon (Anderson & Ullrich, 1982; Korhonen, 1978, 1983) although subcultures from diploidised haploids have proved unstable, reverting to the haploid form.

Group 2 isolates, as well as showing some partial compatibility with those of the temperate *A. mellea*, showed other close similarities to the temperate species including protein and enzyme profiles, basidiome and culture mat morphology, and possibly host specialisation (principally on agricultural not forest hosts). Although group 2 isolates fruited far less readily in culture than those of group 1, some single-spore isolates have been obtained. As for groups 1b and 1c, it would appear from the crustose nature of group 2 single-spore isolates that the African form of *A. mellea* is homothallic not heterothallic as is its temperate counterpart. DNA (RFLP and RAPD) analysis also suggested that there are significant differences between the temperate and tropical forms of *A. mellea*.

Isolates from group 3 and 4 both originated from high altitude Kenya. Although group 3 only comprised 5 isolates and group 4 one isolate, the groups were clearly defined by somatic incompatibility in pairing tests, culture mat morphology and biochemical analysis. In culture these isolates were slow growing and reacted unfavourably to high temperatures. Group 3 and 4 isolates never fruited in culture and so it was not possible to study the sexuality of these two groups. It is intriguing that group 3 was the only group to show intra-specific somatic incompatibility typical of the pairing between diploids belonging to different clones, i.e. the type of intra-specific incompatibility encountered within a temperate *Armillaria* species.

Future perspectives

The two major and widespread groups of African *Armillaria* determined by the authors correspond well with the only two species described in any detail by the morphologists from Africa - *A. mellea sensu stricto* and *A. heimii* Pegler. However, as indicated by the clear distinction of two additional groups in Kenya, other species must be present. As further isolates are collected, the occurrence of species (perhaps of limited distribution) should become clearer.

The group equated by the authors to *A. heimii* is so widespread and variable that it should perhaps be referred to as *A. heimii sensu lato*. Two types of sexual system have been found within this group each associated with different regions of tropical Africa. This plurality of sexual forms is known in a number of basidiomycetes such as *Sistotrema brinkmannii* (Bres.) J. Erikss. (Boidin & Lanquetin, 1984) and *Stereum hirsutum* (Willd. :Fr.) S.F. Gray (Ainsworth *et al.*, 1990). Indeed the rare and non-forest European species of *Armillaria*, *A. ectypa* (Fr.) Lamoure, is homothallic (Guillaumin, 1973). However, more research is needed to establish if a particular sexual form does have some geographical and ecological association, an indication perhaps of the advantage of one or other sexual system to the fungus. It is also important for future taxonomic research of the *A. heimii* group to include isolates originating from South India and Ceylon, so that the proposed synonymy with *A. fuscipes* suggested by the morphologists can be confirmed by techniques based on incompatibility and biochemical and molecular differences. Such research would also investigate the possibility that *A. heimii sensu lato* is a species indigenous to a wide geographical arc - tropical Africa, islands of the Indian Ocean, South India and Sri Lanka, rather than a species introduced from Africa to Sri Lanka.

African *A. mellea* is closely similar to the temperate *A. mellea*. Continued research, especially at the molecular level, must establish whether the African *mellea* should be considered as a homothallic form of the temperate *mellea* or a separate species.

References

Ainsworth, A. M., Rayner, A. D. M., Broxholme, S. J. & Beeching, J. R. (1990). Occurrence of unilateral genetic transfer and genomic replacement between two strains of *Stereum hirsutum* from non-outcrossing and outcrossing populations. *New Phytologist* **115**, 119-128.

Anderson, J. B., Petsche, D. M. & Smith, M. L. (1987). Restriction fragment polymorphisms in biological species of *Armillaria mellea*. *Mycologia* **79**, 69-76.

Anderson, J. B. & Ullrich, R. C. (1979). Biological species of *Armillaria mellea* in North America. *Mycologia* **71**, 402-414.

Anderson, J. B. & Ullrich, R. C. (1982). Diploids of *Armillaria mellea*: synthesis, stability, and mating behaviour. *Canadian Journal of Botany* **60**, 432-439.

Beeli, M. (1927). Contribution à l'étude de la Flore Mycologique II. *Bulletin de la Société royale Botanique Belge* **109**, (Sér. II, 9) 101-112.

Blaha, G. (1978). *Clitocybe* (*Armillariella*) *elegans* Heim un grave pourridié du caféier *Arabica* au Cameroun. *Café, Cacao, Thé* **22**, 203-216.

Boidin, J. & Lanquetin, P. (1984). Répertoire des données utiles pour effectuer les tests d'intercompatibilité chez les Basidiomycètes, I. Introduction. *Cryptogamie Mycologie* **51**, 33-46.

Dadant, M. R. (1963). Contribution à l'étude du pourridié du caféier causé par le *Clitocybe elegans* à Madagascar. Ses relations avec le *Trichoderma viride* Pers. *Revue de Mycologie, Paris* **28**, 94-168.

Gibson, I. A. S. (1960). *Armillaria* root rot in Kenya pine plantations. *Empire Forestry Review* **39**, 94-99.

Gibson, I. A. S. & Corbett, D. C. M. (1964). Variation in isolates from *Armillaria* root disease in Nyasaland. *Phytopathology* **54**, 22-23.

Gibson, I. A. S. & Goodchild, N. A. (1960). *Armillaria mellea* in Kenya forests. *East African Agricultural and Forestry Journal* **26**, 142-143.

Goodchild, N. A. (1960). *Armillaria mellea* in tea plantations. *Journal of the Tea Boards of East Africa* (October 1960), 43-45.

Guillaumin, J. J. (1973). Étude du cycle cytologique de deux espèces appartenant aux genres *Armillariella*. *Annales de Phytopathologie* **5**, 317.

Guillaumin, J. J., Anderson, J. B. & Korhonen, K. (1991). Life cycle, interfertility, and biological species. In *Armillaria Root Disease*, United States Department of Agriculture Forest Service Agriculture handbook no. 691, (ed. C. G. Shaw & G. A. Kile), pp. 10-20. Forest Service, U.S.D.A.: Washington, D.C.

Guillaumin, J. J. & Berthelay S. (1990). Comparaison de deux méthodes d'identification des clones chez le Basidiomycète parasite *Armillaria obscura* (syn. *A.ostoyae*). *European Journal of Forest Pathology* **20**, 257-268.

Guillaumin, J. J., Berthelay, S. & Savin, V. (1983). Étude de la polarité sexuelle des Armillaires du groupe *mellea*. *Cryptogamie Mycologie* **4**, 301-319.

Guillaumin, J. J., Lung, B., Romagnesi, H., Marxmuller H. & Mohammed, C. (1985). Systématique des Armillaires du groupe *mellea*. Consequences phytopathologiques. *European Journal of Forest Pathology* **15**, 268-277.

Guillaumin, J. J., Mohammed, C. & Berthelay, S. (1989). *Armillaria* species in the Northern temperate hemisphere. In *Proceedings of the Seventh International Conference on Root and Butt Rots*, IUFRO Working Party S2.06.01, British

Columbia, August 1988, (ed. D. J. Morrison), pp. 27-44. Forestry Canada, Pacific Forestry Centre: Victoria, B.C.

Heim, R. (1963). L'*Armillaria elegans* Heim. *Revue Mycologique* **28**, 89-94.

Heim, R. (1967). Notes sur la flore mycologique des Terres du Pacifique Sud. IV. Note complementaire sur l'*Armillariella elegans*. *Revue Mycologique* **32**, 9-11.

Hintikka, V. (1973). A note on the polarity of *Armillariella mellea*. *Karstenia* **13**, 32-39.

Hintz, W. E. A., Anderson, J. B. & Horgen, P. A. (1988). Physical mapping of the mitochondrial genome of the cultivated mushroom *Agaricus brunnescens* (= *A. bisporus*). *Current Genetics* **14**, 43-49.

Hood, I. A., Redfern, D. B. & Kile, G. A. (1991). *Armillaria* in planted hosts. In *Armillaria Root Disease*, United States Department of Agriculture Forest Service Agriculture handbook no. 691, (ed. C.G. Shaw & G.A. Kile), pp. 122-149. Forest Service, U.S.D.A.: Washington, D.C.

Ivory, M. H. (1987). In *Diseases and Disorders of Pines in the Tropics - a Field and Laboratory Manual*, pp. 12-16: Overseas Research Publication, Her Majesty's Stationery Office no. 31. Published for the Overseas Development Association by the Oxford Forestry Institute: Oxford

Kile, G. A. (1983). Identification of genotypes and the clonal development of *Armillaria luteobubalina* Watling and Kile in eucalypt forests. *Australian Journal of Botany* **31**, 657-671.

Kile, G. A. & Watling, R. (1983). *Armillaria* species from south-eastern Australia. *Transactions of the British Mycological Society* **81**, 129-140.

Kile, G. A. & Watling, R. (1988). Identification and occurence of Australian *Armillaria* species, including *A. pallidula* sp. nov. and comparative studies between them and non-Australian tropical and Indian *Armillaria*. *Transactions of the British Mycological Society* **91**, 305-315.

Kile, G. A., McDonald, G. I. & Byler J. W. (1991). Ecology and disease in natural forests. In *Armillaria Root Disease*, United States Department of Agriculture Forest Service Agriculture handbook no. 691, (ed. C.G. Shaw & G.A. Kile), pp. 102-121. Forest Service, U.S.D.A.: Washington, D.C.

Korhonen, K. (1978). Infertility and clonal size in the *Armillaria mellea* complex. *Karstenia* **18**, 31-42.

Korhonen, K. (1983). Observations on nuclear migration and heterokaryotization in *Armillaria*. *Cryptogamie Mycologie* **4**, 79-85.

Korhonen, K. & Hintikka, V. (1974). Cytological evidence for somatic diploidization in dikaryotic cells of *Armillariella mellea*. *Archives of Microbiology* **95**, 187-192.

Laemmli, U. K. 1970. Cleavage of stuctural proteins during the assembly of the head of bacteriophage T4. *Nature* **227**, 680-685.

Lockwood, B. C., North, G. J., Scott, K. I., Bremmer, A. F. & Coombs, G. H. (1987). The use of a highly sensitive electrophoretic method to compare the proteinases of trichomonads. *Molecular and Biochemical Parasitology* **24**, 89-95.

Makambila, C. & Bakala, L. (1986). Les pourridiés à *Armillaria* sp., *Sphaerostilbe repens* B. et Br. et *Phaeolus manihotis* Heim sur le manioc (*Manihot esculenta* Crantz). *Agronomie Tropicale* **41**, (3-4), 258-264.

Market, C. L. & Hunter, R. L. (1959). The distribution of esterases in mouse tissue. *Journal of Histochemistry and Cytochemistry* **7**, 42-49.

Mohammed, C. (1985). Croissance et ramification du rhizomorphe des cinq espèces européenes d'Armillaire du group *mellea*. *Agronomie* **5**, 360.

Mohammed, C. (1987). Étude comparée des cinq espèces d'*Armillaria* appartenant au complexe *mellea*. Diplôme de Doctorat. Université Clermont-Ferrand II: France.

Mohammed, C., Guillaumin, J.J. & Berthelay S. (1989). Preliminary investigation about the taxonomy and genetics of African *Armillaria* species. In *Proceedings of the Seventh International Conference on Root and Butt Rots*, IUFRO Working Party S2.06.01, British Columbia, August 1988, (ed. D.J. Morrison), pp. 447-457. Forestry Canada, Pacific Forestry Centre: Victoria, B.C.

Mwangi, L. M., Lin, D. & Hubbes, M. (1989). Identification of Kenyan *Armillaria* isolates by cultural morphology, intersterility tests and analysis of isozyme profiles. *European Journal of Forest Pathology* **19**, 399-406.

Olembo, T. W. (1972). Studies on *Armillaria mellea* in East Africa. Effect of soil leachates on penetration and colonization of *Pinus patula* and *Cupressus lusitanica* wood cylinders by *Armillaria mellea* (Vahl ex Fr.) Kummer. *European Journal of Forest Pathology* **2**, 134-140.

Onsando, J. O. (1988). Tea diseases situation in Kisii district. *Tea* **2**, 47-49.

Peabody, D. C. & Peabody, R. B. (1984). Microspectrophotometric nuclear cycle analyses of *Armillaria mellea*. *Experimental Mycology* **8**, 161-169.

Pegler, D. N. (1977). A preliminary agaric flora of East Africa. *Kew Bulletin Additional Series* **6**, 91-94.

Pegler, D. N. (1986). Agaric flora of Sri Lanka. *Kew Bulletin Additional Series* **12**, 81-86.

Petch, T. (1909). New Ceylon fungi. *Annals of the Royal Botanic Garden of Peradeniya* **4**, 299-307.

Petit-Renaud, D. (1991). Contribution à l'étude du pourridié de l'Hévea (*Hevea brasiliensis*) causé par *Armillaria heimii* au Gabon. Diplôme d'Agronomie Approfondie Sciences et Techniques des Productions Végétales: Protection des cultures et de l'Environment. École Nationale Supérieure Agronomique de Toulouse.

Pichel, R. J. (1956). *Les pourridiés de l'Hévéa dans la cuvette congolaise*. Publication of the Institut National pour l'Étude Agronomique du Congo Belge, Series Techniques No. 49.

Rishbeth, J. (1980). *Armillaria* on cocoa in São Tomé. *Tropical Agriculture (Trinidad)* **57**, 155-165.

Rishbeth, J. (1986). Some characteristics of English *Armillaria* species in culture. *Transactions of the British Mycological Society* **85**, 213-218.

Saccas, A. M. (1975). *Les Pourridiés des Caféiers en Afrique Tropicale*. Publication of the Institut Français du Café et du Cacao: Paris.

Seif, A. A. & Whittle, A. M. (1984). Diseases of citrus in Kenya. *FAO Plant Protection Bulletin* **32**, 122-127.

Swift, M. J. (1968). Inhibition of rhizomorph development by *Armillaria mellea* in Rhodesian forest soils. *Transactions of the British Mycological Society* **51**, 241-247.

Swift, M. J. (1970). *Armillaria mellea* (Vahl ex Fries) Kummer in central Africa: studies on substrate colonisation relating to the mechanism of biological control by ring-barking. In *Root Diseases and Soil-borne Pathogens*, Proceedings of the

Symposium, July 1968, Imperial College, London, (ed. T. A. Toussoun, R. V. Bega & P. E. Nelson), pp. 150-152. University of California Press: Berkeley.

Swift, M. J. (1972). The ecology of *Armillaria mellea* (Vahl ex Fries). *Forestry* **45**, 175-199.

Ullrich, R. C. & Anderson, J. B. (1978). Sex and diploidy in *Armillaria mellea*. *Experimental Mycology* **2**, 119-129.

Wastie, R. L. (1986). Disease resistance in rubber. *FAO Plant Protection Bulletin* **34**, 193-199.

Watling, R. (1992). *Armillaria* Staude in the Cameroon Republic. *Persoonia* **14**, 483-491.

Watling, R., Kile, G. A. & Gregory, N. M. (1982). The genus *Armillaria* - nomenclature, typification and the identification of *Armillaria mellea* and species differentiation. *Transactions of the British Mycological Society* **78**, 271-285.

Williams, J. G. K., Kubelik, A. R., Livak, K. J., Rafalski, J. A. & Tingey, S. V. (1990). DNA polymorphisms amplified by arbitrary primers are useful as genetic markers. *Nucleic Acids Research* **18**, 6531-6535.

[illegible] (1986). [illegible] University of California Press, [illegible]

[illegible]

[illegible] *Experimental Acarology* [illegible]: 119–129.

[illegible] 199.

Walton, [illegible] (1992). [illegible] in the [illegible]

[illegible] (1982). The genus [illegible] nomenclature, reproduction and [illegible]

[illegible] DNA [illegible] markers. [illegible]

Chapter 13

Interactions between the pathogen *Crinipellis perniciosa* and cocoa tissue

Susan Isaac*, Keith Hardwick & Hamish Collin*

*Departments of *Genetics & Microbiology and Environmental & Evolutionary Biology, University of Liverpool, P.O. Box 147, Liverpool, L69 3BX, U.K.*

Introduction

The unusual basidiomycete *Crinipellis perniciosa* (Stahel) Singer causes witches' broom disease of cocoa (*Theobroma cacao* L.). Although cocoa is grown in South and Central America, Africa and Malaysia, outbreaks of the disease have, as yet, only been recorded on plantations in South America and the Caribbean. The pathogen is endemic and highly destructive in the Amazon region of South America where it severely affects cocoa production (Wheeler, 1985), and in consequence has a major influence on the economic status of the countries involved. New (1990) and devastating outbreaks in Bahia, Brazil, have recently resulted in 90 to 95% crop losses. Long term effects on plantations must have serious consequences, since infected trees quickly show excessive, abnormal growth.

A great deal is now known about the development and spread of the disease in the field. The International Witches' Broom Project (IWBP) was set up and supported (1985-91) by finance from the Governments of six cocoa producing countries affected by the disease, and from international organisations concerned with cocoa production worldwide. Under this project an enormous amount of information was collected, concerning the conditions which favour disease establishment and spread in the field. Assessments were made of the efficiency of phytosanitation as a means of control, at various times during disease development and under various environmental conditions in the field. In spite of extensive trials, no effective means of chemical control have yet been achieved. The primary aim of the IWBP was to develop guidelines for farmers whose plantations have already been affected by the disease and education for those whose crops have not yet been affected but may soon be at risk. The data generated by the project are currently being evaluated and prepared for publication (Lass & Rudgard, 1991).

Although the influence of field conditions on disease development has been assessed in some detail, most aspects of the morphological, physiological and biochemical interactions between *C. perniciosa* and *T. cacao* have not yet been satisfactorily described and are still very poorly understood.

In the field, disease spread results from the dispersal of basidiospores (Rocha & Wheeler, 1985). Germ tubes, from recently germinated spores, infect actively growing host tissue, inducing excessive and disorganised growth at apices. Apical dominance is lost, and usually multiple swollen shoots (green brooms) are produced from vegetative buds, and more occasionally from flower cushions (Baker & Holliday, 1957; Thorold, 1975). The infective, biotrophic (primary) phase, mycelium is characterised by quite wide hyphae (5–20 μm) which grow intercellularly within stem apices and give rise to the relatively localised development of symptoms in the host (Cronshaw & Evans, 1978). At some time (several months) after symptom development the fungus dikaryotizes (Evans, 1980) and becomes saprotrophic (secondary phase), penetrating host cells by means of narrow (1·5–3 μm) hyphae. The fungus persists in dead tissues (brown brooms), but the exact timing and cause of broom tissue death is not clear. Eventually, after 6–9 months (Rudgard, 1986), basidiomes are produced on dead host tissue and basidiospores are liberated from them.

Secondary phase, saprotrophic mycelium can be cultured easily in the laboratory (Wheeler, 1985), utilising a range of nutrient substrates, and therefore lends itself well to *in vitro* physiological investigations. However, the biotrophic phase cannot be cultured so easily. Primary mycelium (lacking clamp connections) has been briefly maintained in the presence of cocoa tissue cultures (Evans & Bastos, 1980) and only recently have axenic cultures of this phase been achieved (D. Penman, pers. comm.).

The pathogen exerts a range of important influences on host tissue during the disease cycle, although the physiological basis for many of the interactions is not at all clear. For example, (i) the stimulus to apical growth and the formation of multiple meristems is unknown; (ii) although major morphological changes occur in the host plant, little mycelium can be detected in sections of diseased stems and the distribution of the fungus may be very limited; (iii) dikaryotization of the fungus occurs during disease development, but its significance and what triggers it are unknown; (iv) the causes of broom tissue death have not been determined; (v) the relative proportions of primary and secondary phase mycelium that are present, and the influence either phase has on the progress of the disease are unknown. This chapter will consider recent experiments which have been carried out to provide answers to some of these questions.

The stimulus to apical growth and the formation of multiple meristems

Morphological and physiological changes in infected host tissues

Symptom development in infected plants is diverse. The extent and nature of stem swelling observed and the time course of changes which occur are very variable and unpredictable (Wheeler, 1985; Fonseca & Wheeler, 1990). Observations of gross morphological alterations have been related to more detailed anatomical investigations in infected plants to determine the internal changes which take place as disease progresses. Stages of disease development were identified to standardise the material used as far as possible (Orchard *et al.*, 1989). The early stage of disease development was considered to coincide with the first signs of stem swelling; mid-stage, where stem swelling was well established, and late-stage, when the first necrotic regions were apparent on swollen stems.

The anatomies of healthy and diseased stems have been compared using scanning electron microscopy (Orchard *et al.*, 1989). The radial dimensions of all tissues (particularly the cortex) were increased by the late-stage of infection (Fig. 1). In healthy stems boundaries between tissue types were clear, but in infected stems such boundaries were difficult to determine and tissues were more disorganised. In general, in infected stems individual cells were larger and differentiation was reduced. Vascular tissue in particular was less differentiated, with more parenchymatous cells in the phloem region. Xylem lacked the characteristic, large vessels of healthy stems (Fig. 1); xylem tracheids were less uniform and medullary rays were less distinct. Evidence that vascular function is severely impaired at this stage has also been provided (Orchard, Collin & Hardwick, 1980; 1981). In spite of the swollen nature of the diseased stem the amounts of dry biomass were found to be significantly lower ($P < 0{\cdot}001$) than in uninfected stems, at both mid- and late-stages of infection (Table 1). It is clear therefore, that the massive increase in stem diameter after infection was not due to the accretion of biomass. Estimates of the transverse sectional areas of cells in different tissues were made from scanning electron micrographs. The average cell areas for all tissue types examined were significantly larger ($P < 0{\cdot}001$) and the range of cell sizes was increased, in tissues from diseased stems, at all stages of infection (Fig. 2). Cell number was not increased in infected tissues (Orchard *et al.*, 1989). It is clear

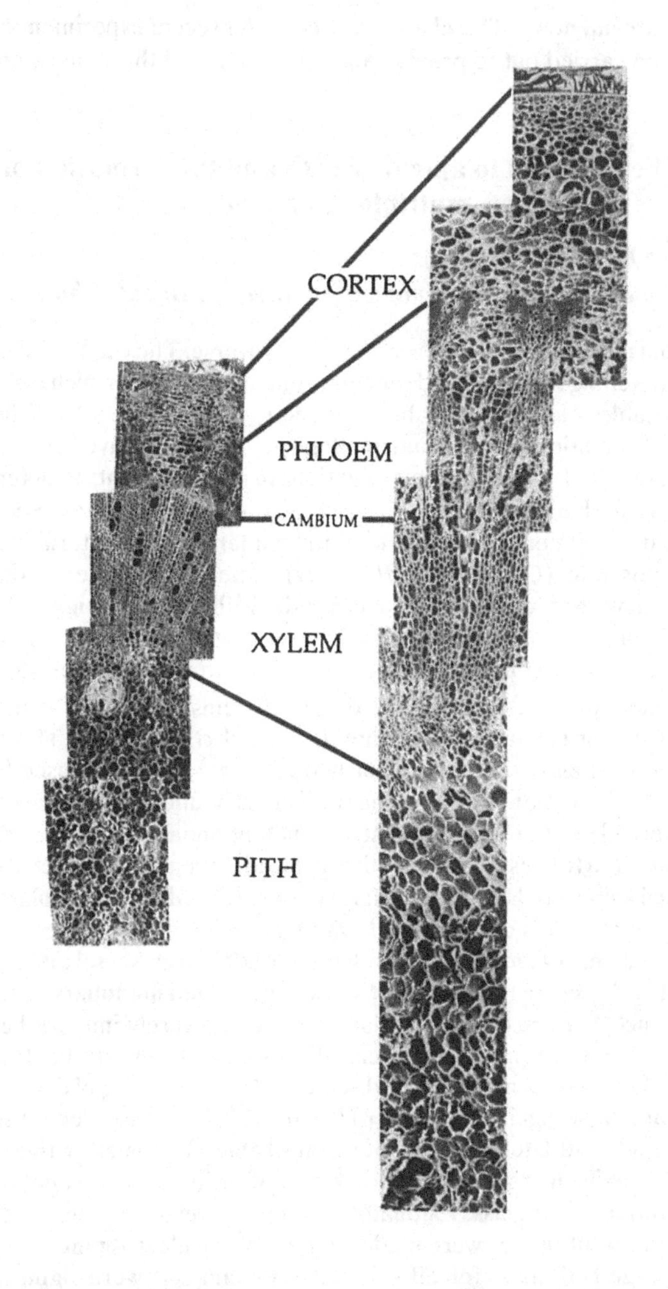

Fig. 1. Composite photomicrographs of transverse sections along a radius through healthy (left) and late-stage, diseased (right) cocoa stems (after Orchard *et al.*, 1989).

Table 1. Dry weight as a percentage of fresh weight for healthy and diseased cocoa stems (data from Orchard *et al.*, 1989).

	Dry weight: Fresh weight ratio x 100 (± SEM)	
Stage of disease	Healthy	Diseased
Mid-stage	26·5±2·2	8·3±1·7
Late-stage	28·2±1·6	15·4±1·4

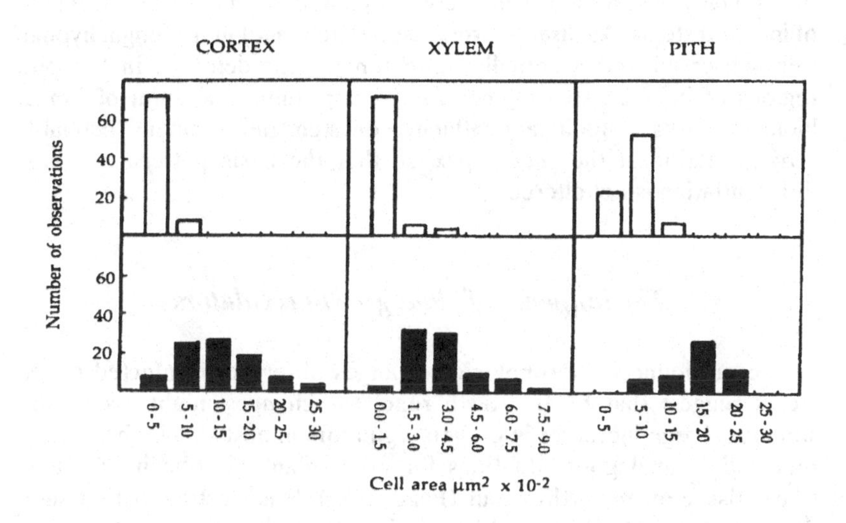

Fig. 2. Range of cell sizes in cortex, xylem and pith tissues in transverse sections of healthy (open bars) and diseased (solid bars) stems at the late-stage of disease development (data replotted from Orchard *et al.*, 1989).

therefore, from this detailed examination, that the increase in stem diameter which occurs during infection is attributable to cell enlargement (hypertrophy) in stem cells but not to cell division (hyperplasia), although both have been suggested previously to occur in broom formation (Evans, 1980).

Pith cells in healthy stems always contained numerous starch grains but none were observed in the pith of diseased stems. Analyses of stem tissue showed that by the late stage of disease development the starch content of infected stems was reduced by over ten-fold from that of healthy tissue. The use of histochemical methods showed (R. Machado, pers. comm.) that starch was present throughout the pith and cortex regions of uninfected stems but occurred only in a narrow zone, just outside the vascular tissue, in infected stems. Clearly, carbohydrate metabolism is altered in infected tissues, and material, normally destined for storage, may be diverted for proliferation of disorganised host tissue.

In general, very limited amounts of fungal tissue were observed in green brooms. In many sections no fungal tissue was seen, but hyphae were occasionally observed growing between plant cells in the cortical region of infected stems. As disease progressed (late-stage) more fungal hyphae were observed and eventually more fungus was detected in the pith regions of infected stems. Therefore, a very limited amount of fungal biomass exerts a significant influence on stem development, probably through action at the shoot apex, so that the basic pattern of tissue differentiation is not altered.

The influence of plant growth regulators

The pronounced morphological changes observed in infected cocoa stems suggest that plant growth regulator changes might act as the stimulus. Large increases in growth regulators in host species have been reported in analagous situations for other plant-microbe interactions where tissue overgrowths occur (Pegg, 1984). It is most likely that such developmental changes would be induced by alterations in cytokinin levels. Cytokinins do have a role in inducing cell division and also affect nutrient mobilisation and transport (Goodman, Kiraly & Wood, 1986).

Immunological methods were used to investigate levels of cytokinins in healthy and diseased cocoa tissues (Orchard *et al.*, 1989). However, no significant differences were detected between zeatin levels in healthy and diseased tissues as disease progressed (Fig. 3). In addition, since only very small elevations in levels of zeatin riboside were detected in infected tissue a direct influence on tissue development from this source is unlikely. Levels of isopentenyl cytokinins were not significantly different in healthy and diseased plants at any stage of the disease.

Evidence to implicate cytokinins is therefore limited, and furthermore the application of kinetin to shoot apices of healthy plants resulted in disruption to the flush cycle, but no stem swelling was observed (Abo Hamed, Hardwick & Collin, 1981). Nevertheless, distortions of host development are distinct and may implicate alterations in the balance

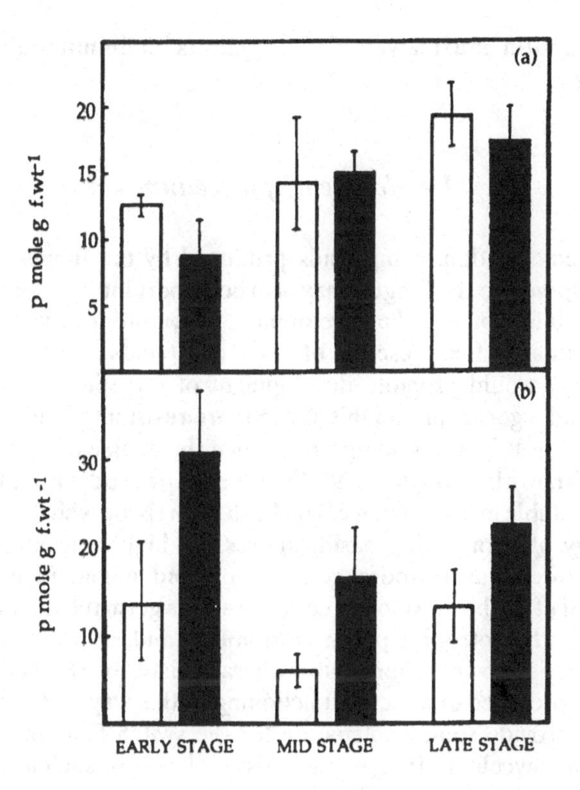

Fig. 3. (a) Zeatin and (b) zeatin riboside contents of healthy (open bars) and diseased (solid bars) stems at each stage of disease development. Bar markers represent standard errors (data replotted from Orchard *et al*., 1989).

---

between different growth regulators during the development of witches' broom disease. In addition variations in host tissues must not be overlooked. It is not easy to identify accurately the stage of the flush cycle of the host plant at the time of infection. Levels of plant growth regulators change significantly during the flush cycle (Orchard, *et al*., 1980; 1981), so that uninfected (control) plants tested at even slightly different developmental stages may contain markedly different levels. All changes in plant growth regulators except the most pronounced would, therefore, be masked. Furthermore, it is possible that any growth regulator changes may be restricted to small meristematic areas at the shoot apex. In analyses of stem sections these changes would, again, not be detected. Any such evaluation must await developments in immunohistological techniques.

Development of mycelium

During colonisation, compounds produced by the fungus, or by the plant in response to the fungus, may also be important in determining the outcome of infection and the development of symptoms. Evans & Bastos (1980) postulated the presence of two compounds in cocoa tissue: a modifier that would promote development of mycelium in susceptible cultivars, and a germination inhibitor in more resistant plants that would influence germ-tube morphology and hence the course of disease. More recently Brownlee *et al.* (1990) have reported the effects of methanol-soluble extracts from cocoa flush stem tissue which affected the morphology of germinating basidiospores. At high concentrations this factor inhibited basidiospore germination and might, *in vivo*, delay colonisation of the host. At lower concentrations germ tubes were formed with apparently biotrophic phase morphology and, at even more dilute levels, gave rise to mycelium more characteristic of the saprotrophic phase. The presence of a factor functioning in this way within host plant tissue may provide some explanation for the switch from biotrophic to saprotrophic mycelium. In dying host tissue, levels of such a factor may decline, promoting the formation of secondary phase mycelium.

Similarly, the presence of such a factor may also provide partial explanation for differences in susceptibility to the fungus shown by different cocoa clones. Some cultivars are extremely susceptible and would not be recommended for planting in areas where witches' broom disease occurs; unfortunately no really resistant cocoa lines have been identified. In addition, some isolates of *C. perniciosa* are more pathogenic then others (Wheeler & Mepstead, 1988; McGeary & Wheeler, 1988). Screening for resistance in cocoa is time-consuming and unreliable. Recently however, it has been shown that it may be possible to use callus cultures to speed up the screening process (Fonseca & Wheeler, 1990).

The cause of host tissue death

Host tissue death does not begin until some weeks after stem swelling and even at this stage little fungal mycelium may be present in infected stems. The relationship between mycelial development and tissue death has not yet been determined. However, it is clear that vascular development is

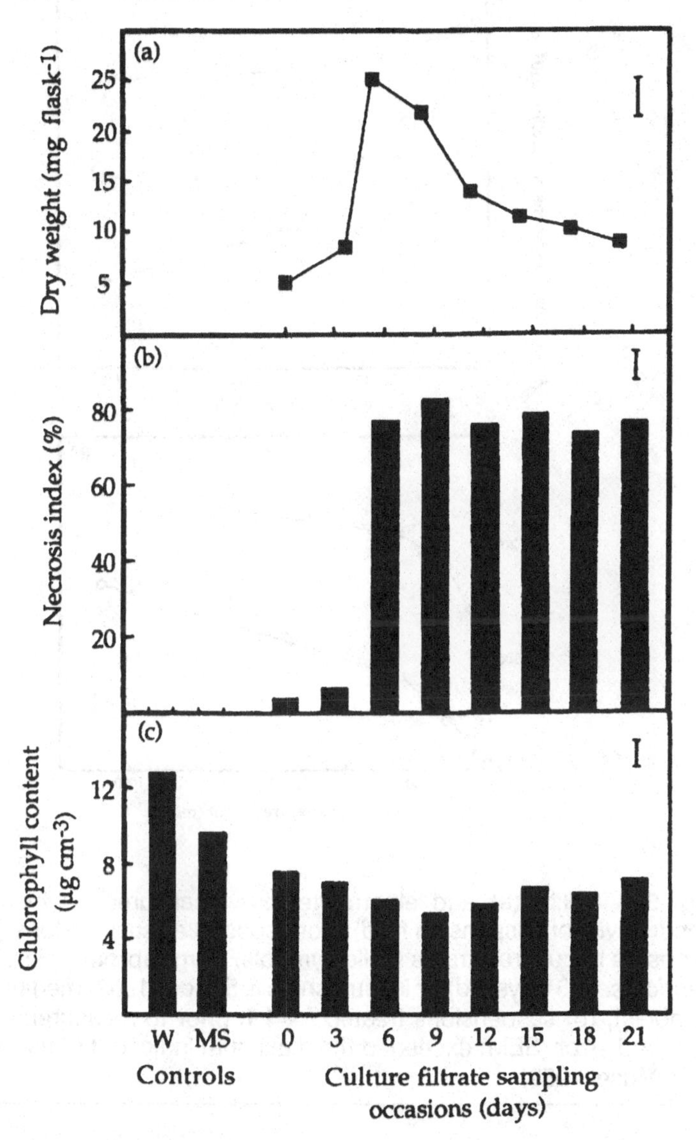

Fig. 4. Change in dry weight of *C. perniciosa* (saprotrophic phase) in stirred Murashige & Skoog (MS) liquid culture medium (Murashige & Skoog, 1962) at 30°C (a); necrosis index (see text) (b) and chlorophyll content (c) of cocoa leaf discs after 6 days incubation in culture filtrates from different stages of growth of *C. perniciosa*. W, water; MS, full strength MS medium. No standard error (SEM) exceeded the maximum indicated (data replotted from Muse *et al.*, 1988).

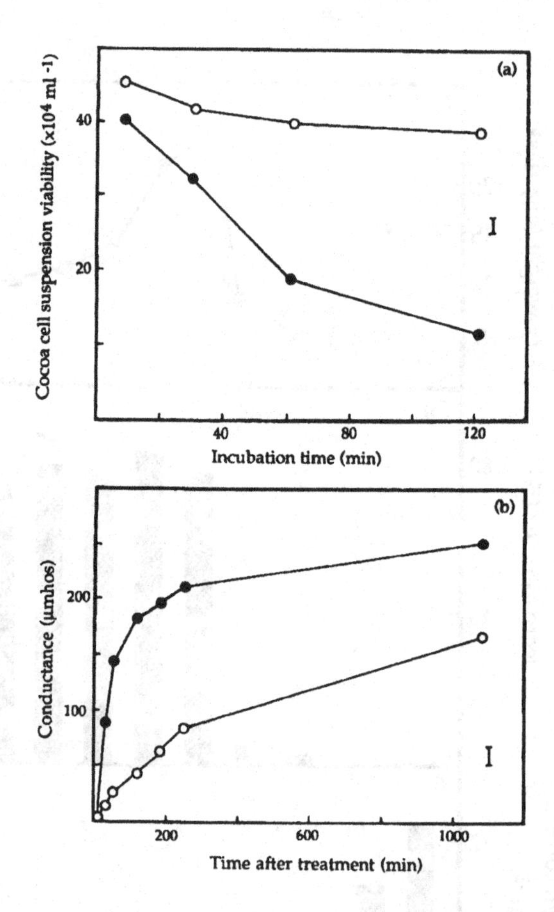

Fig. 5. Viability (a) and electrolyte loss (measured as increase in conductivity of suspension fluid) from cocoa cell suspensions (b) after exposure to culture filtrates (solid symbols) from saprotrophic phase *C. perniciosa* (15 days old) or in Murashige & Skoog (1962) medium (open symbols), (b) suspensions treated for 1 h prior to measurements. No standard error (SEM) exceeded the maximum indicated (data replotted from Muse, 1989).

severely disrupted in infected stems and flush leaves become brown, dehydrated and necrotic although abscission does not occur. It is likely that impaired vascular function contributes to the death of leaf and stem tissue (Orchard & Hardwick, 1988).

Since saprotrophic phase mycelium can be readily grown in culture, physiological investigations have been carried out with this form. It has been shown that a phytotoxic compound was released from saprotrophic mycelium in liquid culture (Muse *et al.*, 1988; Muse, 1989). This compound caused a number of detrimental effects on cocoa tissues although the action was not specific to cocoa. The release of phytotoxin was related to the growth of the fungus in culture and was probably a low molecular weight compound although it has not yet been further characterised.

Cocoa leaf discs were incubated in culture filtrate samples from different stages of growth of saprotrophic phase *C. perniciosa*. The degree of necrosis induced in each disc was scored to give a necrosis index, based on the surface area of the leaf disc showing necrosis after 6 days incubation (Muse *et al.*, 1988). Little effect ($< 12\%$ necrosis) occurred with filtrates taken very early during fungal growth (Fig. 4a, b). However, filtrates taken from fungal cultures more than five-days-old caused 80% necrosis. Control treatments using water or growth medium for leaf disc incubation did not induce necrosis. Chlorophyll levels in leaf discs were less affected by filtrates (Fig. 4c) suggesting that these photosynthetic pigments were relatively stable in the presence of filtrates and were masked by necrotic browning. Filtrates that elicited most response in this non-growing plant tissue were from those cultures in which maximum fungal biomass had been reached.

In similar experiments, using actively growing cocoa cell suspensions (Muse, 1989) it was shown that exposure to fungal culture filtrates (15 day-old cultures) resulted in a rapid decrease in cell viability over a 2 h incubation period (Fig. 5a). Additionally, electrolyte loss was high and occurred very rapidly from cocoa cell suspensions after a 1 h incubation in fungal culture filtrate (15 day-old cultures), compared with control suspensions (Fig. 5b). Very similar results were also obtained using protoplasts from young, rapidly expanding cocoa leaf tissue (Muse, 1989), suggesting that the phytotoxic compound released during growth of *C. perniciosa* detrimentally affected the membranes of cocoa cells.

The degree to which saprotrophic phase mycelium has a role in tissue death *in vivo* is not known. It is likely that as disease progresses there is a gradual switch from the biotropic form to saprotrophic mycelium in host tissue. The biotrophic phase has now been maintained in culture for up to 17 days (D. Penman, unpublished data), and sufficient biomass can be accrued for biochemical and physiological investigations to be carried out. In due course, it should be possible to assess the influence both phases has on the progress of the disease.

Conclusions

It can be seen that the biochemical and physiological interactions between *C. perniciosa* and cocoa tissues are complex, constantly changing with fungal development and as disease progresses.

It is clear that a small amount of fungus exerts a major influence on tissue morphology in infected stems, causing cell enlargement and disruption to tissue organisation. The influence of plant growth regulators, originating either from the plant or from the fungus, on the formation of multiple meristems is still not clear. Changes in cytokinin levels are small and the morphological effects may be the result of very subtle changes in the balance between different growth regulators. The application of new methods, for making such measurements with improved sensitivity, is awaited.

It is likely that broom tissue death occurs as the result of many influences. The impairment of vascular function, mobilisation of nutrient reserves resulting in physiological depletion in infected tissues, and the production of a phytotoxic compound by saprotrophic phase mycelium, undoubtedly contribute to the onset of necrosis. Very little fungal mycelium can be detected during disease development, and the relative proportions of primary and secondary phase mycelium that are present and the influence either phase has on the progress of the disease are unknown.

The growth of *C. perniciosa* and the development of witches' broom disease in cocoa are not easily studied *in vivo* in practical terms and, additionally, it is clear that the interactions between host and pathogen are complex. It may now be possible to use *in vitro* techniques to provide simplified systems to aid investigations, making use of tissue culture methods and laboratory culture of the biotrophic stage. A useful start has been made in understanding the processes involved but many questions remain to be answered.

Acknowledgements The authors wish to thank the Biscuit, Chocolate, Cake and Confectionery Alliance for financial support for much of the work reported here. We are also grateful to Drs J. Orchard, R. Machado, and R. B. Muse, Ms J. Dart and Mr D. Penman for laboratory work, and to Mr C. J. Veltkamp for SEM photography. We also wish to thank Dr B. E. J. Wheeler for his assistance and for helpful discussions during this work.

References

Abo-Hamed, S., Hardwick, K. & Collin, H. A. (1981). Biochemical and physiological aspects of leaf development in cocoa (*Theobroma cacao* L.). VI. Hormonal interactions between mature leaves and the shoot apex. *New Phytologist* **89**, 191-200.

Baker, R. E. D. & Holliday, P. (1957). Witches' Broom Disease of Cocoa (*Marasmius perniciosus* Stahel). *Phytopathological Paper* No. 2, 42pp. Commonwealth Mycological Institute, Kew.

Brownlee, H. E., Hedger, J. N. & Scott, I. M. (1990). Host extracts cause morphological variation in germ-tubes of the cocoa pathogen, *Crinipellis perniciosa*. *Mycological Research* **94**, 543-547.

Cronshaw, D. K. & Evans, H. C. (1978). Witches' Broom disease of cocoa (*Crinipellis perniciosa*) in Ecuador. II. Methods of infection. *Annals of Applied Biology* **89**, 193-200.

Evans, H. C. (1980). Pleomorphism in *Crinipellis perniciosa*, causal agent of witches' broom disease of cocoa. *Transactions of the British Mycological Society* **74**, 515-523.

Evans, H. C. & Bastos, C. N. (1980). Basidiospore germination as a means of assessing resistance to *Crinipellis perniciosa* (witches' broom disease) in cocoa cultivars. *Transactions of the British Mycological Society* **74**, 525-536.

Fonseca, S. E. A. & Wheeler, B. E. J. (1990). Assessing resistance to *Crinipellis perniciosa* using cocoa callus. *Plant Pathology* **39**, 463-471.

Goodman, R. N. Kiraly, Z. & Wood, K. R. (1986). *Biochemistry and Physiology of Plant Disease*. University of Missouri Press: USA.

Lass, R. A. & Rudgard, S. A. (1991). *Managing Witches' Broom Disease of Cocoa*, International Witches' Broom Project 7th Workshop Report. International Office of Cocoa, Chocolate and Sugar Confectionery: Brussels, Belgium.

McGeary, F. M. & Wheeler, B. E. J. (1988). Growth rates of, and mycelial interactions between, isolates of *Crinipellis perniciosa* from cocoa. *Plant Pathology* **37**, 489-498.

Murashige, T. & Skoog, F. (1962) A revised medium for rapid growth and bioassays with tobacco tissue cultures. *Physiologia Plantarum* **15**, 473-497.

Muse, R. B. (1989). Physiology and biochemistry of Witches' Broom disease in cocoa (*Theobroma cacao* L.). Ph. D. Thesis, University of Liverpool.

Muse, R. B., Isaac, S., Collin, H. A. & Hardwick, K. (1988). Phytotoxin production by *Crinipellis perniciosa*. In *Proceedings of the 10th International Cocoa Research Conference, Santo Domingo*, pp. 355-358. Cocoa Producers Alliance: Lagos, Nigeria.

Orchard, J. E. & Hardwick, K. (1988). Photosynthesis, carbohydrate translocation and metabolism of host and fungal tissues in cacao seedlings infected with *Crinipellis perniciosa*. In *Proceedings of the 10th International Cocoa Research Conference, Santo Domingo*. pp. 325-330. Cocoa Producers Alliance: Lagos, Nigeria.

Orchard, J. E., Collin, H. A. & Hardwick, K. (1980). Biochemical and physiological aspects of leaf development in cocoa (*Theobroma cacao* L.). IV. Changes in growth inhibitors. *Plant Science Letters* **18**, 299-305.

Orchard, J. E., Collin, H. A. & Hardwick, K. (1981). Biochemical and physiological aspects of leaf development in cocoa (*Theobroma cacao* L.). V. Changes in growth promotors, auxins and cytokinins. *Café, Cacao et Thé* **25**, 25-29.

Orchard, J. E., Hardwick, K., Collin, H. A., Isaac, S. & Dart, J. (1989). The role of plant growth regulators in the response of cocoa to infection by *Crinipellis perniciosa*. pp. 1-32. Report to the Biscuit, Cake, Chocolate and Confectionery Alliance, Green Street, London, U.K.

Pegg, G. F. (1984). The role of growth regulators in plant disease. In *Plant diseases: Infection, Damage and Loss* (eds J. G. Jellis & R. K. S. Wood), pp. 29-48. Blackwell Scientific Publications: Oxford.

Rocha, H. M. & Wheeler, B. E. J. (1985). Factors influencing the production of basidiocarps and the deposition and germination of basidiospores of *Crinipellis perniciosa*, the causal fungus of witches' broom on cocoa (*Theobroma cacao*). *Plant Pathology* **34**, 319-328.

Rudgard, S. A. (1986). Witches' Broom disease on cocoa in Rondonia, Brazil: basidiocarp production on detached brooms in the field. *Plant Pathology* **35**, 434-442.

Thorold, C. A. (1975). *Diseases of Cocoa*. Clarendon Press: Oxford, U. K.

Wheeler, B. E. J. (1985). The growth of *Crinipellis perniciosa* in living and dead cocoa tissue. In *Developmental Biology of Higher Fungi* (eds D. Moore, L. A. Casselton, D. A. Wood & J. C. Frankland), pp. 103-116. Cambridge University Press: Cambridge, U.K.

Wheeler, B. E. J. & Mepstead, R. (1988). Pathogenic variability amongst isolates of *Crinipellis perniciosa* from cocoa (*Theobroma cacao*). *Plant Pathology* **37**, 475-488.

Chapter 14

Molecular genetics of *Colletotrichum gloeosporioides* infecting *Stylosanthes*

J. M. Manners, A. Masel & J. A. G. Irwin

Cooperative Research Centre for Tropical Plant Pathology, The University of Queensland, Brisbane, Queensland 4072, Australia.

Introduction

The group species *Colletotrichum gloeosporioides* (Penz.) Penz. & Sacc. shows considerable variation in classical morphological taxonomic characters and is considered to be genetically heterogeneous (Sutton, 1980; Cox & Irwin, 1988). Members of this group species infect a wide range of hosts and are of considerable economic importance in the tropics, yet little is known of genetic relationships within the group species. This Chapter will focus on strains of this fungal pathogen which infect the tropical pasture legumes *Stylosanthes* spp. and will discuss the genetic relationships of pathotypes and races of this fungus in Australia. Particular emphasis will be placed on the mechanisms resulting in genetic variation and pathogenic specialisation in the fungus.

Anthracnose of *Stylosanthes* species in Australia

The genus *Stylosanthes* is primarily indigenous to tropical and subtropical South America, but contains several species which have been domesticated for use as pasture legumes in Australia, South East Asia and Africa, as well as South America (Burt *et al.*, 1983). The full commercial potential of *Stylosanthes* spp. in Australia and other regions has not been fully realised because of the development of the anthracnose disease caused by *C. gloeosporioides*. In Australia, the disease was first recognised on *Stylosanthes* spp. in 1973 (Pont & Irwin, 1976). Two *C. gloeosporioides*-incited diseases, designated Types A and B, were subsequently recognised by the different symptoms produced on the same host (*S. guianensis*, (Aubl.) Sw.)(Irwin & Cameron, 1978). Extensive pathogenic specialisation has been reported within both the Type A and

Type B pathogen groups (Irwin *et al.*, 1986). The adaptation of this pathogen to newly deployed resistant cultivars is a major challenge for plant breeders (Chakraborty *et al.*, 1990; Cameron *et al.*, 1989). A total of 12 cultivars of *Stylosanthes* spp. have been released in Australia, but large scale seed production has now ceased for all except three because of anthracnose susceptibility in the other cultivars (Chakraborty *et al.*, 1991). Several mechanisms could have contributed to the extensive adaptation and pathogenic specialisation exhibited by this pathogen in Australia.

Mechanisms contributing to pathogenic specialisation

The processes that could contribute to the generation of a new race or type of *C. gloeosporioides* in Australia are illustrated in Fig. 1. New races or types could arise by genetic variation in strains of *C. gloeosporioides* already present in Australia and these could have been pathogenic on *Stylosanthes* genotypes or perhaps even on other plant genera. Alternatively new strains of *C. gloeosporioides* could arise by introduction from other countries and these could also have originated from *Stylosanthes* or other hosts. It would be expected that if types and races are recently related, then their genomes would be very similar perhaps differing at only a few loci important for pathogenicity and virulence. Such genetic similarity might be less expected if types and races had independent and diverse origins. One aim of recent research has been to distinguish whether genetic change or geographic shift has led to the development of the two types of *C. gloeosporioides* infecting *Stylosanthes* in Australia. To address this question a considerable body of information on pathogenicity, morphology, sexual and parasexual compatibility and genome organisation in Type A and Type B strains has been assembled and these investigations are described briefly below.

Type A and Type B pathogens are genetically distinct

Pathogenicity of the types

The Type A fungus has a wide host range, infecting species of *Aeschynomene, Desmodium, Indigofera* and *Psoralia* in addition to *Stylosanthes*. The fungus produces limited lesions with a light-coloured centre and a dark margin (Vinijsanun, Irwin & Cameron, 1987). The Type B fungus is host specific to *S. guianensis*, producing blight symptoms on

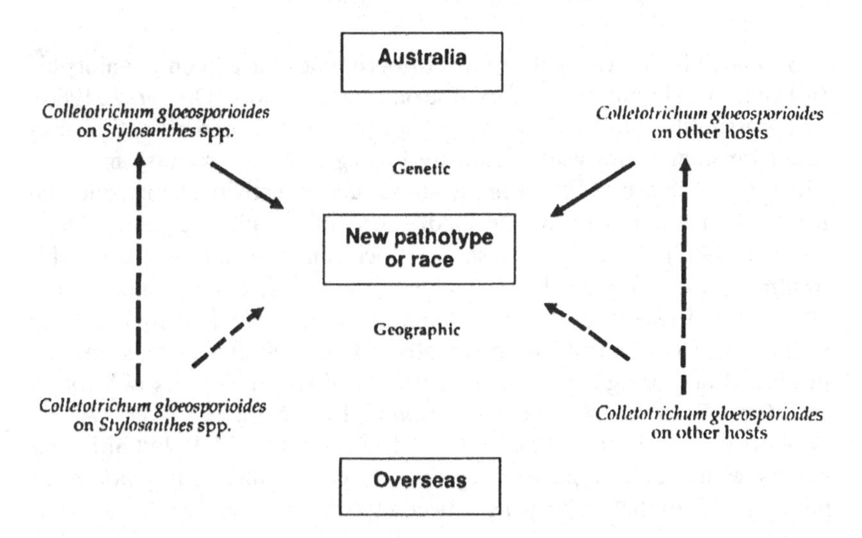

Fig. 1. Possible sources of new types and races of *C. gloeosporioides* pathogenic on specific genotypes of *Stylosanthes* in Australia and the role of geographic migration indicated by dashed lines and genetic alterations of the pathogen indicated by solid lines.

leaves and stems. Histologically, the two pathogen groups invade their hosts in different ways. Both infect 2 to 4 adjacent epidermal cells before spreading intra- and intercellularly into the mesophyll. In the Type B fungus, however, hyphae also emerge from the epidermal cells to grow subcuticularly, or more often superficially, initiating new infections without the formation of appressoria. This fungal growth results ultimately in extensive blight and complete plant death (Trevorrow, Irwin & Cameron, 1988; Ogle, Gowanlock & Irwin, 1990).

Morphology of the types

Conidia of the Type A fungus are relatively uniform in size with rounded ends, whilst those of Type B are more variable and tapered at one end (Ogle, Irwin & Cameron, 1986; Cox & Irwin, 1988). Both groups have mean conidial widths between 3.0 and 4·2 μm but Type B can produce 'giant' conidia on free hyphae (Cox & Irwin, 1988).

Vegetative incompatibility between types

So far all highly virulent isolates of both types have been anamorphic (asexual) precluding any studies of sexual compatibility (Ogle *et al.*, 1986). The possibility that heterokaryosis and parasexual recombination may occur between types was investigated using nitrate reductase mutants. Mutants which are deficient in nitrate reductase activity can be selected for by their ability to grow on media containing chlorate (Cove, 1976; Puhalla, 1985). Mutations at several loci can be selected for by this treatment, these mutants being unable to use nitrate as a nitrogen source efficiently. If two mutants carry mutations at independent loci and are vegetatively compatible then nitrate-utilising heterokaryons can be produced in pairing experiments. Nitrate reductase mutants of Type A and Type B strains of *C. gloeosporioides* have been prepared (A. M. Poplawski, J. M. Manners and J. A. G. Irwin, unpublished), but although strains within each type are vegetatively compatible it has not been possible to form heterokaryons between types.

The distinct morphology, host range, infection processes and vegetative incompatibility of the Type A and Type B strains suggest that they are genetically distinct. However, differences in these characters could be controlled by a relatively small number of genes. To obtain a more comprehensive assessment of genetic relationships between the types, research has concentrated on nucleic acid markers.

Double stranded RNAs in the types

Analysis of cellular protein patterns initially failed to demonstrate any significant differences between the types (Dale, Manners & Irwin, 1988). However, major differences in double stranded RNA components were detected between the types (Dale *et al.*, 1988). Double stranded RNA is thought to be cytoplasmic and the major differences in these components suggest that Type A and Type B do not readily recombine in the field. This observation is consistent with the apparent somatic incompatibility of the two types. Double stranded RNA is potentially infectious and may not reflect true genetic relationships between organisms. Analysis of genetic differences is best achieved by studies of the nuclear DNA.

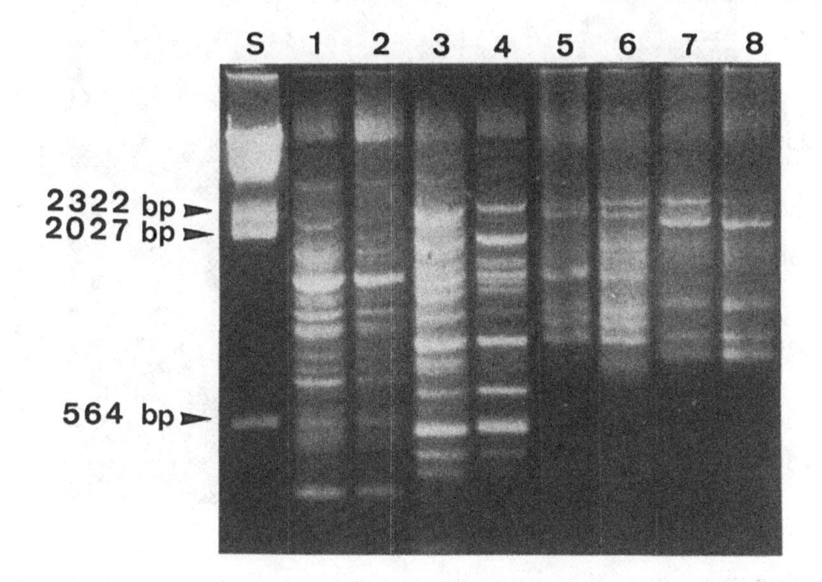

Fig. 2. Random amplified polymorphic DNA markers generated by the method of Williams *et al.*, (1990) distinguish strains of the Type A pathogen (lanes 1, 2, 5 and 6) and the Type B pathogen (lanes 3, 4, 7 and 8). Lanes 1 to 4 were with one primer (random 10mer) and lanes 5 to 8 were with a different primer. Lane S shows DNA size markers.

RFLPs and RAPDs distinguish types

True genetic comparisons of organisms have to be carried out at the DNA level to avoid physiological and environmental ambiguities. Restriction fragment length polymorphisms (RFLPs) and the more recently developed random amplified polymorphic DNA (RAPD, Williams *et al.*, 1990) markers permit a rapid comparison of the genomes of organisms.

The genetic relationships between Type A and Type B pathogens of *Stylosanthes* spp. have been studied in most detail using RFLPs (Braithwaite & Manners, 1989; Braithwaite, Irwin & Manners, 1990 a,b). The easiest RFLP markers to use are sequences from the ribosomal DNA

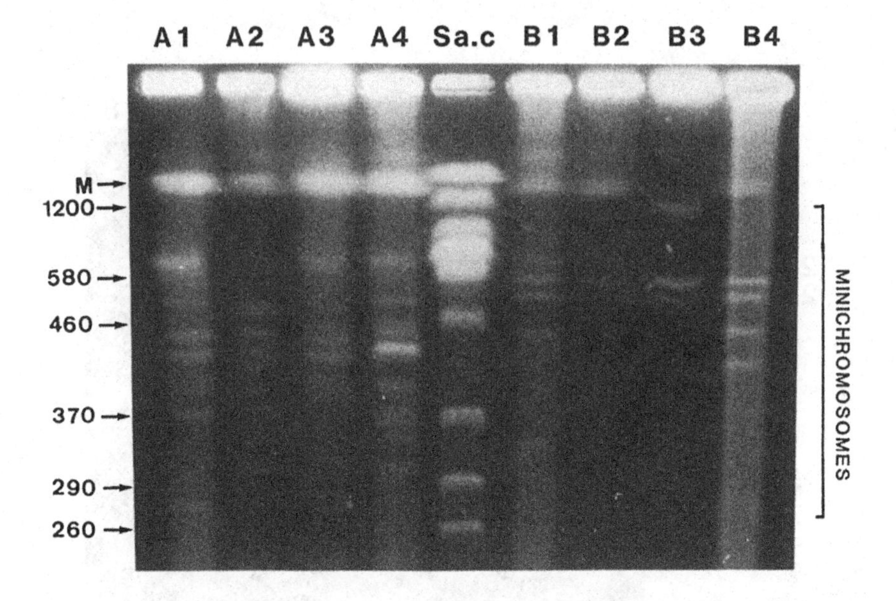

Fig. 3. Pulse-field electrophoresis of mini-chromosomes of four races of each of the Type A pathogen (A1–A4) and the Type B pathogen (B1–B4). *S. cerevisiae* chromosomes are shown as standards with their size in kb on the left (Sa.c). M refers to the maxi-chromosomes. Reproduced from Masel *et al.* (1990) with permission.

repeats (rDNA) which are readily detected because their tandem arrangement leads to intense banding patterns. Using a probe representing the entire ribosomal tandem repeat unit of *Aspergillus nidulans* (Eidam) Winter, it has been shown that the ribosomal repeat of Type A and Type B strains differs in length and contains a readily detectable RFLP using the restriction enzyme *Xba*1. In order to survey a more comprehensive amount of the genome, dispersed repeats, low copy sequences isolated from the pathogen, and heterologous human mini-satellite DNA probes were used to compare isolates representative of 4 physiological races from each of the types. These data demonstrated that the two types are genetically distinct and shared only 43% of the total number of hybridising bands revealed by these probes. Variation between strains and races within the two types was very slight (95% similarity) and this suggests that two distinct clonal populations of *C. gloeosporioides* infecting *Stylosanthes* spp. exist in Australia. The large genetic distance

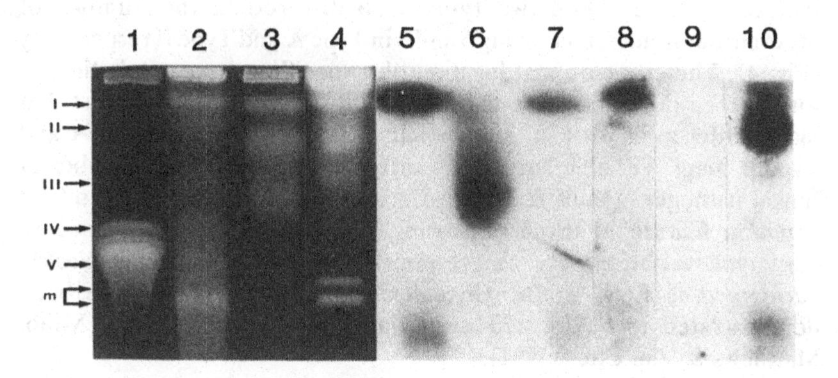

Fig. 4. Pulse-field electrophoresis of maxi-chromosomes of the Type A pathogen (lanes 2,5 and 8) and the Type B pathogen (lanes 4,7 and 10). Lanes 3,6 and 9 are chromosomes of *Schizosaccharomyces pombe* Lind and lane 1 shows *Saccharomyces cerevisiae* Meyer ex Hansen yeast standards. Lanes 1–4 are ethidium bromide stained DNA. Lanes 5–7 and 8–10 represent hybridisation studies using ribosomal and tubulin gene probes respectively. The position of the maxi-chromosomes is arrowed on the left for Type A and on the right for Type B and m refers to the position of mini-chromosomes. Reproduced from Masel *et al.*, (1990) with permission.

observed between the two types by RFLPs has been confirmed using RAPD markers (Fig. 2). These markers are much easier to use than RFLPs and permit the screening of larger numbers of isolates and may be of greater use in future, more detailed, population studies.

Electrophoretic karyotypes of the types

Chromosomes of *C. gloeosporioides* can be separated on agarose gels by pulse-field electrophoresis. When the chromosomes of isolates representing the races of Type A and Type B were compared by this method, considerable differences in karyotype were observed (Masel *et al.*, 1990). The chromosomes of the two types could be separated into two size classes termed mini- (< 1·3 Mb in size, e.g. Fig. 3) and maxi- (> 1·3 Mb, Fig. 4) chromosomes. The size and number of the mini-chromosomes were extremely variable between isolates, but the Type A isolates contained 8–11 of these small chromosomes compared to 2–5 in Type B isolates (Fig. 3). The two types also differed in the number of maxi-chromosomes, containing 5 and 3 in Type A and Type B respectively (Fig. 4). The chromosomal location of a specific gene e.g. tubulin was found to be different in the two types (Fig. 4). These data suggest that major differences exist in the genomic organisation of the Type A and Type B fungi. Variable karyotypes have been reported for a number of fungal pathogens (Mills & McCluskey, 1990) and would appear to be a common feature of these organisms. The functional significance of karyotype variations are as yet poorly defined, although in *Nectria haematococca* Berk. & Br. phytoalexin inactivating genes have been demonstrated to be located on variable mini-chromosomes (Miao, Matthews & Van Etten, 1991).

The karyotype analysis suggests that this method may be a readily used diagnostic technique capable of distinguishing specific pathotypes of *C. gloeosporioides*. However, a wider survey of the mini-chromosome complement of 60 isolates from a number of species of *Stylosanthes* from West Africa, South-East Asia and South America indicated that there was no correlation between the general karyotype pattern of an isolate and the host species and country of origin (Masel, 1992). The association of a particular karyotype pattern with each of the types in Australia may simply reflect the narrow genetic base of the pathogen in Australia. Such a narrow genetic base could have resulted from introduction of only a limited number of strains into the country.

Origin of the two types

The data described above demonstrate unequivocally that the Type A and Type B fungi are genetically distinct. The distinction between these two pathogen groups is sufficiently large to rule out the possibility that they are recently related. It thus seems likely that they have separate origins in Australia either by introduction from overseas or by adaptation from another host. To investigate these possibilities we have examined isolates from other countries and from other hosts by DNA fingerprinting (Braithwaite *et al*., 1990b; Manners *et al*., 1992). Some isolates from *S. guianensis* in South America appear to resemble closely the Type B pathogen, but as yet no isolate closely resembling a Type A pathogen has been identified from South America (Manners *et al*., 1992). Our inability to identify an isolate which is similar to the Type A in our collection of South American isolates could be explained by the small sample size of only 16 isolates and further broader surveys are necessary. Analysis of Australian isolates of *C. gloeosporioides* from the alternate hosts of the

Type A fungus, *viz. Aeschynomene falcata* Desv.and *Centrosema pubescens* Benth., has also failed to reveal one identifying with the Type A pathogen. Seed transmission of anthracnose-inciting *C. gloeosporioides* has been reported (Irwin & Cameron, 1978) and imported seed is a likely route of introduction into Australia.

Isolates of *C. gloeosporioides* from *Stylosanthes* which resemble both Type A and Type B strains by DNA fingerprinting criteria have been identified in South East Asia (Manners *et al.*, 1992). These strains may have been derived from Australia, since seed grown in Australia has been utilised extensively in South-East Asia. Importantly though, in Thailand and the Philippines unique strains unrelated to Australian isolates have been identified (Manners *et al.*, 1992). These results indicate that Australia may have only a limited amount of the potential genetic diversity of this pathogen and quarantine should be strictly maintained.

Race specialisation in the Type B pathogen

The low amount of genetic variation between races within each type detected by RFLPs (Braithwaite *et al.*, 1990b) indicates that race specificity has most probably arisen by simple variation in loci affecting virulence rather than by either independent introduction or transfer from another host. We have focused on the Type B host-pathogen interaction for a detailed analysis of the molecular basis of race specificity in *C. gloeosporioides*.

Physiology of race-cultivar specificity

In the Type B interaction four distinct races are very clearly distinguished on 4 differential cultivars (Table 1) and resistance in each of these cultivars is monogenically inherited (D. F. Cameron, unpublished data). This resistance is visually expressed as an almost complete immunity, but histologically fungal growth is arrested at different growth stages during early stages of infection of each resistant cultivar (Ogle *et al.*, 1990). Cytochemical and biochemical studies of the interaction of *S. guianensis* cultivar Graham with races 1 (incompatible) and 3 (compatible) have revealed that incompatibility is associated with callose deposition at the site of penetration (Sharp *et al.*, 1990). Callose deposition coincided with the cessation of fungal growth between 1 and 2 days after inoculation. Incompatibility in this interaction did not appear to be associated with enhanced levels of phytoalexins, lignification,

Table 1. Race-cultivar specificity in Type B C. gloeosporioides on *S. guianensis*

Race	Disease reaction on differential *S. guianensis* cultivars			
	Endeavour	18750*	Graham	Cook
1	S	R	R	R
2	S	S	R	R
3	S	R	S	R
4	S	R	R	S

R = resistant reaction, S = susceptible reaction
* = Commonwealth Plant Introduction number

peroxidase, chitinase or other pathogenesis-related proteins (Sharp *et al.*, 1990). This information implies that host recognition processes which determine race specificity occur either during or prior to the penetration phase in this interaction.

Neither the biochemical or genetic basis of race specificity is understood in this interaction and the approach we have taken has been to develop a molecular genetic approach for the analysis of race specialisation. Particular emphasis has been placed initially on the analysis of fungal chromosomes and the identification, cloning and characterisation of race-specific DNA fragments.

A race-specific mini-chromosome

Individual isolates of the Type B pathogen differ in size and number of their mini-chromosomes (Fig. 5A, Masel *et al.*, 1990). This was a surprising observation in view of the conclusion from RFLP analysis that these isolates were essentially monomorphic and clonal (Braithwaite *et al.*, 1990b). Importantly, the mini-chromosomes appear to be transcriptionally active linear DNA molecules (Manners *et al.*, 1992; Masel, 1992).

There was no overall correlation of the mini-chromosome complement with race, but it was observed that two independent race 3 isolates contained a 1·2 Mb mini-chromosome (Fig. 5A) which was completely absent from duplicate isolates of the other three physiological races. Within the range of isolates representing each race this mini-chromosome was defined as being race specific. To investigate this specificity of the

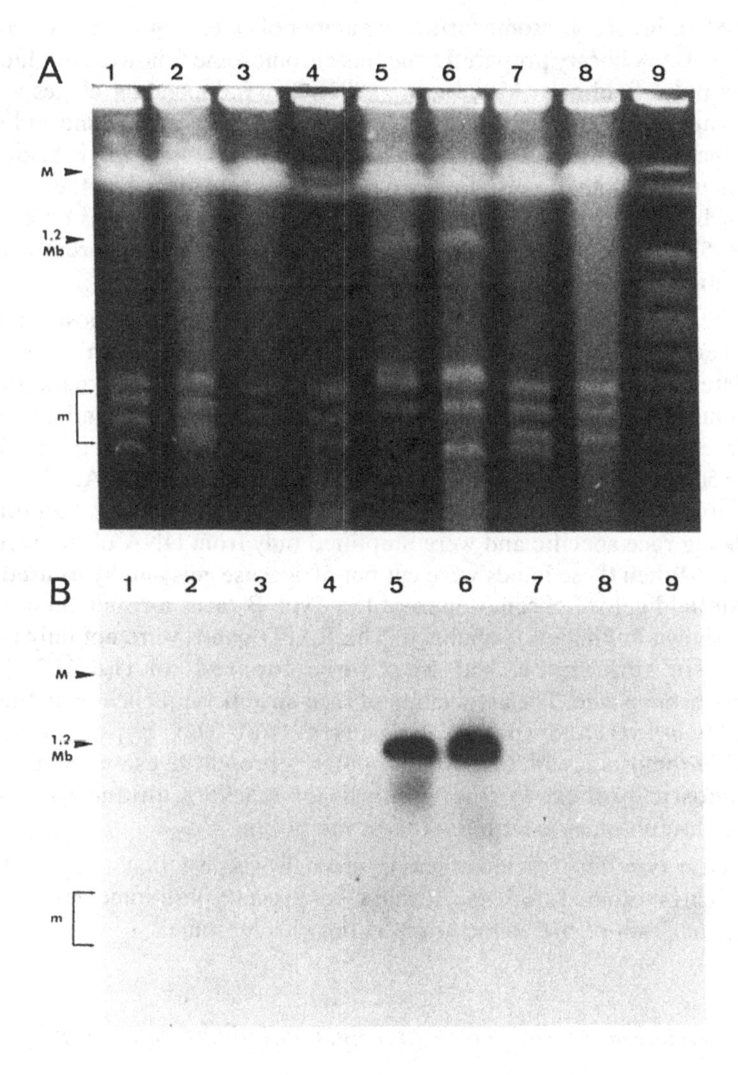

Fig. 5. Evidence for a race-specific 1·2 Mb mini-chromosome in the Type B pathogen. Two independent isolates (a and b) of races 1–4 (B1–B4) were analysed by pulse-field electrophoresis and the stained gel (panel A) revealed a 1·2 Mb mini-chromosome only in the race 3 isolates (lanes 5 & 6). Southern blot of the above gel (panel B) was hybridised to one of ten DNA probes specific to the 1·2 Mb mini-chromosome. M refers to the maxi-chromosomes, m to the mini-chromosomes and the 1·2 Mb mini-chromosome is as indicated.

1·2 Mb mini-chromosome further, a number of DNA clones were isolated from a DNA library prepared from this chromosome following its elution from pulse-field gels (Masel *et al.*, 1992). Ten independent clones were obtained which only hybridised to the 1·2 Mb mini-chromosome and not to other mini- or maxi-chromosomes. When these probes were hybridised to Southern blots of pulse-field gels of isolates representing other races, no hybridisation to any chromosome in these other races was observed (Fig. 5B). Thus, these chromosome-specific probes were apparently race specific.

The unique race-specific nature of the 1·2 Mb mini-chromosome was further demonstrated in an independent experiment which aimed to isolate race-specific sequences and then subsequently determine their chromosomal location. Replicate DNA samples of a race 2 and a race 3 isolate were compared by RAPD analysis using 90 primers generating over 500 RAPD bands (A. M. Poplawski, J. M. Manners & J. A. G. Irwin, unpublished). From this large number of bands only two were identified as being race specific and were amplified only from DNA of the race 3 isolate. When these bands were cut out of agarose gels and hybridised to pulse-field gels of chromosomes of the Type B races a result similar to that shown in Fig. 5 was obtained. The RAPD bands were not only race specific sequences, but also were located on the 1·2 Mb mini-chromosome. The association of race-specific sequences with the 1·2 Mb mini-chromosome suggests that the hypervariable mini-chromosomes of *C. gloeosporioides* represent a useful source of diagnostic probes. Further analysis of markers unique to other mini-chromosomes is required to test this notion.

These two lines of investigation strongly suggest that the 1·2 Mb mini-chromosome is race specific and these results pose some interesting questions with regard to the origin of this chromosome.

Deletion, addition or rearrangement of chromosomes?

At present it is not known how the variation in the size and number of mini-chromosomes arises. If the genome is undergoing rearrangements then sequences in one chromosome may be present on either one or several differently sized chromosomes in another strain. However, it is possible that DNA exchange between isolates via parasexuality is occurring, leading to strains with additional chromosomes or other aneuploid strains lacking particular chromosomes. Chromosomes may also be lost during somatic growth if they are intrinsically unstable and physiologically redundant. DNA addition or deletion would mean that

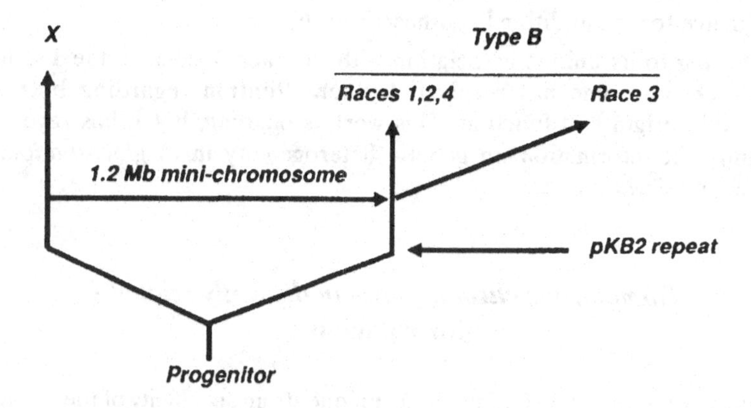

Fig. 6. Possible evolutionary path for the 1·2 Mb mini-chromosome and its transfer into the Type B pathogen from a less related strain. This model is consistent with the race-specific occurrence of this chromosome and its lack of the pKB2 repeat which is present on other chromosomes.

sequences present in a chromosome of one strain may be completely absent from another strain of the fungus. To test between these mechanisms generating variation, it is necessary to clone marker sequences for particular variable chromosomes and trace their occurrence amongst a range of independent isolates with variable karyotypes. These results of Masel *et al.* (1992) described above, which demonstrated the race specific nature of all clones specific to the 1·2 Mb mini-chromosome, indicate that this chromosome did not arise by rearrangement of the genome of another race but that DNA addition or deletion must play some role in its variable occurrence in the pathogen. It is possible that initially the race 3 strains were introduced into Australia and that the chromosome was then lost. Alternatively the chromosome may have been acquired via horizontal transfer by parasexual interaction with another distinct strain of *C. gloeosporioides* or another fungus.

Masel *et al.* (1992) also investigated the occurrence of repeat sequences on the chromosomes of a race 3 isolate of Type B *C. gloeosporioides*. A repeat sequence termed pKB2 previously cloned from this isolate (Braithwaite *et al.*, 1990b) was present on all chromosomes except the 1·2 Mb mini-chromosome. This also indicates that the 1·2 Mb mini-chromosome may have been added to the genome following the elaboration of this repeat sequence and provides some circumstantial evidence for the addition hypothesis (Fig. 6).

Owing to its unique association with the race 3 isolates, the 1·2 Mb mini-chromosome has received further attention regarding both its possible origin and function. This work is ongoing, but it has revealed significant information on genetic heterogeneity in *C. gloeosporioides* infecting *Stylosanthes*.

Homologous chromosomes in distantly related C. gloeosporioides

It is currently not known whether the unique strain specificity of the 1·2 Mb mini-chromosome is because these strains resemble the initial introduction into Australia and the mini-chromosome was subsequently lost, or whether the 1·2 Mb mini-chromosome was acquired by horizontal transfer from another genetically distinct fungal strain *via* parasexuality. We have conducted a survey using pulse-field electrophoresis to determine the distribution of chromosomes homologous to the 1·2 Mb mini-chromosome in a variety of isolates of *C. gloeosporioides* and other species of *Colletotrichum*. Southern blots of these pulse-field gels were prepared and hybridised with each of the 10 DNA clones specific to the 1·2 Mb mini-chromosome to detect homologous chromosomes which might be of different size (Masel, 1992). Surprisingly each of the 10 chromosome-specific DNA probes hybridised to a chromosome present in each of the races of the Type A pathogen in Australia. The homologous chromosome in the Type A pathogen was larger, being about 2 Mb in size (Fig. 7). This suggested that the 1·2 Mb mini-chromosome may have originated from the Type A pathogen and then been transferred to Type B by a rare parasexual event. The Type A chromosome and the 1·2 Mb mini-chromosome of the race 3 Type B pathogen were compared by DNA fingerprinting methods to determine whether they were closely related. The results indicated that these two chromosomes were only distantly related and showed only 35% similarity. This degree of similarity is about equal to that between Type A and Type B strains over other regions of the genome (Braithwaite *et al*., 1990b) and indicates that the 2.0 Mb and 1·2 Mb homologous chromosomes are not more recently related. Thus, it would appear that the Type A pathogen is not a donor of the 1·2 Mb mini-chromosome to the Type B pathogen and it is also unlikely to be a progenitor of the race 3 isolates in Australia. More recently, we have observed sequences homologous to those specific to the 1·2 Mb mini-chromosome in a number of other isolates of *C.*

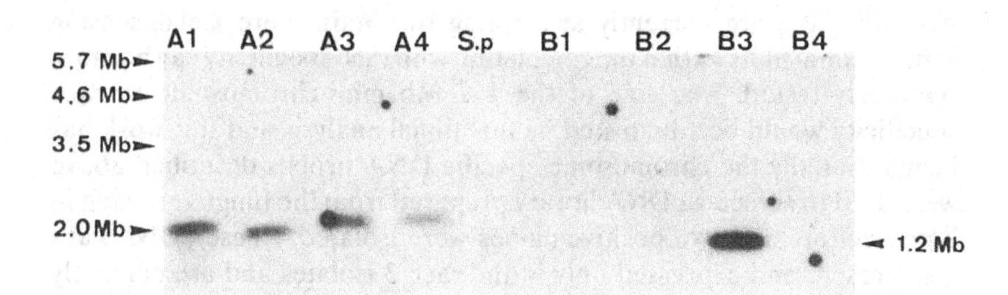

Fig. 7. Evidence that the Type A pathogen and race 3 of the Type B pathogen share a homologous chromosome. One of the 10 DNA probes specific to the 1·2 Mb mini-chromosome of the race 3 Type B pathogen was hybridised to a Southern blot of a pulse-field gel of chromosomes of each of the 4 races of the Type A (A1–A4) and Type B (B1–B4) pathogen. Note the non-race-specific presence of the chromosome in the Type A pathogen and its race-specificity in the Type B pathogen. S.p refers to the *S. pombe* size standards which are indicated on the left.

gloeosporioides. These include isolates from *Stylosanthes* from South East Asia, and Australian isolates from *Carica papaya*, *Centrosema* spp. and *Aeschynomene* spp. However, DNA fingerprint analysis of these isolates with the chromosome-specific DNA probes has also failed to reveal a possible source of the mini-chromosome.

It may not be possible to answer the question of whether deletion or addition gave rise to the race-specific nature of the 1·2 Mb mini-chromosome by the analysis of isolates randomly sampled from field populations, and it may be necessary to simulate these events in the laboratory. Such experiments would also provide some insights into the possible role of this chromosome in determining specificity in this pathogen.

Functional analysis of the 1·2 Mb mini-chromosome

At present the evidence that the 1·2 Mb mini-chromosome has a role in determining race specificity is only correlative within a limited sample of isolates. Unfortunately *S. guianensis* cv Graham which is specifically infected by the race 3 pathogen has not been sown for over 10 years in Australia. We are currently attempting to obtain more isolates using experimental plots so that the correlation with race specificity can be more rigorously tested. The role of the 1·2 Mb mini-chromosome in race specificity would best be tested by functional analysis and this work has begun. Initially the chromosome-specific DNA probes described above were used to screen a cDNA library prepared from the fungus growing in liquid culture and two positive clones were isolated. These cDNAs are also present and expressed only in the race 3 isolates and are currently being sequenced. At present the DNA probes that we have isolated represent less than 5% of the total DNA of this chromosome and more representative cosmid clones are being prepared. Cosmid clones may permit the isolation of a larger number of race-specific cDNA clones and their corresponding genes. It is hoped that DNA sequencing of the cDNA clones and searching for homologies in the sequence data banks will provide some insights as to the possible functions of gene products encoded on this chromosome. It is possible to transform this pathogen at low frequency using plasmid DNA carrying fungal selectable marker genes e.g. hygromycin resistance (Manners *et al.*, 1992; Masel, 1992). It should therefore be possible to transfer either cloned genes or cosmid clones of the 1·2 Mb mini-chromosome to a race 1 isolate and test for effects on race specificity.

Conclusions

Our understanding of the genetic mechanisms leading to pathogenic specialisation in *C. gloeosporioides* is still very rudimentary. It would appear that the two types are very dissimilar genetically and probably have separate origins. Although relatively homogeneous genetically, individual isolates within a type can show large variations in karyotype. We have shown that one variable chromosome, possibly associated with a particular race of the Type B, is highly strain specific and this strain specificity is the result of either chromosome addition or deletion during the evolution of this pathogen. These observations demonstrate that the biology of adaptation of *C. gloeosporioides* is extremely dynamic and involves the production of distinct pathogen populations by geographic displacement and also novel genetic mechanisms which alter the genome organisation and composition between otherwise closely related strains. Future work will focus on identifying genes that determine specificity and pathogenicity. This should permit a more refined analysis of the role that these genes play in determining pathotype and race. Ultimately it may be possible to use information on the genes that the pathogen requires to infect particular host cultivars in order to design molecular-based control strategies. For example, the production of either toxins or plant defence suppressors might be countered by the transfer of genes encoding enzymes for their degradation into the host. The genetic transformation of *Stylosanthes* spp. has been reported (Manners, 1987, 1988; Manners & Way, 1989) and this plant system should be amenable to genetic engineering for improved disease resistance.

References

Braithwaite, K. S., Irwin, J. A. G. & Manners, J. M. (1990a). Ribosomal DNA as a molecular taxonomic marker for the group species *Colletotrichum gloeosporioides*. *Australian Systematic Botany* **3**, 733-738.

Braithwaite, K. S., Irwin, J. A. G. & Manners, J. M. (1990b). Restriction fragment length polymorphisms in *Colletotrichum gloeosporioides* infecting *Stylosanthes* spp. in Australia. *Mycological Research* **94**, 1129-1137.

Braithwaite, K. S. & Manners, J. M. (1989). Human hypervariable minisatellite probes detect DNA polymorphisms in the fungus *Colletotrichum gloeosporioides*. *Current Genetics* **16**, 473-475.

Burt, R. L., Cameron, D. G., Cameron, D. F., 't Mannetje, L. & Lenne, J. (1983). *Stylosanthes*. In *The Role of Centrosema, Desmodium and Stylosanthes in Improving Tropical Pastures*, (ed. R. L. Burt, P. P. Rotar, J. L. Walker, & M. W. Silvey), pp. 141-185. Westview Press: Boulder.

Cameron, D. F., Chakraborty, S., Davis, R. D., Edye, L. A., Irwin, J. A. G., Manners, J. M. & Staples, I. B. (1989). A multidisciplinary approach to anthracnose diseases of *Stylosanthes* in Australia. *Proceedings, XVI International Grassland Congress, Nice*, pp. 719-720. Association Française pour la Production Fouragere, INRA: France.

Chakraborty, S., Pettitt, A. N., Boland, R. M. & Cameron, D. F. (1990). Field evaluation of quantitative resistance to anthracnose in *Stylosanthes scabra*. *Phytopathology* **80**, 1147-1154.

Chakraborty, S., Pettitt, A. N., Cameron, D. F., Irwin, J. A. G. & Davis, R. D. (1991). Anthracnose development in pure and mixed stands of the pasture legume *Stylosanthes scabra*. *Phytopathology* **81**, 788-793.

Cove, D. J. (1976). Chlorate toxicity in *Aspergillus nidulans*: the selection and characterisation of chlorate resistant mutants. *Heredity* **36**, 191-203.

Cox, M. L. & Irwin, J. A. G. (1988). Conidium and appressorium variation in Australian isolates of the *Colletotrichum gloeosporioides* group and closely related species. *Australian Systematic Botany* **1**, 139-149.

Dale, J. L., Manners, J. M. & Irwin, J. A. G. (1988). *Colletotrichum gloeosporioides* isolates causing different anthracnose diseases on *Stylosanthes* in Australia carry distinct double-stranded RNAs. *Transactions of the British Mycological Society* **91**, 671-676.

Irwin, J. A. G. & Cameron, D. F. (1978). Two diseases in *Stylosanthes* spp. caused by *Colletotrichum gloeosporioides* in Australia, and pathogenic specialisation within one of the causal organisms. *Australian Journal of Agricultural Research* **29**, 305-317.

Irwin, J. A. G., Cameron, D. F., Davis, R. D. & Lenne, J. (1986). Anthracnose problems with *Stylosanthes*. *Tropical Grasslands Society Occasional Publication* **3**, 38-46.

Manners, J. M. (1987). Transformation of *Stylosanthes* spp. using *Agrobacterium tumefaciens*. *Plant Cell Reports* **6**, 204-207.

Manners, J. M. (1988). Transgenic plants of the tropical pasture legume *Stylosanthes humilis*. *Plant Science* **55**, 61-68.

Manners, J. M. & Way, H. (1989). Efficient transformation with regeneration of the tropical pasture legume *Stylosanthes humilis* using *Agrobacterium rhizogenes* and a Ti plasmid binary vector system. *Plant Cell Reports* **8**, 341-345.

Manners, J. M., Masel, A., Braithwaite, K. S. & Irwin, J. A. G. (1992). Molecular analysis of *Colletotrichum gloeosporioides* pathogenic on the tropical pasture legume *Stylosanthes*. In *Colletotrichum, Biology, Pathology and Control*, (ed. J. A. Bailey & M. Jeger), pp. 250-268. CAB International: Oxford.

Masel, A. (1992). *Molecular analysis in the plant pathogenic fungus Colletotrichum gloeosporioides*. Ph.D. Thesis. The University of Queensland.

Masel, A., Braithwaite, K. S., Irwin, J. A. G. & Manners, J. M. (1990). Highly variable molecular karyotypes in the plant pathogen *Colletotrichum gloeosporioides*. *Current Genetics* **18**, 81-86.

Masel, A., Irwin, J. A. G. & Manners, J. M. (1992). DNA deletion or addition is associated with a major karyotype polymorphism in the phytopathogen *Colletotrichum gloeosporioides*. *Molecular and General Genetics*, in press.

Miao, V. P. W., Matthews, D. E. & Van Etten, H. D. (1991). Identification and chromosomal locations of a family of cytochrome P-450 genes for pisatin detoxification in the fungus *Nectria haematococca*. *Molecular and General Genetics* **226**, 214-223.

Mills, D. & McCluskey K. (1990). Electrophoretic karyotypes of fungi: The new cytology. *Molecular Plant Microbe Interactions* **3**, 351-357.

Ogle, H. J., Gowanlock, D. H. & Irwin, J. A. G. (1990). Infection of *Stylosanthes guianensis* and *S. scabra* by *Colletotrichum gloeosporioides*. *Phytopathology* **80**, 837-842.

Ogle, H. J., Irwin, J. A. G. & Cameron, D. F. (1986). Biology of *Colletotrichum gloeosporioides* isolates from tropical legumes. *Australian Journal of Botany* **34**, 281-292.

Pont, W. & Irwin, J. A. G. (1976). *Colletotrichum* leaf spot and stem canker of *Stylosanthes* spp. in Queensland. *Australian Plant Pathology* **5**, No. 1 (supplement: abstract no. 35).

Puhalla, J. E. (1985). Classification of strains of *Fusarium oxysporum* on the basis of vegetative compatibility. *Canadian Journal of Botany* **62**, 546-550.

Sharp, D., Braithwaite, K. S., Irwin, J. A. G. & Manners, J. M. (1990). Biochemical and cytochemical responses of *Stylosanthes guianensis* to infection by *Colletotrichum gloeosporioides*: association of callose deposition with resistance. *Canadian Journal of Botany* **68**, 505-511.

Sutton, B. C. (1980). *The Coelomycetes*. Commonwealth Mycological Institute: Kew, U.K.

Trevorrow, P. R., Irwin, J. A. G. & Cameron, D. F. (1988). Histopathology of compatible and incompatible interactions between *Colletotrichum gloeosporioides* and *Stylosanthes scabra*. *Transactions of the British Mycological Society* **90**, 421-429.

Vinijsanun, T., Irwin, J. A. G. & Cameron, D. F. (1987). Host range of three strains of *Colletotrichum gloeosporioides* from tropical pasture legumes and comparative histological studies of interactions between Type B disease-producing strains and *Stylosanthes scabra* (non-host) and *S. guianensis* (host). *Australian Journal of Botany* **35**, 665-677.

Williams, J. G. K., Kubelik, A. R., Livak, K. J., Rafalski, J. A. & Tingey, S. V. (1990). DNA polymorphisms amplified by arbitrary primers are useful as genetic markers. *Nucleic Acids Research* **18**, 6531-6536.

Chapter 15

Tropical fungi: their commercial potential

F.M. Fox

Xenova Ltd., 545 Ipswich Road, Slough, SL1 4EQ UK

Introduction

Fungal biotechnology has wide ranging commercial potential and this Chapter describes some of the ways fungi and their products (secondary metabolites) can be used commercially in industries for the direct benefit of users. Undoubtedly, tropical habitats house the greatest diversity of organisms, so it can be expected that tropical fungi will tend to be more abundant and more species-diverse than in other geographical regions of the world (see Chapter 16). In terms of commercial potential, however, tropical fungi are no different from those found elsewhere, but the most appropriate application for fungal biotechnology may well vary considerably between tropical and non-tropical habitats depending on the priorities of the user and the availability of resources.

In this Chapter, the commercial potential of fungi will be firstly illustrated with some examples (Fig. 1). Natural product screening for the discovery of molecules with biological activity for application in the pharmaceutical industry is the first example. This type of industry generally requires high capital investment, sophisticated equipment and multidisciplinary teams of workers. Some examples will also be given of the technology available for the production of food and energy from bioconversion of lignocellulosic and other solid resources. These do not require the same degree of capital investment, but they offer promising opportuntities for sustainable, environmentally sound industries and, in the immediate future at least, they may be more applicable to tropical countries which have a rich natural fungal resource but limited access to technology.

Secondly, some issues related to the management and conservation of biodiversity will be addressed. This is fundamental to the success of biotechnology and is critical if the commercial potential of fungi is to be fully realised. This topic is especially relevant and sensitive at present. Indeed, the recent United Nations Conference on Environment and

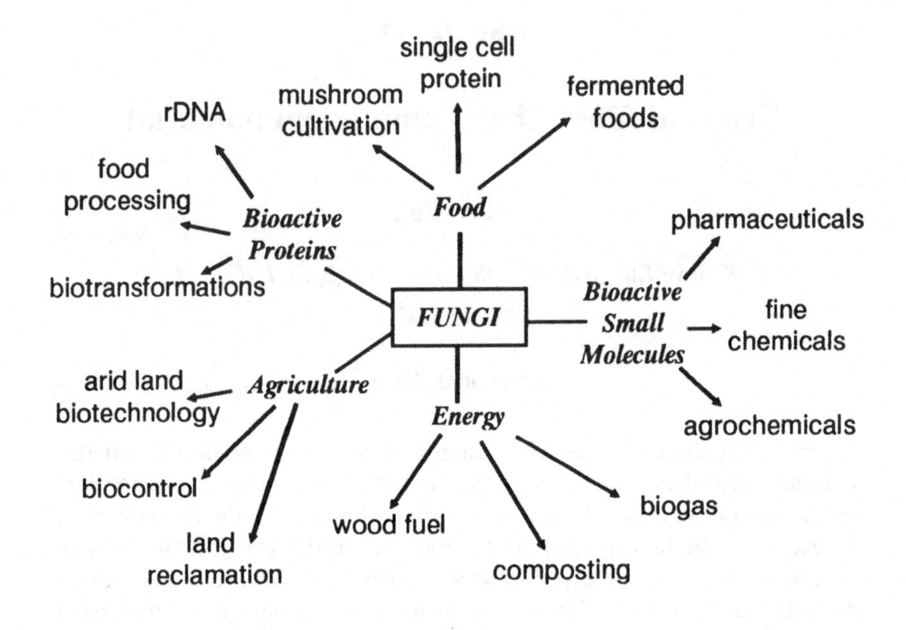

Fig. 1. Biotechnological application of fungi.

Development held in Rio de Janeiro considered biodiversity as a major part of the agenda.

Commercial potential of fungi

Microbial screening for the discovery of biologically active molecules

The discovery of penicillin from *Penicillium notatum* Westling in 1928 and its application during the second World War laid the foundation of many of today's pharmaceutical companies in that it started an intensive search for antibiotics and other therapeutic chemicals from microorganisms. In recent years, advances in molecular biology, automation, and computer science have changed the direction of screening for novel molecules. It is no longer random. Sensitive, target-specific, mode-of-action screens have been designed to search for small molecular weight molecules that can be developed as new drugs by intervening at critical points in biochemical pathways.

There is evidence that fungi, being simple but multicellular eukaryotic systems, have similar biochemical mechanisms to those found in mammals. It has been suggested (Roth *et al.*, 1986) that because some vertebrate tissues, such as glands and the immune system, have developed relatively recently in evolutionary terms, some of the molecules through which these cells communicate (hormones and neuropeptides) may have appeared much earlier in evolution in unicellular or simple multicellular organisms such as bacteria and fungi. In fact, there is evidence that some fungal biochemical control mechanisms have been conserved through evolution in higher eukaryotes (man included), and the regulatory molecules involved may function in mammalian systems by interfering with receptor-ligand interactions and intracellular signalling. *Achyla bisexualis* Coker, for example, produces the sex hormones antheridiol and oogoniol which are steroidal structures directly analagous to sex hormones found in mammals (Williams *et al.*, 1989). Ergot alkaloids produced by *Claviceps purpurea* (Fr.) Tulasne are known to be active at mammalian cell receptors (Nisbet & Porter, 1989) and act as agonists or antagonists of dopamine. These compounds are now used clinically for the treatment of postpartum haemorrhage and migraine. Such evidence provides increasing support for the hypothesis that receptor-ligand recognition systems in vertebrates may have evolved from similar systems in microorganisms and thus the screening of microbial products may be a fruitful approach towards the discovery of new therapeutics for medicine (Nisbet & Porter, 1989).

Fungi are a good source of molecules for other reasons too. Over the last 30 years, screening for useful microbial products has focused on bacteria and actinomycetes as a source of novel molecules so the search for novel molecules from fungi could be rewarding. Despite the outstanding success of penicillin, cyclosporin A, an immunosuppressive agent originally isolated from *Trichoderma polysporum* (Link: Pers.) Rifai, and mevinolin, a cholesterol-reducing agent from *Aspergillus terreus* Thom, many groups of fungi, particularly the higher fungi, have not been systematically investigated for biologically active molecules. Most fungi can be reliably cultured in liquid or solid phase fermentation systems. Although our knowledge is limited, they appear to be genetically manipulable. The synthesis of secondary metabolites is often associated with cellular differentiation and the complex life cycles of many fungi indicate a considerable degree of differentiation. So it is quite probable that highly differentiated fungi such as the Basidiomycotina will produce a rich array of secondary metabolites. Finally, fungi are exceedingly diverse and abundant, a fact often overlooked by conservationists interested in global biodiversity.

It has been estimated recently (Hawksworth, 1991) that there are at least 800,000 species of fungi, a high proportion of which has not yet been discovered, and less than 1·0% of which are held in culture collections. Hawksworth (Chapter 16) now estimates 1·5 million fungal species as a conservative figure. If each fungal species contains between 100,000 and 400,000 genes (Wilson, 1988), the fungal kingdom provides molecular biologists with a playground of 80 to 600 thousand million genes from which to discover and develop new products. The potential for discovery of new medicines can best be illustrated by the statistic that more than 25% of prescriptive drugs in current use originate from plant sources. And yet fungi are ten times more abundant than plants and produce at least as rich an array of secondary metabolites.

Screening for novel pharmaceutically useful microbial secondary metabolites is broadly a three stage process. Firstly, a large collection of organisms are needed as a source of material, and appropriate fermentation systems must be operated in such a way as to optimise the synthesis and production of metabolites. Secondly, appropriate screens must be designed to detect biological activity. Sophisticated equipment is then needed to screen large numbers of samples and to elucidate the chemical structures of detected active compounds. Thirdly, biologically active molecules have to be turned into fully formulated, marketable drugs, a process which is hugely expensive and time consuming. This type of technology is not available in many countries and is perhaps not appropriate for many tropical developing countries hoping to realise the potential of their fungal resource. So, the question to be asked is: what technology is appropriate for tropical countries, especially the poorest where production of food and energy are pressing priorities?

Production of food, protein and energy by solid substrate fermentations

Solid substrate fermentation is an example of fungal biotechnology which holds promise for sustainable development. Processes are now available for the production of food and energy using environmentally sound, low-capital processes involving the bioconversion of various solid substrates by fungi into useful products. Some examples are given below.

Table 1. Cultivation of mushrooms on solid substrates

Species	Level of Environmental Control Required*	Substrate
Agaricus bisporus	++++	Composted horse manure, rice straw
Auricularia spp.	+++	Sawdust + rice bran
Lentinula edodes	++	Logs, sawdust + rice bran
Coprinus cinereus	+	Straw
Pleurotus ostreatus	+	Straw, paper, sawdust
Volvariella volvacea	+	Straw, cotton waste

* Required level of environmental control: + = low, ++++ = high.

Mushroom cultivation

Mushrooms such as Shiitake (*Lentinula edodes* (Berk.) Pegler) have been cultivated for food in the east for 2000 years and this represents one of the oldest known examples of fungal biotechnology based on fermentation of solid substrates (DaSilva & Taguchi, 1987). Indeed, mushroom cultivation is probably the most economic bioconversion process known. *Agaricus bisporus* (Lge) Imbach currently occupies 75% of the world market in terms of commercial production of edible mushrooms. However, there is a wide range of basidiomycetes which can be cultivated either commercially or at the family or village level which are not currently exploited. Edible mushrooms provide a valuable source of nutritious food, containing 16-20% dry weight protein and they are also rich in minerals and vitamins but low in fat.

The West could learn much from the East where mushrooms are cultivated for food using systems that do not require a high degree of environmental control and which utilise readily available, renewable and cheap substrates such as agricultural waste (Table 1). An enormous amount of agricultural waste is produced each year. Europe alone produces 25 billion tonnes of cereal waste annually, most of which is burnt but which could be converted into food.

Table 2. Global mycoprotein production from industrial and agricultural waste materials

Waste material	Organism	Current Production (tonnes y^{-1})
Molasses	*Candida utilis*	80,000
Starch hydrolysate	*Fusarium graminearum*	50-100
Corn Waste	*Trichoderma viride*	Pilot
Confectionery	*Candida utilis*	500
Cellulosic	*Chaetomium cellulolyticum*	Pilot

Mycoprotein

According to an FAO estimate in 1987 (Senez, 1987), more than one thousand million people are suffering from protein malnutrition in the developing world and, because of demographic growth, this situation is rapidly getting worse. For many tropical countries, food production is the key to development but is often restricted by available energy resources and environmental problems such as drought.

Research into the utilisation of mycoprotein (more accurately called microbial biomass protein, MBP) dates back to the start of modern biotechnology. In fact it was the recognition of world protein shortages in the 1950's that led to rapid developments in the conversion of microbial biomass into protein for human consumption and as a supplement for animal feeds.

Quorn mycoprotein, made from *Fusarium graminearum* Schwäbe, is the best known example of MBP used for human consumption though the production process requires sophisticated technology (Trinci, 1992). Recently, a number of similar systems have been designed which use lignocellulosic and other agricultural residues as substrates for a range of fungi (Table 2). Some of these, such as the Waterloo process using *Chaetomium cellulolyticum* Chahal & Hawksworth (Moo-Young *et al.*, 1979), are especially attractive because the fermentation process does not require complex pre-treatments or sterile conditions. However, most of these processes are still at the experimental or pilot stage and therefore most products are only suitable for animal feeds. Research is needed to overcome the toxicological barriers for human nutrition, to find new strains of fungi that will improve the bioconversion process and to find ways of simplifying the fermentations. The attraction of this type of biotechnology is very great. Production of protein is critical on a global scale. A process such as the Waterloo process makes use of readily available, renewable resources and is environmentally sound. It can be cheap to build and maintain and could be appropriate for rural areas. Also, as has been demonstrated with Quorn which now has an annual market value of £25 million, this type of technology can be very profitable in addition to providing good overall potential for the improvement of world food production.

Energy: developments in biogas technology

Fermentation of solid substrates can be used for the production of energy. Biogas systems, using anaerobic fermentations for the generation of methane and fertiliser, have been established with great success in China, especially in rural areas where alternative forms of energy are not readily available. Recently, Caro (1988) developed a two step system for biogas production using *Aspergillus niger* van Tieghem for aerobic fermentation of cellulosic beet pulp. This type of technology based on renewable waste resources is environmentally sound and offers enormous potential as a sustainable, alternative source of energy that could be readily established in both developing and developed countries.

This section has briefly outlined some of the ways by which fungi can be used to make products that are commercially valuable and offer opportunities for sustainable economies. Biotechnology relies on a diverse source of materials whether it is from whole microorganisms as used in mushroom cultivation or biocontrol strategies, parts of organisms as in MBP production, or genetic material as used in strain improvement and genetic engineering. Conservation of natural resources is fundamental to successful biotechnology and the second part of this Chapter will address some of the issues of biodiversity and its management which are particularly topical at this time.

Value of natural resources and conservation of biodiversity

Biodiversity can be defined as the variability of life: variation between species, genetic diversity within species and variation between ecosystems. Its estimated magnitude is vast; there are between 5 and 30 million species of organism on earth (May, 1988), less than 1.4% of which have been described. There are few available data on the diversity of tropical fungi, primarily due to the lack of mycologists working in appropriate geographical regions. During an ecological study in Puerto

Rico, Lodge (1988) estimated that 20-30% of the agarics she found were new species. Dennis (1970) listed 2412 species of macrofungi in his fungal biota of Venezuela but believed it to be a 20-fold underestimate of the probable species total for the region.

If the magnitude of biodiversity is great, so is the estimated rate of loss. Tropical rain forest covers about 14% of the world's land surface, but may house over 70% of the world's species (Flint, 1991). At present rates of destruction, over half of the remaining tropical rain forests will have been destroyed within 50 years (May, 1988). Twenty thousand species of organism become extinct each year (Wilson, 1988). Once a species is extinct it is lost to society together with its potential contribution to world economy and sustainable development. Undoubtedly, amongst those species being lost are some which contain potentially valuable germplasm for new food crops and pharmacologically important chemical molecules. But what are the main causes of this rapid rate of loss? Habitat destruction, because of the pressure of development, is usually blamed as the main culprit. Perhaps loss of habitat is inevitable with the pressures of a still rapidly rising population (expected to increase by 1 billion in the 1990's) on limited, non-renewable resources. However, it now seems apparent that inappropriate government policies, in both developed and developing countries, are causing an unprecedented acceleration of the rate of of loss.

There are two important issues worth considering in respect of habitat loss: market value and valuation of the unknown. With the environment now recognised as an economic issue, it has become necessary for governments to assign values to natural resources so that environmental effects of economic policies can be measured. This has revealed an inadequacy in conventional national income accounts, which record changes in wealth only when they pass through market. Thus, a forest providing food and employment for its inhabitants and acting as a buffer against soil erosion and flooding is assigned no value whatsoever. However, if it is cut down and sold as timber, the country appears to grow wealthier in terms of market value even though the forest is not easily renewable, and hence a valuable resource may be lost. Similarly, there is no incentive to conserve a tropical mushroom growing wild which has not yet been shown to have potential commercial application. Partly as a consequence, measures of *in situ* conservation and *ex situ* conservation (such as microbial culture collections, germplasm banks and botanical and zoological gardens) are not adequately supported at government level, even when for instance wild relatives of important food crops are threatened.

The second issue concerns the way that assumed biodiversity is estimated in economic terms. In the fungal kingdom, only a fraction of the estimated number of species believed to exist have been described (Hawksworth, 1991, Chapter 16 this volume) and this scale of ignorance is reflected throughout the plant and animal kingdom. So, given that we do not know what we have now, it is almost impossible to measure the cost of habitat destruction for the future.

There is an urgent need for taxonomic surveys of the mycobiota of the tropics and the work of mycologists should be strongly supported. For countries wanting to conserve and make scientific and commercial use of their fungal resource, an important first step lies in the establishment of culture collections. These can be used to support the infrastructure of biotechnology, agriculture and healthcare. In the same way as botanical gardens and seed banks are used for the conservation of plant germplasm, microbial culture collections can be used to assess biodiversity, preserve genetic material and support research in ecology and taxonomy. MIRCEN (Microbial Resources Centres), MSDN (Microbial Strain Data Network) and the WFCC (World Federation for Culture Collections) have made valuable contributions to the establishment of global networks for culture collections providing training and technology transfer, and there is no doubt this work will become increasingly important in the future.

In the coming decades, it is very likely that both developed and developing countries will rely heavily on fungal biotechnology to meet the needs of food and energy production, healthcare and environmental problems. Indeed, advances in biotechnology have been heralded as providing the same kind of boost to world economic growth as the green revolution heralded the development of new plant varieties in the 1970s. It is up to academia, industry and an educated and aware public to make this a reality and to avoid the pitfalls and conflict of interests that occurred as a consequence of the green revolution.

Even with the commercial potential of fungi directly in mind, the conservation and management of biodiversity is not solely an economic issue. Scientifically, biodiversity is essential for the healthy balance of ecosystems (May, 1988). Also, on ethical grounds, it can be argued that we have an obligation to conserve the world's natural resources for the benefit of future generations, and also that other species have inherent value (Zell, 1991).

Conclusions

In conclusion, there is no doubt that fungi have enormous potential commercial value. How their exploitation contributes to sustainable, long term benefit to humans will depend on the type of technology and end product for which they are used. This Chapter has outlined some of the applications of fungal technology including those that require high financial investment such as pharmaceuticals, and also those that can make use of renewable waste materials for the production of food and energy and which do not necessarily require high capital investment and availability of sophisticated equipment. The success of biotechnology depends on the availability of a diverse genetic resource and continued research into appropriate technologies. Tropical countries may best realise the value of their fungal resource, firstly through the establishment of culture collections and, secondly, by the development of low capital technologies based on renewable waste materials adapted to meet the needs of specific users. Finally to reiterate, conservation and sustainable management of biodiversity are fundamental to the success of biotechnology.

References

Caro, T. (1988). Bioreactor for the cultivation of microorganisms of lignocellulosic materials. In *Treatment of Lignocellulosics with White Rot Fungi,* (ed. F. Zadrazil & P. Reiniger), pp 90-99. Elsevier Applied Science.

Da Silva, E.J. & Taguchi, H. (1987). An international network exercise: the MIRCEN programme. In *Microbial Technology in the Developing World*, (ed. E.J. Da Silva., Y.R. Dommergues., E.J. Nyns & C. Ratledge), pp 313-335. Oxford University Press: Oxford, U.K.

Davies, J. (1990). What are antibiotics? Archaic functions for modern activities. *Molecular Microbiology* **4**, 1227-1232.

Dennis, R.W.G. (1970). Fungal flora of Venezuela and adjacent countries. Kew Bulletin, additional series **3 i-xxxiv**, 1-531.

Flint, M. (1991). Biological Diversity and Developing Countries. Issues and Options. pp13-45, Overseas Development Administration: London.

Hawksworth, D.L. (1991). The fungal dimension of biodiversity: magnitude, significance, and conservation. *Mycological Research* **95**, 441-452.

Lodge, J. (1988). Three new species (Basidiomycotina, Tricholomataceae) from Puerto Rico. *Transactions of the British Mycological Society* **91**, 109-116.

May, R.M. (1988). How many species are there on Earth? Science **24**, 1441-1449

Moo-Young, M., Dangulis, A., Choral, D. & Macdonald, D.G. (1979). The Waterloo process for SCP production from waste biomass. *Processes in Biochemistry* **14**, 38-45.

Nisbet, L.J. & Porter, N. (1989). The impact of pharmacology and molecular biology on the exploitation of microbial products. In *Symposium for Society of General Microbiology*, **vol 44** (ed. S. Baumberg, I. Hunter & M. Rhodes), pp. 309-342, Cambridge University Press: Cambridge, U.K.

Roth, J., LeRoith, D., Collier, E.S., Watkinson, A & Lesnisk, M.A. (1986). The evolutionary origins of intercellular communication and the Maginot Lines of the mind. *Annals of the New York Academy of Sciences* **436**, 1-11.

Senez, J.C. (1987). Single cell protein: past and present developments. In *Microbial Technology in the Developing World*, (ed. E.J. DaSilva., Y.R. Dommergues., E.J. Nyns & C. Ratledge), pp 238-260. Oxford University Press: Oxford, U.K.

Smith, J. E., Fermor, T. R. & Zadrazil, F. (1988). Pretreatment of lignocellulosics for edible fungi. In *Treatment of Lignocellulosics with White Rot Fungi,* (ed. F. Zadrazil & P. Reiniger), pp 3-13. Elsevier Applied Science.

Trinci, A.P.J. (1992) Mycoprotein: a twenty-year overnight success story. *Mycological Research* **96**, 1-13.

Williams, D.H., Stone, M.J., Hauck, P.R. & Rahman, S.K. (1989). Why are secondary metabolites (natural products) biosynthesised? *Journal of Natural Products* **52**, 118-120.

Wilson, E.O. (1988). The current state of biological diversity. In *Biodiversity*, (ed. E.O. Wilson), pp. 3-18. National Academy Press: Washington DC.

Zell, R.A. (1991). Where bioscience could take us in the future. In *Bioscience Society*, (ed. D.J. Roy, B.E. Lynne & R.W. Old), pp 97-108. John Wiley & Sons: Chichester.

Chapter 16

The tropical fungal biota: census, pertinence, prophylaxis, and prognosis

D. L. Hawksworth

International Mycological Institute, Bakeham Lane, Egham, Surrey TW20 9TY, UK

Introduction

The tropics are generally recognized as embracing the greatest genetic variation on Earth. In the case of plants, about two thirds (180,000 species) are believed to occur in the tropics (Raven, 1988). What do we really know about tropical fungi? Here, an assessment is made of the scale of the problem, the extent of the knowledge base, and its pertinence to issues of contemporary concern. Means by which some of the lacunae might be or are being addressed are considered, with reflections on the prospects for the subject.

Census

The extent of the fungal biota

Estimates of the number of fungi in the world, primarily based on extrapolations from three independent data sets, suggest a total of about 1·5 million species (Hawksworth, 1991*a*). While wide confidence limits must be applied to this figure, it is probably an underestimate for reasons presented in that paper.

Smith & Waller (1992) considered 1·5 million too low; they suggested that there are probably 1 million undescribed fungi on tropical plants alone. In the case of Australia, by extrapolation from the number of fungi on selected hosts, Pascoe (1990) estimated that there are 'at least ten times as many fungi as vascular plants', implying 250,000 for the country, of which fewer than 5% are known, and 2·7 million for the world.

Caution is essential in making extrapolations. Bisby (1933) pointed out that 'the smaller the area surveyed the more the species of fungi outnumber those of flowering plant'. This is consequent upon the tendency of fungi, particularly non-host specific saprobes, to have wider distributions than vascular plants. While such a statement cannot apply to host-specific fungi, it also holds for myxomycetes, lichen-forming species, and some polypores.

May (1991) was sceptical as to the large numbers of fungi postulated, as the percentages of new species described in various tropical studies was typically around 15 to 30%. For newly studied sites, that figure would be expected to approach 95% if a 1·5 million total is correct. However, most tropical mycological studies are based on short-term visits by scientists from Europe or North America, frequently do not access the least disturbed sites, and do not involve exhaustive habitat or systematic group sampling. Species encountered in cursory visits will inevitably be the most conspicuous and/or widespread. What is remarkable is that brief and non-exhaustive collecting in the tropics invariably yields a substantial proportion of species which prove to be new, especially of microfungi. For example, on a visit by Dr B.C. Sutton to Malawi in April 1991, 221 species of microfungi were collected over a period of 40 h, of which *at least* 33 (15%) were undescribed. In the case of Tremellales, Lowy (1971) noted that eight new species had been discovered in six weeks spent in the field in Guatemala. While he considered that to be evidence of richness, the frequency, as to be expected in this group, was much lower than for the less conspicuous micromycetes.

It is pertinent that in Corner's extended studies on macromycetes in the Malaysian region, including 15 years *in situ*, 66% of the species collected were new to science (Table 1). Further, I suspect Professor Corner would agree that he had probably not collected all species in these groups present in the region, notwithstanding the fact that he discovered some 3000 species of macromycetes within one mile of the Mount Kinabulu field station in North Borneo (E. J. H. Corner, pers. comm., January 1992).

An indication that Corner's Malaysian data may not be atypical comes from D. J. Lodge's ongoing studies in Puerto Rico (pers. comm., May 1992). About 15 to 25% of each collection of agarics she makes appears to be undescribed and, more pertinently, this range *still* holds for collections being made in the area. The percentage varies according to the group, and she considers that in the Entolomataceae the percentage of undescribed taxa in Puerto Rico could be as high as 50%. This suggests that ultimately the overall percentage of novel agaric taxa in this area could reach 50%, approaching the 66% that emerged from Corner's studies, and reinforces the need for long-term *in situ* studies rather than

Table 1. Numbers of new macromycetes described by Professor E. J. H. Corner from the Malaysian region in monographs.

	Total species	New species†	% new	Source
cantharelloids	50	44	88%	Corner (1966)
Pleurotus	19	15	79%	Corner (1981*a*)
Amanita	37	28	78%	Corner & Bas (1962)
clavarioids	54	39	72%	Corner (1950)
boletes	140	100	71%	Corner (1972)
Thelephora	18	12	67%	Corner (1968)
polypores	450	282	62%	Corner (1983, 1984, 1987, 1989a & b, 1991)
Panus	12	6	50%	Corner (1981*a*)
Lentinus	6	1	17%	Corner (1981*a*)
Totals	786	517	66%	

†Totals include some species not formally named and denoted by numerals.

short-term sporadic trips to determine the full richness of an area's mycobiota.

If long-term studies of macromycete fungi can yield so many novelties in a single region, it seems not unreasonable to suggest that similarly high or even higher percentages might be expected among, for example, soil and litter microfungi, endophytic fungi, entomogenous species, caulicolous and foliicolous microfungi, and associated fungicolous species.

An alternative ecological approach to the estimation of species numbers has been to extrapolate on the basis of body size through from well-known larger mammals to insects (May, 1988). This model should also be applicable to certain fungal orders, for example, Leotiales or Pezizales, and could provide new insights into this question. Nevertheless, as stressed by May (1991), intensive site studies encompassing both vascular plants and fungi are prerequisite for further progress towards a refinement of the estimated number of fungi on Earth. Such studies will be costly as a consequence of the number of specialists required, time-consuming isolations into pure culture, preparation of leaf-surface strips, and length of time over which collections must be made.

Table 2. Numbers of new species described from the tropics in 1981-90 as catalogued in the *Index of Fungi*.

Region	New species described	% of world total
Africa*	1225	8%
Asia*	3523	22%
Australasia & Pacific*	950	6%
Central America & West Indies	811	5%
South America*	1369	8%
All tropical countries	7878	49%
Rest of world	8135	51%
World	16013	100%

* Excluding wholly non-tropical countries.

The closest approximation to an in-depth study of a single tropical site involving microfungi is that of the Gombe Stream Reserve in north-east Tanzania (Pirozynski, 1972); more than 700 species of vascular plants were collected and 'the preliminary sorting indicates that the species ratio of microfungi alone to phanerogams is at least 3:1 and may even be as high as 5:1'. As Pirozynski's collections were made over only five months, and as he did not collect macromycetes, entomogenous fungi, lichenized fungi, or undertake isolations from any habitat, his data are not incompatible with the postulated 6:1 ratio used in the extrapolations of Hawksworth (1991*a*).

Whatever future research establishes as to the true number, it is indisputable that there is a tremendous number of undescribed fungi in tropical regions. If only 5% of the world's species (i.e. 70,000) have now been described, 1·43 million must remain unrecognized. Even if the estimated world figure of 1·5 million should prove an overestimate by a factor of two, the prospect of 680,000 undescribed species would still be daunting.

Incontrovertible evidence that there are large numbers of fungi to be described was also obtained from the *Index of Fungi*. The world rate of description of new species has accelerated over the last 70 years, and currently stands at about 1700 per year (Hawksworth, 1991*a*). An analysis of the provenance of the 16,013 fungi (including lichen-forming species) newly described in the ten years 1981-90 revealed that 7878 (49%) were

Table 3. Countries from which 1% or more of the new fungi (including fossil and lichen-forming fungi) catalogued in the *Index of Fungi* were described in 1981-90.

Country	Species	Country	Species
USA	1623 (10·1[†])	Canada	326 (2·0)
India	1554 (9·7)	Cuba	292 (1·8)
France	1230 (7·7)	South Africa	281 (1·8)
PR China	919 (5·7)	Taiwan	260 (1·5)
Japan	736 (4·6)	Germany	249 (1·6)
Australia	634 (4·0)	Czechoslovakia	213 (1·3)
Brazil	569 (3·6)	Switzerland	204 (1·3)
Former USSR	524 (3·3)	Spain	198 (1·2)
British Isles[‡]	459 (2·9)	Netherlands	195 (1·2)
New Zealand	437 (2·7)	Papua New Guinea	173 (1·1)
Malaysia	426 (2·7)	Mexico	162 (1·0)
Rest of World	4349 (27·1)		

[†] Percentage of world species described 1981-90.

discovered in tropical countries (Table 2). Bearing in mind the shortage of taxonomic mycologists working in the tropics (see below), this may appear to be a not inconsiderable achievement, but it has to be remembered that perhaps three-quarters of that number were recognised by specialists based outside the tropics. Further, the effort varies enormously from country to country, 22 contributing 73% of the novel taxa, and a mere nine 52% (Table 3). It is sobering that 12 of the 22 countries contributing more than 1% to the ten-year total are temperate ones. That the USA leads the list is not surprising as extrapolative studies suggest that only about one sixth of even that country's fungi have been documented (Hawksworth, 1991*a*).

In summary, our knowledge of the tropical mycobiota remains in the 'pioneer (or exploratory) phase' of the four stages of understanding the species present in a region distinguished by Davis & Heywood (1963). This phase represents only the first portion of *alpha*-taxonomy, as defined by those authors, in that consolidated knowledge of the species present and their variability will be unattainable in the foreseeable future.

The resource base

Recognizing that large numbers of undocumented tropical fungi exist, it is appropriate to assess the extent of the resources available in the tropics to address the problem, prior to considering actions that might be initiated to redress the situation.

The countries wholly or partially situated between the Tropics of Cancer and Capricorn are listed in Table 4. Information on the numbers of mycologists and reference collections is provided, together with an indication of the level of documentation that exists, and the number of species described from each in 1981-90.

Reviews of the state of mycology in India, Indonesia, Malaysia, The Philippines, Singapore, Sri Lanka, Taiwan, and Thailand have been given in Subramanian (1986*c*).

Human resources

The key resource is the number of experienced mycologists. Some indication of this number can be derived from membership lists of appropriate international and national societies. While the number who belong to the major British and North American societies in many tropical countries is disappointing, to some extent this reflects the ability to pay dues in hard currency. Whereas for the four societies analysed in Table 4 the percentage of tropical members may not at first sight appear disheartening at 4 to 12%, these figures are skewed by significant memberships in Australia and South Africa in particular.

If national society memberships or listings from tropical countries are examined, substantial numbers can be identified in certain countries, notably Argentina, Brazil, India, Mexico, and Taiwan. While such larger totals might be taken as encouraging, closer scrutiny reveals that most of those included are not taxonomists but rather plant pathologists, medical mycologists, food microbiologists, industrial mycologists, and in some cases lichenologists.

Experience at the International Mycological Institute suggests that authoritative expertise across all systematic and ecological groups of fungi is unlikely to be found in a team of less than 25 specialists. Most tropical countries lack *any* active taxonomic mycologists, and only in Argentina, Australia, Brazil, Egypt, India, Mexico, South Africa, the People's Republic of China, and Taiwan do their numbers approach the critical 25 needed for in-depth surveys of tropical mycobiotas.

The shortage of mycological taxonomists has been of concern to the British Mycological Society for almost 50 years (British Mycological Society, 1944), and is currently also recognized as a problem in North America (Burdsall, 1990). In relation to the numbers of taxa they have to contend with, mycologists tend to be disproportionately under-represented amongst taxonomists as a whole. Pascoe (1990) noted that in Australia fewer than 15 fungal taxonomists have to grapple with perhaps 250,000 native fungi, whereas 19 fern taxonomists work on a group with only about 350 Australian species, i.e. 16,667 *vs.* 18 species per taxonomist.

Further, mycologists resident in the tropics rarely have the opportunity of attending pertinent meetings to discuss their problems and formulate collaborative proposals with colleagues in other regions. Swinscow (1991) remarked, of an international symposium on tropical lichens held in London in 1989, 'that, of about 60 people attending . . . only two live in the tropics.' The situation was not so extreme at this Symposium, but was not dissimilar. Fortunately, meetings in the tropics between tropical mycologists are now actively encouraged by the International Mycological Association (IMA; see below).

Information resources

Each country, scored for the level of information available on selected aspects of its fungi, is given in Table 4. The first tool required for progress in the documentation of the mycobiota of a region is a checklist of what has already been reported, together with its associated bibliography, and ideally locations of type or authenticating reference collections. Mycological literature is generally so scattered that unless this homework has been done, a person interested in a particular family or genus is unlikely to have the time or resources to make the necessary bibliographic searches. Regrettably, many tropical countries lack such basal checklists, even for economically important crop pathogens.

The vision of open-ended regional monograph series such as the *Flore Illustrée des Champignons d'Afrique Centrale* initiated in 1972 by the Jardin Botanique National de Belgique, its predecessor the *Flora Iconographique des Champignons du Congo* (1935-72), and *Flora Neotropica*, is to be applauded. The reality is, however, far from comforting. Between 1968 and 1986, *Flora Neotropica* considered 949 fungi; Prance & Campbell (1988) noted that 'this amounted to only 1.8% of the estimated 50,000 fungi of the region, and that it would take 948 years to complete the fungi'. However, even that statement is optimistic as the figure quoted for the fungi is a gross underestimate. At least for the

Table 4. Indicators of the resource base for mycological investigations *within* tropical countries.

	New fungi described 1981/90[1]	Human resources					Reference collection resources		Information resources			
		National[2] Societies	BMS[3]	MSA[4]	IAL[5]	BLS[6]	Dried[7]	Living[8]	Plant Diseases[9]	Lichens[10]	Macro-mycetes[11]	Micro-mycetes[12]
Africa	1226	-	33	20	15	1	15	5	-	-	-	-
Algeria*	22	-	1	-	1	-	-	-	2	1	1	1
Angola	4	-	-	-	-	-	-	-	-	1	1	1
Benin	-	-	-	-	-	-	-	-	-	-	1	1
Botswana	1	-	-	-	-	-	-	-	-	-	1	1
Burkina Faso	1	-	-	1	-	-	-	-	-	-	-	-
Burundi	30	-	-	-	-	-	-	-	2	2	-	-
Cameroon	56	-	-	-	-	-	-	-	-	-	1	-
Cape Verde Islands	-	-	-	-	-	-	-	-	-	2	-	-
Central African Republic	94	-	-	-	-	-	-	-	-	-	3	-
Chad	1	-	-	-	-	-	-	-	-	1	1	-
Comoros	-	-	-	-	-	-	-	-	-	1	-	-
Congo	11	-	-	-	-	-	-	-	2	-	3	1
Cote d'Ivoire	43	-	-	-	-	-	-	-	1	1	1	1
Djibouti	-	-	-	-	-	-	-	-	-	-	-	-
Egypt*	17	-	8	3	-	-	-	1	2	1	1	1
Equatorial Guinea	1	-	-	-	-	-	-	-	-	-	-	-

Ethiopia	23	-	1	-	1	-	-	-	2	1	-	2
Gabon	39	-	-	-	-	-	-	-	1	-	-	-
Gambia	-	-	-	-	-	-	-	-	-	-	-	-
Ghana	21	-	-	-	-	-	2	-	2	-	1	2
Guinea	-	-	-	-	-	-	-	-	-	1	-	1
Guinea Bissau	-	-	-	-	-	-	-	-	-	-	-	-
Kenya	79	-	-	-	-	-	2	-	2	3	3	1
Lesotho	17	-	-	-	-	-	-	-	-	-	-	-
Liberia	6	-	-	-	-	-	-	-	-	-	-	-
Libya*	5	-	-	1	-	-	-	-	1	1	-	-
Madagascar	14	-	1	-	-	-	1	-	2	1	3	-
Malawi	14	-	-	-	-	-	-	-	2	-	3	-
Mali*	-	-	-	-	-	-	-	-	-	-	-	-
Mauritania*	7	-	-	-	-	-	-	-	-	1	-	-
Mauritius	8	-	1	-	-	-	1	-	2	1	1	1
Mayotte	-	-	-	-	-	-	-	-	-	-	-	-
Mozambique	3	-	-	-	-	-	-	-	2	1	-	-
Namibia	43	-	-	-	-	-	-	-	-	1	-	-
Niger	-	-	-	-	-	-	-	-	1	1	-	-
Nigeria	21	-	3	6	-	1	2	1	2	1	1	1
Réunion	27	-	-	-	-	-	-	-	1	1	-	-
Rwanda	7	-	-	-	-	-	-	-	2	2	1	-
Saint Helena & Ascension	-	-	-	-	-	-	-	-	-	-	-	-
Sao Tome & Principe	-	-	-	-	-	-	-	-	1	1	-	-

Senegal	6	·	·	·	·	·	·	·	2	·	·	·
Seychelles	·	·	·	·	·	·	·	·	1	·	·	·
Sierra Leone	26	·	1	1	·	·	1	·	2	2	1	2
Somalia	2	·	·	·	·	·	·	·	2	1	·	·
South Africa*	282	·	14	7	13	·	4	1	3	2	3	3
Sudan, The	8	·	1	1	·	·	·	·	2	·	·	1
Swaziland	2	·	·	·	·	·	·	·	·	·	·	·
Tanzania	72	·	·	·	·	·	·	·	2	3	3	·
Togo	1	·	·	·	·	·	·	·	1	·	·	1
Uganda	31	·	·	·	·	·	1	·	1	3	3	2
Zaire	112	·	·	·	·	·	·	·	2	2	1	1
Zambia	19	·	·	·	·	·	·	·	2	1	1	·
Zimbabwe	13	·	2	·	·	·	1	2	2	1	1	1
Asia	3523		74	72	15	3	39	41	·	·	·	·
Bangladesh*	2	·	·	·	·	·	·	·	1	2	·	1
Brunei	24	·	·	·	·	·	·	·	2	·	1	·
Cambodia	·	·	·	·	·	·	·	·	2	·	·	·
Hong Kong	6	·	4	2	·	·	·	1	2	3	1	·
India*	1554	200	38	27	8	3	11	12	3	3	2	3
Indonesia	67	·	3	5	·	·	1	3	2	1	1	1
Laos	1	·	·	·	·	·	·	·	1	·	·	·
Malaysia	432	·	5	·	·	·	2	2	3	1	1	1
Maldives	·	·	·	·	·	·	·	·	·	·	·	1
Myanmar (Burma)	·	·	·	·	·	·	·	·	·	·	·	·

Oman	2	-	2	-	-	-	1	-	1	1	1	3
Peoples Republic of China*	919	-	4	5	5	-	15	5	3	3	1	3
Philippines	33	-	1	5	1	-	3	5	2	2	2	2
Saudi Arabia*	2	-	5	4	-	-	-	-	1	2	-	-
Singapore	82	-	4	2	-	-	1	2	1	1	-	-
Sri Lanka	107	-	1	-	-	-	1	4	2	3	3	2
Taiwan* (Republic of China)	260	225	5	21	1	-	3	1	3	3	1	2
Thailand	31	-	1	1	-	-	-	6	2	1	1	-
United Arab Emirates*	-	-	1	-	-	-	-	-	-	-	-	-
Vietnam	-	-	-	-	-	-	1	-	2	1	-	-
Yemen, Republic of	1	-	-	-	-	-	-	-	2	1	-	-
Australasia & Pacific	947	-	74	11	9	11	24	32	-	-	-	-
Australia*	634	-	72	11	9	10	20	32	2	3	2	2
Caroline Islands	16	-	-	-	-	-	-	-	-	-	-	-
Cayman Islands	1	-	-	-	-	-	-	-	-	-	-	-
Cook Islands	2	-	-	-	-	-	-	-	-	1	-	-
Fiji	4	-	-	-	-	-	1	-	2	1	1	-
French Polynesia	-	-	-	-	-	-	-	-	1	1	-	-
Guam	4	-	-	-	-	-	-	-	-	-	-	-
Hawaii	10	-	-	-	-	1	-	-	2	2	-	1
Kiribati	-	-	-	-	-	-	-	-	-	-	-	-
Macau	-	-	-	-	-	-	-	-	-	-	-	-

Micronesia, Fed. States of	-	-	-	-	-	-	-	-	1	1	-	-
Nauru	-	-	-	-	-	-	-	-	-	-	-	-
New Caledonia	18	-	-	-	-	-	-	-	1	1	1	1
Norfolk Island	2	-	-	-	-	-	-	-	-	-	1	1
Papua New Guinea	174	-	2	-	-	-	3	-	2	2	2	2
Pitcairn Island	2	-	-	-	-	-	-	-	-	-	-	-
Solomon Islands	75	-	-	-	-	-	-	-	2	1	2	2
Tokelau	-	-	-	-	-	-	-	-	-	-	-	-
Tonga	1	-	-	-	-	-	-	-	2	-	-	-
Tuvalu	-	-	-	-	-	-	-	-	-	-	-	-
Vanuatu	-	-	-	-	-	-	-	-	2	-	1	-
Wallis & Futuna	-	-	-	-	-	-	-	-	-	-	-	-
Western Samoa	3	-	-	-	-	-	-	-	2	-	-	-
Central America & West Indies	811	-	14	30	2	1	19	10	-	-	-	-
Anguilla	-	-	-	-	-	-	-	-	-	-	-	-
Antigua & Barbuda	1	-	-	-	-	-	-	-	2	-	-	-
Aruba	-	-	-	-	-	-	-	-	-	-	-	-
Bahamas, The	-	-	-	-	-	-	-	-	-	2	-	-
Barbados	-	-	1	-	-	-	-	-	2	-	-	-
Belize	23	-	-	-	-	-	-	-	1	1	-	-
Bermuda	5	-	-	-	-	-	-	-	2	1	-	-
Costa Rica	89	4	2	1	-	-	1	-	-	1	1	1

Cuba	291	99	-	-	-	-	2	[1]	2	3	3	2
Dominica	5	-	-	-	-	-	-	-	2	3	3	-
Dominican Republic	5	-	1	1	-	-	-	-	2	1	2	2
El Salvador	-	14	1	1	-	-	-	-	2	1	-	-
Grenada	2	-	-	-	-	-	-	-	-	-	1	-
Guadeloupe	23	-	-	-	-	-	-	-	-	2	3	-
Guatemala	12	4	-	-	-	-	-	1	1	1	-	-
Haiti	-	-	-	-	-	-	-	-	2	1	-	2
Honduras	12	-	1	-	-	-	-	-	-	1	-	-
Jamaica	29	-	-	-	-	-	1	-	2	2	2	-
Martinique	59	-	-	-	-	-	-	-	-	-	3	-
Mexico	165	201	5	11	2	1	12	8	2	3	3	-
Montserrat	1	-	-	-	-	-	-	-	-	-	-	-
Netherlands Antilles	1	-	-	-	-	-	-	-	-	-	-	-
Nicaragua	5	-	-	-	-	-	-	-	1	-	-	-
Panama	29	1	-	-	-	-	1	-	2	1	1	-
Puerto Rico	27	1	1	16	-	-	1	-	2	2	2	2
Saint Kitts & Nevis	-	-	-	-	-	-	-	-	-	-	-	-
Saint Lucia	2	-	1	-	-	-	-	-	-	-	-	-
Saint Vincent & The Grenadines	1	-	-	-	-	-	-	-	2	-	-	-
Trinidad & Tobago	8	-	1	-	-	-	[1]	-	2	2	2	-
Virgin Islands	7	-	-	-	-	-	-	-	2	1	2	-

South America	1369	-	19	14	6	7	35	10	-	-	-	-
Argentina	256	189	7	3	1	4	6	3	2	3	3	2
Bolivia	21	-	-	-	-	-	-	-	1	1	-	-
Brazil*	569	297	4	6	4	2	15	6	3	3	3	2
Chile*	152	12	2	2	-	-	3	1	2	3	1	2
Colombia	101	37	1	2	-	-	1	-	2	1	1	-
Ecuador	50	-	1	-	-	-	1	-	1	1	1	-
French Guiana	24	-	-	-	-	-	1	-	-	2	2	-
Guyana	-	-	-	-	-	-	1	-	2	2	-	-
Paraguay*	5	-	-	-	-	-	-	-	1	1	-	2
Peru	52	-	-	-	-	-	2	-	2	2	-	2
Suriname	1	-	-	-	-	-	-	-	1	-	1	-
Venezuela	135	-	4	1	1	1	5	-	2	2	3	3

Notes: The names of countries mainly follow Daume (1991); * countries the boundaries of which are not all located between the Tropics of Cancer and Capricorn (i.e. between latitudes 24°N and 24°S); the information resources are scored on a subjective 1-5 point scale (1 = scattered mainly short papers; 2 = checklists or other extensive compilations; 3 = monographs with descriptions and keys available for some groups; 4 = monographs available for most groups; 5 = exhaustive monographs available for all groups). Numbers in square brackets are based on information additional to that in the sources cited.

Sources: [1]*Index of Fungi* (1981-90), compiled from information extracted from the database by P.M. Kirk - some continental totals exceed those from the summation of the country figures due to the inclusion of species only localized to a region; [2]Hall & Hawksworth (1990); [3]British Mycological Society (G. W. Beakes, pers. comm., February 1992), total membership 2000 (11% tropical); [4]Mycological Society of America (Blackwell, 1988), total membership 1620 (9% tropical); [5]International Association for Lichenology (H.J. Sipman, pers. comm., February 1992), total membership 387 (12% tropical); [6]British Lichen Society (Gray (1991), total membership 535 (4% tropical); [7]Holmgren *et al.* (1990); [8]Takishima *et al.* (1989); [9]Johnston & Booth (1983), M. Holderness (pers. comm., March 1992); [10]Hawksworth & Ahti (1990); [11]D. L. Hawksworth and D. N. Pegler (pers. comm., March 1992); [12]D. L. Hawksworth and B. C. Sutton (pers. comm., March 1992).

lichen-forming fungi, 14 families are now scheduled, and most should appear by 1996 (Gradstein, 1991). *Flora Fungorum Sinicorum*, the first volume of which appeared in 1987 (Chen *et al.*, 1987), plans to provide monographic treatments in Chinese of the mycobiota of the People's Republic of China. A ten-year work programme to prepare a *Fungal Flora of Taiwan* has also been drawn up (Chen, 1987). In the case of Australia, plans for lichen volumes in the mammoth *Flora of Australia Project* are advancing well, and three fungal groups have been initiated; a grant to produce a checklist of the macrofungi was secured in 1991 (I. Pascoe, pers. comm.).

Even where substantial publications exist, they have to be seen in context. For example, in the case of Venezuela and adjacent countries, Dennis (1970) treats 3996 species, but notes '... rather optimistically, that perhaps one-half of the Myxomycetes, Polystigmataceae, Uredinales and Polypores have been collected and anything between one-fifth and one-tenth of those of other groups of fungi'.

In the absence of authoritative checklists and national and regional monographic treatments, it is scarcely feasible to produce the meaningful identification manuals and teaching aids that are required by teachers, students, ecologists, conservationists, pathologists, and other applied biologists. We are failing potential customers (Hawksworth & Bisby, 1988).

A major constraint to the production of basal checklists is the inadequacy of library facilities in tropical countries. Much of the exploration of tropical biotas during the last 200 years has been by the scientists of former colonial powers located in developed countries, particularly in Europe. The results of their studies are generally included in journals and books published in their own countries at dates when university and institute libraries had not been founded in many tropical countries. Where sound libraries did exist, such as that in Bogor prior to World War II, they have sometimes been destroyed. Those in Singapore narrowly escaped a similar fate (Corner, 1981*b*).

The large number of journals in which mycologists publish systematic work is in itself a problem. The IMI Library finds it necessary to receive 692 current serial titles, and the *Index of Fungi* cited names published in 851 titles (674 journals, 142 books, and 35 collective titles) in the years 1981-90. *Mycotaxon*, which was founded in 1974, is now a valuable focus for such descriptive work, but no mandatory requirement to use it exists.

When a relevant reference is located, obtaining a photocopy, let alone an original of a scarce out-of-print or recently published book, can be impractical without foreign exchange. Corner (1946) pragmatically advised young tropical botanists to ignore, of necessity, what they could not possibly obtain from distant libraries. Improved technologies have yet to fully alleviate the situation, and, in practice, mycologists in developing countries rely on the generosity of overseas colleagues or donors.

Reference collection resources

The resource value of reference collections is generally underestimated; they are rarely granted support commensurate with a crucial component of the information transfer system of all biology (Hawksworth & Mound, 1991). They constitute vouchers for published reports in the scientific literature, provide the basic working material a taxonomist needs to understand the characters, limits and variation of a species, contain information on geographical and host distributions, and are essential in the identification process itself.

The numbers of collections of dried reference specimens (herbaria) located in each tropical country that are stated to include or consist entirely of fungi in *Index Herbariorum* (Holmgren *et al.*, 1990) have been abstracted (Table 4). Of the total of 132, 77 (58%) are located in six countries (Australia, Brazil, India, Mexico, the People's Republic of China, and South Africa). None to few exist in most tropical countries, and when present they tend to be modest in size, to have been founded during the present century, to have few or no mycologically trained staff, and not to encompass all fungal groups.

The situation with living genetic resource collections (culture collections) of fungi parallels that of dried collections. While 107 such tropical collections are now listed in the *World Directory* (Takishima *et al.*, 1989), 44 (41%) are located within two countries, Australia and India. Most are broad microbiological collections including a few fungi, but a significant proportion are restricted to medical mycology or plant pathogens. Living collections are almost invariably held in different institutions from those with dried collections, and generally lack access to the range of expertise necessary to verify the identity of strains.

Collections of living cultures, in addition to having comparable roles to those of dried reference collections as vouchers, data sources, and in identification, have a unique value. They are the mechanism by which the fungal genetic resource is made available for research, screening, and exploitation.

The practical difficulties of maintaining both dried and living reference collections in the tropics are daunting. High and fluctuating humidities encourage the growth of unwanted spoilage fungi, and may also lead to some of the stored material itself starting to grow. The supply of electricity in many countries is also frequently erratic, causing problems with any air-conditioning and dehumidifying equipment, and alsowith refrigerators and cold rooms for live culture storage. In addition, paper materials are subject to destruction by termites, silver fish, beetles, and other arthropods, which can also severely damage specimens, especially of macromycetes. In the case of cultures, liquid nitrogen storage is ideal provided that local supplies can be secured; this is achieved at the Institute of Microbiology in Beijing by a liquid nitrogen generation plant being installed alongside the refrigeration flasks.

The long-term security of reference collections is a major concern in many tropical countries, both with respect to funding and to the threat of destruction through fire, looting, and military action. Some important mycological collections have been lost in East Africa, and several are currently a major cause for concern both elsewhere in Africa and in India.

Pertinence

In order to redress the current lack of resources available for exploring the tropical mycobiota, it is a prerequisite that pragmatic reasons so to do are voiced. Academic enquiry alone is unlikely to sway the apportionment of limited resources in the tropics. Attention needs to be focused on the case for integration in, and relevance to, programmes aimed at improvements in sustainable development, such as those advocated by Swaminathan (1991). Such a case can be founded on both the role of fungi in the maintenance of the ecosystems on which local communities may depend, and the potential benefits to human welfare, development, and sustainability.

Mutualisms involving fungi are of especial significance in the maintenance of tropical ecosystems. Mycorrhizas surely play a crucial role in the transport of minerals otherwise limiting tree growth in the tropics, as elsewhere (Read, 1991), but firm data on their significance in nutrition and rainforest regeneration are required (Janos, 1983; Herrera *et al.*, 1991). Lichens act as food and camouflage for an amazing range of organisms (Seaward, 1988), and species with cyanobacterial partners are nitrogen-fixing. The significance and widespread occurrence of mycophyllas (involving fungal endophytes) is also starting to be recognized. Arthropod gut symbionts are vital to the existence of the myriads of insects feeding on plant materials which require fungi as a source of enzymes (Martin, 1987). Fungi are also important as a food source for numerous arthropods, including substantial numbers of beetles and flies (Hammond & Lawrence, 1989).

The biocontrol of insect pests and weeds using fungi is a key aspect of environmentally friendly and sustainable crop production. However, it is scarcely appreciated that in a natural system the dynamic balance between a pathogen and its host may be crucial to its maintenance. Fungi may limit the vigour of pernicious shrubby or tree species that otherwise could assume dominance, or of populations of insect pests that otherwise might explode and defoliate dominant trees.

Fungi also occupy a key role in the biodegradation of the massive amounts of lignocellulose debris that otherwise would accumulate on forest floors, and without which scarce nutrients would not be both distributed and recycled. These functions are also significant in the formation and maintenance of soil structure, although knowledge of soil fungi in undisturbed tropical ecosystems is meagre.

Additionally, the tropics are of importance to humans as an immense source of novel organisms for screening for beneficial pharmaceutical and agrochemical applications (Nisbet & Fox, 1991), as potential mycopesticides, mycoherbicides and biocontrol agents (Evans & Ellison, 1990), and as sources of industrially important enzymes (see also Chapter 15, this volume). While the first benefactors of novel products arising from tropical fungi may be companies and individuals from the Northern Hemisphere, that need not always be the case. Questions on the property rights of countries from whom exploitable organisms are obtained, and 'debt-for-nature' cash stimuli for the conservation of genetic diversity are currently under active discussion (e.g., McNeely *et al.*, 1990).

The contribution that fungi can make to improved sustainability at the small farmer level in the tropics awaits realisation. Notwithstanding applications in biocontrol, there remains immense scope for the employment of fungi in waste utilisation and as sources of food. These desiderata can sometimes be combined, as in the utilisation of agricultural wastes (e.g. bagasse, coire, paddy straw) for the cultivation of edible mushrooms – including local species.

Prophylaxis

The census presented above demonstrates that the resources available for basic mycological work in the tropics are grossly inadequate. Further, single comprehensive, regional or national dried and living reference collections, with associated libraries and other information resources, should be favoured over the predominating pattern of dispersed and under-resourced small collections. National and regional nodes are required to make the most effective use of limited resources, and also to develop a cadre of staff with complementary skills. To this end, a strong plea for the establishment of a National Culture Collection and Identification Centre for Fungi in India has been made (Subramanian, 1982, 1986*a*).

Almost without exception, the crucial publications and reference collections relating to particular tropical countries, including type material of new species described from them, are located in institutions in Europe, and to a lesser extent North America. Consequently, the most substantial tropical revisions of fungal groups continue to be authored in the Northern Hemisphere (e.g. Dennis, 1970; Lowy, 1971; Corner, 1972; Singer, 1976; Pegler, 1977, 1983; Ryvarden & Johansen, 1980; Swinscow & Krog, 1988). The colonial histories that led to this circumstance render it almost impossible for individual mycologists located in tropical countries to make major contributions to the description of their own mycobiotas without substantial and costly periods overseas. The situation is exacerbated by a reluctance of some Northern Hemisphere reference collections to mail irreplaceable type material to the tropics, not because of doubts that it will be treated appropriately on arrival but rather as a result of the uncertainty of the mail services. Most major institutions have tales of such collections being irretrievably lost or damaged. As a safeguard, IMI loans only a portion of the specimen or preparations requested at any one time.

Accepting the mismatch of limited resources and the immense amount yet to be accomplished to attain even a modest knowledge of tropical fungi, their biologies, and their ecological roles, the only remedy to alleviate if not cure this condition is for mycologists internationally to prioritize, rationalize, mobilize, organize, and publicize to a degree hitherto not contemplated.

Prioritize

It is first necessary to identify the knowledge gaps and priorities for advancing tropical mycology. Otherwise, limited resources may be inappropriately deployed.

Key scientific questions relating to biodiversity and its importance for ecosystem function have already been identified (Solbrig, 1991). However, in addition to intellectually challenging issues, the need to improve the level of mycological knowledge in the tropics is desperate in relation to the damage fungi cause to humans, crops, food, and materials. In developing tropical countries, practical problems with prospects of short-term benefits will be accorded the highest priority. Subramanian

(1986*b*) considered the principal justification for developing a centre of expertise in tropical mycology in south-east Asia to be that 'tropical mycology impinges on every aspect of human health and development and to be able to harness fungal power for human welfare and needs, it is necessary to strengthen the base of mycology in the tropics'.

The common limiting factors to advances in mycology, on both the scientific enquiry and human welfare fronts, are the inadequacy of resources located in the tropics, the development and strengthening of the resources dependent on the supply of sufficient numbers of appropriately trained scientists, and the reference and information resources they require. Institution-building must be recognized as the first priority.

With such daunting numbers of undescribed species (see above) it also has to be recognized that all cannot be documented in the foreseeable future. As a general principle Soulé (1990) considered that for speciose groups systematists should consider: (1) concentrating on 'representative' taxa; (2) focusing on ecologically keystone and indicator taxa and their mutualists; (3) focusing on phylogenetic relicts; (4) focusing on endemics; and (5) developing improved methods for rapid classification. These principles can be applied to mycology, but to them I would add: (6) emphasizing groups most likely to be of importance to human development and sustainability.

Rationalise

When resources are scarce, the prospect of rationalisation has to be faced. A single institution, or even country, cannot expect to cover all aspects of mycology in depth in the 1990s. In the case of systematic biology in the U.K., the Select Committee on Science and Technology (1992) recommended that major institutions start 'discussions on rationalisation of holdings, staff appointments and areas of specialisation'. If this is perceived as a need in a developed country, it is surely a requirement where resources are even more restricted.

Rationalisation in a parallel manner of the numbers of journals in which data on tropical mycology appear also merits further consideration (see above), bearing in mind the expense of the current system and often declining library budgets. In the 1950s, IMI planned (but later had to abort) the reprinting of descriptions and illustrations of newly described fungi on microcards. Optical disc technology could now make that goal realisable.

Mobilise

As the relevant resources for the study of tropical mycology are almost exclusively in North Temperate countries, the means of mobilizing these so that they are available in the tropics must be found.

In principle, the immeasurable amount of information hidden in reference collections throughout Europe and North America can be made accessible by computer technology. At IMI, computerisation means that all material accessed since 1989 from 155 countries can be output on a country basis, or in other formats, as hard-copy or computer disks. If resources to computerise data from the 320,000 specimens accessed in 1920-89 were secured, this would immediately facilitate its use by tropical mycologists. Sutton (1993) noted that IMI held microfungi from 46 countries in Africa alone, and that they were particularly extensive for ten. Similar and complementary patterns exist in many major collections, but only in a few instances has computerisation been initiated.

An important exception is the U.S. National Fungus Collection in the U.S. Department of Agriculture's Agricultural Research Service, Beltsville, Maryland. A comprehensive programme of computerisation started in 1980, and details from 477,000 specimens, about half of their holdings, had been keyed in by February 1992 (A. Y. Rossman, pers. comm.).

The last decade has seen major advances in computerized access to the information on living fungal strains held in the world's culture collections (Krichevsky *et al.*, 1988). Direct dialling into databases held at individual collections or on-line database hosts (e.g. MINE [Microbial Information Network Europe] on DIMDI from late 1992) is becoming increasingly feasible. Such data are becoming available on CD-ROM discs for use in individual laboratories (e.g. CD-STRAIN by Hitachi Ltd).

The world literature on systematic mycology is covered in IMI's twice yearly *Bibliography of Systematic Mycology* and *Index of Fungi*. The *Bibliography* will become available as a component of the CAB ABSTRACTS database shortly, and as such will also be included in updates of the CAB ABSTRACTS CD-ROM.

Organize

An imaginative system of Microbial Resource Centres (MIRCENs) was established by UNESCO in 1975. The scheme currently comprises 24 centres, 10 of which are located in tropical countries. These foster cooperation in applied microbiology and technology, encourage self-reliance and sustainable development, and disseminate appropriate skills (Da Silva, 1991). While, with the exception of IMI, the focus of almost all MIRCENs is on bacteriological applications, this pioneering initiative could be strengthened mycologically if appropriate funding could be secured.

In 1977 the International Mycological Association (IMA) established Committees for Asia, Latin America, and from 1983 Africa, each comprising representatives from the various countries in their regions. Those for Africa and Asia have been organising symposia since 1986 (e.g. Tubaki & Yokoyama, 1987), and further considering how to advance the study of mycology in their regions. In both cases strengthening of regional capabilities by the establishment of formal or informal networks is required. The same conclusion was reached at a UNESCO Workshop on Applied Mycological Research in the Tropics held in Singapore in 1985, at least until regional or national centres had been established (Subramanian, 1986*b*).

The need for national and regional centres with appropriate expertise and reference collections, especially for organisms of actual or potential importance in agriculture and allied fields, has been identified (Harris & Scott, 1989). Subramanian (1982) had the vision of an Institute of Tropical Mycology stating 'tropical fungi must be studied by us in the tropics so that we become completely self reliant'.

Institution-building is necessarily long term and expensive. Haskell & Morgan (1988) identified networking as the best way forward to utilize limited resources, and an IUBS/SCOPE/UNESCO workshop saw it as a key component of a research agenda for biodiversity (Solbrig, 1991). A formal global Network of Biosystematic Centres was advocated by an international workshop on the significance of biodiversity in microorganisms and invertebrates (Hawksworth, 1991*b*). It was recognized that the establishment of such a system would involve both the strengthening of existing centres and development of new ones through training, infrastructure building, and information transfer. In 1991 CAB International initiated discussions with other major centres and possible donor organisations as to how such a network, provisionally termed BIONET, might be established. A system of regional loops is under discussion, linked into centres of excellence with complementary expertise in both temperate regions and the tropics. To date, discussions on BIONET have focused on arthropods, but the model could be adapted for microorganisms, including fungi. Regional networks of systematic resources and expertise have also been advocated by IUBS and IUMS (Hawksworth & Colwell, 1992).

Raven (1988) stressed '.. the absolute necessity of forming partnerships with our colleagues in the tropical countries themselves'. I concur. This requires centres in developed countries with appropriate skills to actively assist those in tropical countries to develop groups of effective, well-trained, and well-supported scientists. A variety of bilateral, inter-institute links already exist and can make significant progress even over a modest time span. For example, the number of fungal cultures maintained in the Instituto de Investigaciones Fundamentales en Agricultura Tropical (INIFAT) in Cuba increased from 401 to 2952 from 1979-89 through collaborative work with mycologists in the University of Jena, Germany (Castañeda Ruiz *et al.*, 1990).

IMI and its sister CAB International Institutes are according increasing emphasis to institution-building and training, and expect to progress into the position of final points of reference rather than simply undertaking large numbers of routine identifications (CAB International, 1990). IMI has formal collaborative agreements with several institutes in the tropics, including the Fundacao Tropical de Pesquisa e Tecnologia 'Andre Tosello' (Campinas, Brazil), and the Systematic Mycology and Lichenology Laboratory (Beijing) which involve training elements. Since 1985, 568 mycologists from 76 countries (including 45 tropical) have trained at IMI. The Institute has also run or contributed to 19 courses outside the UK during this period including ones in Argentina, Chile, Egypt, India, Kenya, Malaysia, and the People's Republic of China in which a further 222 mycologists have participated.

In September 1991, IUBS and IUMS developed a joint Action Statement, MICROBIAL DIVERSITY 21, aimed at addressing issues related to the biodiversity of microorganisms (Hawksworth & Colwell, 1992). A proposal for an international network on Inventorying and Monitoring of Biodiversity, taking note of fungi (including lichens), has subsequently been developed (Di Castri *et al.*, 1992). Funding for this initiative is currently being sought from the Global Environment Facility (GEF).

Consequently, it is now opportune for tropical mycologists to consider in earnest how best to contribute to world networking and inventorying initiatives. The prospects for concrete action are more promising than at any previous period in history. The current political concerns reflected in the convening of the United Nations Conference on the Environment (UNCED) held in Rio de Janeiro in 1992 are unlikely to recur at such a scale in the foreseeable future. To exploit this circumstance, the International Mycological Association (IMA) has contacted mycological organisations adhering to it, is holding *ad hoc* discussions here and during meetings of its Committees for Africa and Asia during 1992, and is being represented at meetings planning for international projects on biodiversity.

Publicise

However cogent arguments may be or acute the needs, in the information explosion that characterises the 1990s, the reality is that those charged with decision-making are unlikely to note items not inescapably brought to their attention. Astronomers and particle physicists have been particularly adept in securing major funding for primarily academic enquiry. They have generated a perceived image of exciting discovery at the lay level, notably through the media and semi-popular but authoritative, best-selling books. Biologists concerned with systematics and biodiversity need to emulate these peer groups.

To increase awareness of tropical mycology we must ensure that: (1) fungi no longer repeatedly continue to be overlooked in the development of policies related to conservation and biodiversity; (2) fungi are included in multidisciplinary research projects and surveys; (3) mycological teaching is made exciting and relevant with 'user friendly' well-illustrated books focusing on contemporary topics and issued in local languages; (4) opportunities to address general audiences in all media are sought and grasped; (5) opportunities are created to meet with key senior scientists, policy makers, and politicians; and (6) appropriate ranges of non-technical narratives targeted at the general reader are both available at competitive prices and well-marketed.

Prognosis

I consider the prognosis for substantial progress in tropical mycology to be good, provided that the prerequisite organisation and planning can be effected by 1995. If we are lackadaisical we could find that the currently receptive international audience has either allocated all its resources or found new topics to support before we can prepare a united case. The present window of opportunity will soon be passed.

This will not be an easy task, as it involves a reprioritizing and reconsideration of both institutional and individual objectives. Increasingly effective methods of utilizing the limiting scarce human resources have to be addressed, and will inevitably involve modifications in working practices. There is a need to: reduce the time currently devoted to sterile nomenclatural investigations (Hawksworth, 1992), avoid unnecessary duplication, ensure the range of systematic coverage required exists, restrict the numbers of hard copy publication outlets, utilize non-print modern technology for the cataloguing of descriptive and taxonomic information, and work as parts of clearly focused transnational teams. The latter approach has been advocated by the International Commission on the Taxonomy of Fungi (ICTF), established in 1982, through subcommissions concerned with *Aspergillus*, *Fusarium*, *Penicillium*, and *Trichoderma*.

The retail products of taxonomy must become as optimally user friendly as is compatible with good science, and be tailored to the needs of those concerned with biodiversity surveys as well as applied aspects. Corner (1946) criticised most tropical floras then existing as being fundamentally unsuitable and useless for the local user. His sentiments have been strongly echoed by Heywood (1983). Cullen (1984) estimated that 90% of the work produced by taxonomists is for the consumption of other taxonomists, and viewed the subject as characterized by introversion and ritualization.

If we default, and fail to grasp the opportunity to contribute to current world concerns and programmes on biodiversity, mycology will become increasingly marginalised. Further, as time passes, the objects of our study will increasingly be irrevocably lost as forests are burned and bulldozed. The solution is in *our* hands, and failure to act will jeopardize the future of our subject. We need to seek *solutions* and not *problems*.

Acknowledgements I am indebted to Professor E. J. H. Corner whose *Life of Plants* (1964) convinced me of the need for a tropical orientation in systematics as an undergraduate almost 30 years ago, and for stimulating and entertaining discussions in relation to the topics addressed. The following are thanked for kindly making available data cited in the text: Drs G. W. Beakes, M. Holderness, P. M. Kirk, D. J. Lodge, I. Pascoe, D. N. Pegler, A. Y. Rossman, H. J. Sipman, and B. C. Sutton.

References

Bisby, F. A. & Hawksworth, D. L. (1991). What must be done to save systematics? In *Prospects in Systematics*, (ed. D. L. Hawksworth), pp. 323-336. Koeltz Scientific Books: Königstein, Germany.

Bisby, G. R. (1933). The distribution of fungi as compared with that of phanerogams. *American Journal of Botany* **20**, 246-254.

Blackwell, M. (ed.) (1988). *Mycological Society of America. Membership Directory and Handbook*. Mycological Society of America: Baton Rouge, U.S.A.

British Mycological Society (1944). *The Need for Encouraging the Study of Systematic Mycology in England and Wales.* British Mycological Society: London, U.K. [Subsequently published in *Transactions of the British Mycological Society* **32**, 104-112 (1949).]

Burdsall, H. H. (1990). Taxonomic mycology: concerns about the present; optimism for the future. *Mycologia* **82**, 1-8.

CAB International (1990). *Eleventh Review Conference, London 1990. Report of Proceedings*. CAB International: Wallingford, U.K.

Castañeda Ruiz, R. F., Arnold, G. R. W., Rodríguez de la Rosa, N. & González Merchán, M. (1990). Resultados de 10 anos de trabajo en la micoteca del INIFAT. In *25 Años de Colaboración Cientifico-Tecnica Cuba-RDA 1965-1990*, (ed. R. Martinez Viera, M. Born & P. Löser), pp. 1-4. Ministerium für Forschung und Technologie: Jena, Germany.

Chen, G.-q., Han, S.-j., Lai, Y.-q., Yu, Y.-n., Zheng, R.-y. & Zhao, Z.-y. (1987). *Flora Fungorum Sinicorum.* **Vol. 1.** *Erysiphales*. Science Press: Beijing, People's Republic of China.

Chen, Z.-c. (1987). Mycological survey in Taiwan: history and present status. *Transactions of the Mycological Society of Japan* **28**, 89-94.

Corner, E. J. H. (1946). Suggestions for botanical progress. *New Phytologist* **45**, 185-192.

Corner, E. J. H. (1950). *A Monograph of Clavaria and allied Genera.* [Annals of Botany Memoirs No. 1.] Oxford University Press: London, U.K.

Corner, E. J. H. (1964). *The Life of Plants.* Weidenfeld and Nicolson: London, U.K.

Corner, E. J. H. (1966). *A Monograph of the Cantharelloid Fungi.* Oxford University Press: London, U.K.

Corner, E. J. H. (1968). A monograph of *Thelephora* (Basidiomycetes). *Beihefte zur Nova Hedwigia* **27**, 1-110.

Corner, E. J. H. (1972). *Boletus in Malaysia*. Government Printing Office: Singapore, Malaysia.

Corner, E. J. H. (1981*a*). The agaric genera *Lentinus, Panus,* and *Pleurotus* with particular reference to Malaysian species. *Beihefte zur Nova Hedwigia* **69**, 1-169.

Corner, E. J. H. (1981*b*). *The Marquis. A tale of Syonan-to*. Heinemann Asia: Singapore.

Corner, E. J. H. (1983-91). Ad polyporaceae I-VII. *Beihefte zur Nova Hedwigia* **75**, 1-182 (1983) [I]; **78**, 1-222 (1984) [II & III]; **86**, 1-265 (1987) [IV]; **96**, 1-218 (1989a) [V]; **97**, 1-197 (1989b)[VI]; **101**, 1-175 (1991)[VII].

Corner, E. J. H. & Bas, C. (1962). The genus *Amanita* in Singapore and Malaya. *Persoonia* **2**, 241-304.

Cullen, J. (1984). Libraries and herbaria. In *Current Concepts in Plant Taxonomy*, (ed. V.H. Heywood & D.M. Moore), pp. 25-38. Academic Press: London, U.K.

Da Silva, E. J. (1991). Biotechnologies, microbes and the environment. *Nature & Resources* **27**, 23-29.

Daume, P. (1991). *1991 Britannica Book of the Year.* Encyclopaedia Britannica: Chicago, U.S.A.

Davis, P. H. & Heywood, V. H. (1963). *Principles of Angiosperm Taxonomy*. Oliver & Boyd: Edinburgh and London, U.K.

Dennis, R. W. G. (1970). *Fungus Flora of Venezuela and Adjacent Countries*. [Kew Bulletin, Additional Series No. 3.] Her Majesty's Stationery Office: London, U.K.

Di Castri, F., Robertson-Vernhes, J. & Younès, T. (1992). A proposal for an international network on inventorying and monitoring of biodiversity. *Biology International, Special Issue* **27**, 1-27.

Evans, H. C. & Ellison, C. A. (1990). Classical biological control of weeds with microorganisms: past and present prospects. *Aspects of Applied Biology* **24**, 39-49.

Gray, J. M. (1991). *British Lichen Society Membership List*. British Lichen Society: London, U.K.

Gradstein, S. R. (1991). Flora neotropica news: lichens. *International Lichenological Newsletter* **24**, 60-61.

Hall, G. S. & Hawksworth, D. L. (1990). *International Mycological Directory*, 2nd edition. CAB International: Wallingford, U.K.

Hammond, P. M. & Lawrence, J. F. (1989). Mycophagy in insects: a summary. In *Insect-Fungal Interactions*, (ed. N. Wilding, N. M. Collins, P. M. Hammond & J. F. Webber), pp. 275-325. Academic Press: London, U.K.

Harris, K. M. & Scott, P. R. (eds) (1989). *Crop Protection Information: An International Perspective.* CAB International: Wallingford, U.K.

Haskell, P. T. & Morgan, P. J. (1988). User needs in systematics and obstacles to their fulfilment. In *Prospects in Systematics*, (ed. D. L. Hawksworth), pp. 399-413. Clarendon Press: Oxford, U.K.

Hawksworth, D. L. (1991*a*). The fungal dimension of biodiversity: magnitude, significance, and conservation. *Mycological Research* **95**, 641-655.

Hawksworth, D. L. (ed.) (1991*b*). *The Biodiversity of Microorganisms and Invertebrates: Its Role in Sustainable Agriculture.* CAB International: Wallingford, U.K.

Hawksworth, D. L. (1992). The need for a more effective biological nomenclature for the 21st century. *Botanical Journal of the Linnean Society*, in press.

Hawksworth, D. L. & Ahti, T. (1990). A bibliographic guide to the lichen floras of the world . 2nd edition. *Lichenologist* **22**, 1-78.

Hawksworth, D. L. & Bisby, F. A. (1988). Systematics: the keystone of biology. In *Prospects in Systematics*, (ed. D. L. Hawksworth), pp. 3-30. Clarendon Press: Oxford, U.K.

Hawksworth, D. L. & Colwell, R. R. (1992). Biodiversity amongst microorganisms and its relevance. *Biology International* **24**, 11-15.

Hawksworth, D. L. & Mound, L. A. (1991). Biodiversity databases: the crucial significance of collections. In *The Biodiversity of Microorganisms and Invertebrates: Its Role in Sustainable Agriculture*, (ed. D. L. Hawksworth), pp. 17-29. CAB International: Wallingford, U.K.

Herrera, R. A., Capote, R. P., Menéndez, L. & Rodriguez, M. E. (1991). Silvigenesis stages and the role of mycorrhiza in natural regeneration in Sierra del Rosario, Cuba. In *Rain Forest Regeneration and Management*, (ed. A. Gómez-Pompa, T. C. Whitmore & M. Hadley), pp. 211-233. United Nations Educational, Scientific and Cultural Organisation: Paris, France.

Heywood, V. H. (1983). The mythology of taxonomy. *Transactions of the Botanical Society of Edinburgh* **44**, 79-94.

Holmgren, P. K., Holmgren, N. H. & Barnett, L. C. (1990). *Index Herbariorum. Part I: The Herbaria of the World.* 8th edition. [Regnum Vegetabile No. 120.]. New York Botanical Garden: New York, U.S.A.

Janos, D. P. (1983). Tropical mycorrhizas, nutrient cycles, and plant growth. In *Tropical Rain Forest Ecology and Management*, (ed. S. L. Sutton, T. C. Whitmore & A. C. Chadwick), pp. 327-345. Blackwell Scientific Publications: Oxford, U.K.

Johnston, A. & Booth, C. (eds) (1983). *Plant Pathologist's Pocketbook*. 2nd edition. Commonwealth Agricultural Bureaux: Slough, U.K.

Krichevsky, M. I., Fabricus, B.-O. & Sugawara, H. (1988). Information resources. In *Living Resources for Biotechnology: Filamentous Fungi*, (ed. D. L. Hawksworth), pp. 31-53. Cambridge University Press: Cambridge, U.K.

Lowy, B. (1971). Tremellales. *Flora Neotropica Monographs* **6**, 1-153.

McNeely, J. A., Miller, K. R., Reid, W. V., Mittermeir, R. A. & Werner, T. B. (1990). *Conserving the World's Biological Diversity*. International Union for the Conservation of Nature: Gland, Switzerland.

Martin, M. M. (1987). *Invertebrate-Microbial Interactions. Ingested Fungal Enzymes in Arthropod Biology.* Comstock Publishing Associates: Ithaca, New York.

May, R. M. (1988). How many species are there on Earth? *Science* **241**, 1441-1449.

May, R. M. (1991). A fondness for fungi. *Nature* **352**, 475-476.

Nisbet, L. J. & Fox, F. M. (1991). The importance of microbial biodiversity to biotechnology. In *The Biodiversity of Microorganisms and Invertebrates: Its Role in Sustainable Agriculture*, (ed. D. L. Hawksworth), pp. 229-244. CAB International: Wallingford, U.K.

Pascoe, I. G. (1990). History of systematic mycology in Australia. In *History of Systematic Botany in Australia*, (ed. P. S. Short), pp. 259-264. Australian Systematic Botany Society: South Yarra, Australia.

Pegler, D. N. (1977). *A Preliminary Agaric Flora of East Africa.* [Kew Bulletin, Additional Series No. 6.] Her Majesty's Stationery Office: London, U.K.

Pegler, D. N. (1983). *Agaric Flora of the Lesser Antilles.* [Kew Bulletin, Additional Series No. 9.] Her Majesty's Stationery Office: London, U.K.

Pirozynski, K. A. (1972). Microfungi of Tanzania. I. Miscellaneous fungi on oil palm. II. New Hyphomycetes. *Mycological Papers* **129**, 1-64.

Prance, G. T. & Campbell, D. G. (1988). The present state of tropical floristics. *Taxon* **37**, 519-548.

Raven, P. H. (1988). Tropical floristics tomorrow. *Taxon* **37**, 549-560.

Read, D. J. (1991). Mycorrhizas in ecosystems - nature's response to the 'Law of the Minimum'. In *Frontiers in Mycology*, (ed. D. L. Hawksworth), pp. 101-130. CAB International: Wallingford, U.K.

Ryvarden, L. & Johansen, I. (1980). *A Preliminary Polypore Flora of East Africa.* Fungiflora: Oslo, Norway.

Seaward, M. R. D. (1988). Contribution of lichens to ecosystems. In *Handbook of Lichenology*, (ed. M. Galun), **vol. 2**, pp. 107-129. CRC Press: Boca Raton, U.S.A.

Select Committee on Science and Technology (1992). *Systematic Biology Research.* **vol. 1.** *Report.* Her Majesty's Stationery Office: London, U.K.

Singer, R. (1976). Marasmieae (Basidiomycetes-Tricholomataceae). *Flora Neotropica Monographs* **17**, 1-347.

Smith, D. & Waller, J. M. (1992). Culture collections of microorganisms: their importance in tropical plant pathology. *Fitopatologia Brasileira*, in press.

Solbrig, O. T. (ed.) (1991). *From Genes to Ecosystems: A Research Agenda for Biodiversity.* International Union of Biological Sciences: Cambridge, Mass., U.S.A.

Soulé, M. E. (1990). The real work of systematics. *Annals of the Missouri Botanical Garden* **77**, 4-12.

Subramanian, C. V. (1982). Tropical mycology: future needs and development. *Current Science* **51**, 321-325.

Subramanian, C. V. (1986*a*). The progress and status of mycology in India. *Proceedings of the Indian Academy of Sciences, Plant Sciences* **96**, 379-392.

Subramanian, C. V. (1986*b*). Foreword. *Proceedings of the Indian Academy of Sciences, Plant Sciences* **96**, 333-334.

Subramanian, C. V. (ed.) (1986*c*). UNESCO Workshop on Progress on Applied Mycological Research in the Tropics, University of Singapore, Singapore, May 1985. *Proceedings of the Indian Academy of Sciences, Plant Sciences* **96**, 333-392.

Sutton, B.C. (1993). Contribution of the International Mycological Institute to African mycology. In *Proceedings of the 13th AETFAT Congress, Zomba*, (ed. J. H. Sayani), in press. National Herbarium and Botanical Garden: Zomba, Malawi.

Swaminathan, M. S. (1991). *From Stockholm to Rio de Janeiro. The Road to Sustainable Agriculture.* [Monograph No. 4.] M.S. Swaminathan Research Foundation: Madras, India.

Swinscow, T. D. V. (1991). Epilogue. In *Tropical Lichens: Their Systematics, Conservation, and Ecology*, (ed. D. J. Galloway), pp. 275-277. Clarendon Press: Oxford, U.K.

Swinscow, T. D. V. & Krog, H. (1988). *Macrolichens of East Africa.* British Museum (Natural History): London, U.K.

Takishima, Y., Shimura, J., Udagawa, Y. & Sugawara, H. (1989). *Guide to World Data Center on Microorganisms with a List of Culture Collections in the World.* World Data Center on Microorganisms: Saitama, Japan.

Tubaki, K. & Yokoyama, T. (1987). Reports from the Business Meeting of the Committee for the Development of Mycology in Asian Countries, IMA, held in Malaysia, March 1986. *Transactions of the Mycological Society of Japan* **28**, 75-104.

ABSTRACTS

INVITED PAPERS

The agaricoid fungi from the Guianas

Régis Courtecuisse

Département de Botanique, Faculté des Sciences Biologiques et Pharmaceutiques, B.P. 83; F.59006 Lille Cedex, France

High endemism is related to the geological history of the Guianese area and an almost unique mycological biota can be expected there. However, there have been few serious collections and we have only a rudimentary knowledge of the fungal biota. In French Guiana, the most recent mycological book is dated 1854. A recent provisional check-list includes only about 400 species, of which 152 are agaricoid. Other lists include 110 species (11 agaricoid) for Surinam and 379 species for Guyana (19 agaricoid).

We can give a first and very tentative account of the biogeographical elements of the agaricoid mycobiota. The main groups are: cosmopolitan and subcosmopolitan: 4 species; pantropical: 12 species; neotropical: 67 species; species with a special distribution pattern: Guiano-Caribbean: 12 species; Guiano-Brazilian pattern: 5 species; species present through the neotropics and S. E. Asia: 9; species present through the neotropics and tropical Africa: 5; endemic or recently described species: 23 taxa from French Guiana. No recent data have been collected from Surinam or Guyana. Very few basidiomes from potentially mycorrhizal species have been collected and saprobes seem to dominate the mycological spectrum. The scarcity of basic information raises the question of reliability in these respects and a review of current plans and perspectives was discussed.

Fungi as biological control agents of arthropods in the tropics

H. C. Evans

International Institute of Biological Control, Silwood Park, Ascot, Berks. SL5 7TA, U.K.

The role of entomopathogenic fungi in the population dynamics of arthropods in natural tropical ecosystems, with particular reference to rainforests, was discussed. The clavicipitaceous genera *Cordyceps, Torrubiella* and *Hypocrella* are well represented in these habitats and are considered to exert a significant (natural) control of certain key arthropod groups, notably ants (Formicidae), scale insects (Coccidae), whiteflies (Aleyrodidae) and spiders (Araneida). It is concluded that the biology and epizootiology of these fungal pathogens have been inadequately studied. In sharp contrast, in annual agricultural ecosystems, such as rice, ascomycete genera are poorly represented, while zygomycetes, (viz. *Entomophthora, Erynia, Entomophaga, Zoophthora)* and hyphomycetes (viz. *Beauveria, Hirsutella, Metarhizium, Nomuraea)* predominate. As a direct consequence of their agricultural importance, much more is known about their modes of action and life histories. The occurrence of natural epizootics in crop systems was reported and the possibilities of manipulating these fungi as biological control agents discussed. Problems concerning the use of mycoinsecticides to control arthropod pests were outlined and the feasibility of applying these products against target pests, such as migratory locusts and rice brown planthoppers, assessed. In tropical tree crops with multi-storeyed canopies, the entomopathogenic mycobiota is less subject to seasonal fluctuations and may approach that of a forest ecosystem in its diversity. Because of their unique ecological niche, entomopathogenic fungi could represent an important resource for both the pharmaceutical and agrochemical industries.

Tropical hyphomycetes from submerged litter in freshwater streams

A. J. Kuthubutheen

Department of Botany, University of Malaya, 59000 Kuala Lumpur, Malaysia

Our knowledge of the taxonomy, distribution, ecology, anamorph/teleomorph relationships, and biochemistry of metabolites of the hyphomycetes and particularly the 'aquatic' forms from the temperate regions far outweighs that of hyphomycetes from the tropics. This situation, due more to the later and slower development of hyphomycete taxonomy in the tropics, has often made it difficult to draw definite conclusions as to whether a hyphomycete is indeed temperate in distribution. From studies of the anamorphic fungi in freshwater habitats it is apparent that species in genera such as *Actinospora*, *Alatospora*, *Anguillospora*, *Articulospora*, *Brachiosphaera*, *Calcariospora*, *Camposporium*, *Campylospora*, *Ceratosporella*, *Clathrosphaerina*, *Clavariopsis*, *Colispora*, *Dendrosporium*, *Flagellospora*, *Helicoma* and related genera, *Heliscus*, *Isthmotricladia*, *Isthmolongispora*, *Laridospora*, *Lateriramulosa*, *Lunulospora*, *Scorpiosporium*, *Speiropsis*, *Subulispora*, *Taeniospora*, *Tetracladium*, *Tricladium*, *Trinacrium*, *Tripospermum*, *Triscelophorus*, *Troposporina*, *Vargamyces*, and *Varicosporium* are evidently cosmopolitan. Species in the anamorph genera such as *Bahusutrabeeja*, *Beverwykiella*, *Camposporidium*, *Cancellidium*, *Ceratosporium*, *Clathrosporium*, *Condylospora*, *Crucispora*, *Cryptophialoidea*, *Diplocladiella*, *Dendrosporomyces*, *Flabellospora*, *Fusticeps*, *Ingoldiella*, *Isthmophragmospora*, *Iyengarina*, *Nawawia*, *Nidulispora*, *Obeliospora*, *Phalangispora*, *Polysynnema*, *Polytretophora*, *Quadricladium*, *Setosynnema*, *Tetrabrachium*, *Tricladiomyces*, *Tricladiospora*, *Tridentaria*, *Triramulispora*, *Triscelosporium*, and *Wiesneriomyces* have hitherto been reported mostly or exclusively from freshwater habitats in the tropics. Several of these fungi are known only from their type localities. As the hyphomycetes from the tropics continue to receive more attention, many of those species that are now known almost exclusively only in freshwater habitats in the temperate regions may well turn out to be cosmopolitan in distribution.

Submerged leaves and twigs when placed in moist chambers incubated at 28±2°C yield mostly hyphomycetes with conidia that are either setulate or appendiculate, branched, sigmoid, or that form bubble traps of various sorts.

Many of the taxa listed above either as cosmopolitan or as tropical are evidently amphibious and readily grow out from and sporulate on submerged twigs and leaves incubated in moist chambers. There is often a fair representation of typical terrestrial hyphomycetes on submerged twigs although this may depend on how long the litter had remained submerged. Marked seasonal variations have not been seen with hyphomycetes in streams particularly in the equatorial rainforests.

Although the large majority of the fungi from submerged litter can grow on 2% malt extract agar or cornmeal agar, sporulation does not occur readily on these media. Sporulation in these fungi may be induced either by floating mycelial plugs in sterile water or in many cases by merely cutting out wells in the colony. *Selenosporella aristata* isolated from submerged litter is capable of xylanase production. Other hyphomycete taxa from submerged litter show good cellulolytic or amylolytic activities.

Soil-borne plant diseases in South East Asia

Leka Manoch

Department of Plant Pathology, Faculty of Agriculture, Kasetsart University, Bangkok 10900, Thailand

Environment and farming practices make soil-borne diseases economically important in South East Asia with income loss to farmers through plant death and yield reduction in the Philippines, Indonesia, Singapore, Malaysia and Thailand. Little is known about their occurrence in Vietnam, Cambodia, Laos and Burma. *Fusarium moniliforme* Sheld., *Magnaporthe salvinii* (Cattaneo) Krause & Webster, *Sclerotium rolfsii* Sacc., and *Thanatephorus cucumeris* (Frank) Donk cause important rice diseases. *F. moniliforme* also infects sorghum and sugarcane and on maize causes stalk rots with *F. graminearum* Shwabe. Chili, citrus, onion, soybean, wheat and wingbean are also affected by fusaria.

Pythium aphanidermatum (Edson) Fitzp. infects maize and papaya, *P. vexans* de Bary infects Cruciferae and mango seedlings and *P. splendens* Braun infects betel, black pepper and Chinese cabbage. Other *Pythium* spp. cause damping-off in mungbean, pine, tobacco and wheat. *Phytophthora* includes many important pathogens. Avocado is affected by *Ph. cinnamomi* Rands, *Ph. nicotiana* van Breda de Haan and *Ph. palmivora* Butl.; the latter also infects black pepper, durian and rubber. Others are *Ph. botryosa* Chee (rubber), *Ph. infestans* (Mont.) de Bary (tomato) and *Ph. parasitica* Dast. (betel and black pepper). Other pathogens include *Rhizoctonia solani* Kühn (betel, black pepper, maize, Siratro and wheat), *S. rolfsii* (betel, black pepper, cassava, castor, cotton, groundnut, onion, soybean, tomato, wheat and wingbean) and *Colletotrichum* spp. (betel, maize, onion, sugarcane and black pepper). Black pepper is also infected with *Corticium salmonicolor* (Berk. & Br.) Burdsall, *Ganoderma lucidum* (Leyss.: Fr.) Karst. and *Fomes lignosus* (Klotzsch) Bres.

Tropical myxomycetes

J. Rammeloo

National Botanic Garden of Belgium, Domein van Bouchout, B-1860 Meise, Belgium

Plasmodial myxomycetes are known to have very large distribution areas, often being cosmopolitan or nearly so. Few species are typically tropical or subtropical; however, many have been collected only occasionally. In cases of intensive search, especially in remote mountainous areas in the tropics, a certain degree of endemism has been demonstrated or at least it has been shown that our knowledge of the tropical myxomycete biota is incomplete. Certain ecological niches, such as soil, have not yet been exploited. Some species from temperate regions do not fruit on their typical substrates when occurring occasionally in the tropics, just as some tropical species prefer an atypical substratum when occasionally occurring in temperate regions.

The notion of being cosmopolitan is based on a morphological species concept. From genetic research it is known that complex species exhibit several mating types. More research will be necessary to elucidate whether cosmopolitan species really behave as one biological species.

Microfungi in diverse tropical habitats

Gary J. Samuels & Amy Y. Rossman

United States Department of Agriculture, Agriculture Research Service, Systematic Botany and Mycology Laboratory, Beltsville, Maryland, U.S.A.

Few microfungi are strictly tropical (i.e. between latitudes 22°N and 22 °S) compared with many primarily tropical species that extend into subtropical and temperate zones. Most genera are cosmopolitan although many are known only from single collections, making it difficult to assess their distribution. There are many examples of pantropical fungi, but some genera have a tropical South American/African distribution and are not found in the Asian or Australasian tropics. In general a greater diversity of species and forms occurs in tropical and subtropical zones than in temperate or boreal zones.

A clear distinction exists between the mycobiota of Andean tropical areas at high elevation (over 2000 m) and Amazonian areas at low elevation (700 m or less). Areas of high elevation that are cool and moist may be more species rich than those at lower elevation, and species found in high elevation sites may have larger spores. Mycological endemism may occur only among parasites of plants or animals that have a limited distribution. Evaluation of the distribution of ascomycetes must always be done in consideration of the full life cycle as some teleomorphs are known only from one or a few tropical sites while their anamorphs are known to be cosmopolitan. Careful evaluation of ascomycetes thought to have broad distribution can reveal a complex of more than one species, each of which may have a limited geographic distribution. Microsites (e.g. savannas, shrub piles, rosette leaves, palms, bamboos, exposed logs) support predictable fungi. The distribution of fungi in litter is probably not haphazard but may reflect an endophytic relationship.

Population dynamics of vesicular-arbuscular mycorrhizal (VAM) fungi in secondary moist forest in West Africa

Julia Wilson & Philip A. Mason

Institute of Terrestrial Ecology, Bush Estate, Penicuik, Midlothian EH26 0QB, U.K.

Studies of VAM spore populations in secondary moist forest in West Africa (Cameroon and Côte d'Ivoire) have demonstrated the sensitivity of VAM populations to site disturbance. When forest was cleared and the land prepared for subsequent planting with indigenous hardwood species, the method of site preparation had considerable effects upon the number and species of mycorrhizal spores present in proximity to trees of *Terminalia spp.* Intensive studies in Cameroon indicated that complete clearance of sites before replanting resulted in a short-term reduction in spore numbers of 65%, while spore populations were damaged to a lesser extent by partial site clearance.

In both Côte d'Ivoire and Cameroon, spore numbers increased dramatically after the sites were planted and the mycorrhizal species distribution changed. Evidence from both countries indicated that the changes in VAM species composition were associated with changes in the weed flora at the sites. In particular, in Cameroon *Glomus occultum* Walker and *Acaulospora scrobiculata* Trappe proliferated after site clearance and planting. Both appeared to be associated with the invasive weed *Chromolaena odorata* (L.) R. M. King & H. Robinson. Studies on a chronosequence of plantations in Côte d'Ivoire indicated that site clearance and replanting caused changes in spore populations which were still apparent 23 years after planting.

OFFERED PAPERS

Observations on tropical fungi

Gunter R.W. Arnold

Friedrich-Schiller-Universität Jena, Pilzkulturensammlung, Freiherr-vom-Stein-Alee 2, D-O-5300 Weimar, Germany

Largely unknown fungi of Cuba were studied over a six year period in the field and in laboratory culture. Many interesting observations relating to ecological niches, biochemical activities and taxonomic position were made. Basidiomes of *Oudemansiella canarii* (Jung.) v. Hoenel, *Lentinus crinitus* (L.: Fr.) Fr., and *Coprinus clastophyllus* Maniotis obtained in the laboratory are better formed than those in their natural habitat; tropical microfungi behave in the laboratory in Europe as they do in Cuba. *Paronia*, an ascomycete genus very rare in Europe, was frequently found in various parts of Cuba. *Exobasidium fawcetti* Massee, causing flower-like hypertrophied leaves of *Lyonia* spp. (Ericaceae), produces a powerful metabolite which can be used in agriculture. Fungal communities on palm leaves need more detailed investigation. Why do the biotechnologists neglect the extremely slow growing imperfect fungus *Hermatomyces tucumanensis* Speg. which inhibits the almost invincible *Trichoderma*? Is the genus *Koorchaloma* distributed worldwide? How can we use tropical agarics and boletes as food? Are there ties between the mycobiota of Europe and the New World?

The decay of leaf litter in a mangrove at Morib, Malaysia; the involvement of lower marine fungi in colonisation and decay.

Graham Bremer & Chang May Hing*

School of Biological Sciences, University of Portsmouth, Portsmouth, U.K. and
**Institute of Advanced Studies, University of Malaya, Kuala Lumpur, Malaysia*

Nylon mesh bags containing irradiated discs of *Sonneratia* and *Rhizophora* leaves were exposed at an intertidal site in a mangrove at Morib Malaysia. The discs were recovered and assessed for colonisation by thraustochytrid and labyrinthulid organisms, and for weight loss over a 7 week exposure period. Leaf discs of *Sonneratia* lost about 40% of wet weight during the first two weeks of exposure. The decline in weight of *Rhizophora* discs was slower, but by the end of the exposure period both species showed extensive decay and mean weight losses of 70-90%. Leaf discs of both species were colonised by *Labyrinthula* and zoosporic thraustochytrid-like organisms during the first 24 hours of exposure and these were consistently isolated from the decaying leaf discs during the first two weeks of exposure.

In vitro studies using axenic type cultures of thraustochytrids from temperate waters and those obtained from mangrove showed that these organisms caused extensive decay of mangrove leaves. Isolates of *Labyrinthula* tested for decay activity in culture, grew luxuriantly on the leaf but caused no significant decay or weight loss. Freeze fracture SEM studies of *Sonneratia* leaves inoculated with cultures of a thraustochytrid and *Labyrinthula* confirmed the ability to invade the leaf tissue within 48 h of inoculation.

A systematic survey of insect fungi from natural, tropical forest in Thailand

Nigel L. Hywel-Jones

National Biological Research Centre, Kasetsart University, PO Box 9-52, Bangkhen, Bangkok, 10900, Thailand

Recent estimates of fungal biodiversity indicate at least $1{\cdot}5 \times 10^6$ insect fungi await discovery. The few surveys of tropical insect fungi have mostly been in agro-ecosystems. In Thailand, 5–10 species (recorded from agro-ecosystems) are pan-global. A three year survey of insect fungi on the underside of leaves in natural forest produced a list of 66 species. Partial surveys of other micro-habitats list 80+ species (14 new to science) compared with 100 species in 200 years for the U.K. Of 66 species from leaves 38 are in culture. The dominant genus was *Aschersonia* (13 species, 10 isolated) and its teleomorph *Hypocrella* (10 species, 6 isolated). *Aschersonia* spp. are recognized as potential biocontrol agents of whitefly but only two species (common in tree crops) have been considered. Not all genera have potential as biocontrol agents. *Gibellula* are specific to spiders (beneficial arthropods). *Cordyceps* do not tolerate disturbed habitats and have restricted host ranges.

With increasing interest in tropical fungi as sources of metabolites only commonly isolated insect fungi have come to the attention of industry. Specialized genera such as *Gibellula* and *Cordyceps* which are able to enter invertebrates and overcome them should be of particular interest to the agro-chemical industry in their search for new pesticides.

Laboratory studies on aerial rhizomorphs from the tropics

Peter Lewis & John Hedger

Department of Biological Sciences, The University College of Wales, Aberystwyth, Dyfed, SY23 3DA, U.K.

Several fungal species from moist tropical forests extend as a rhizomorph network in aerial habitats (see Chapter 2). Successful isolations were made from rhizomorph networks in Cuyabeno Reserve, Ecuador, and Varirata National Park, Papua New Guinea. Some of these isolates produced rhizomorphs in culture, and were used to investigate their response to light, gravity, water availability and carbon dioxide. The results support the suggestion that aerial rhizomorphs are modified stipes with suppressed pileal apices.

Aeroallergens from an industrial belt of India

Sudhendu Mandal

Department of Botany, Visva-Bharati University, Santiniketan -731235, India

Allergic reactions are being caused by the allergens of nonpathogenic fungi which come into contact with a usually predisposed person by inhalation, ingestion, or immediate contact. The incidence of major allergic disorders in India is estimated to be about 15 percent of the total population.

The nature of aeroallergens from different parts of Durgapur, a highly industrialised area of eastern India, in terms of their role as organic environmental pollutants was investigated. The air, water and soil of this area carry various types of pollutants. The survey of aerollallergens has been carried out using a Gravity Slide Sampler and Rotorod Sampler. The most common and dominant fungal spores were *Cladosporium, Aspergillus, Penicillium, Curvularia, Alternaria,* and *Fusarium*. Mycotoxic and pathogenic moulds (*Aspergillus flavus* Link., *A. fumigatus* Fres., *A. niger* v. Tiegh., *Rhizopus nigricans* (Ehrenb.) now *R. stolonifer* (Ehrenb.) Lind., etc.) were found to occur in a higher concentration than several other phytotoxic molds, e.g. *Cladosporium, Alternaria, Curvularia, Fusarium* and *Trichoderma*. Numbers and types of fungal spores in the air change with the time of day, weather, season, geographical location, and the presence of local spore sources. The study was performed in order to identify suspected aeroallergens from this hitherto unexplored area of India.

Eccrinales in the guts of tropical millipedes

Stephen T. Moss

School of Biological Sciences, University of Portsmouth, Portsmouth, Hampshire, PO1 2DY, U.K.

Eccrinales (Trichomycetes) infesting the hindguts of millipedes (Diplopoda) from southern India, South Africa and Malaysia are reported and represent new geographic records for this group of fungi. All species found have been assigned tentatively to the genus *Enterobryus* Leidy. Thalli are unbranched, determinate, and attached to the host cuticle by a basal secreted holdfast. The holdfast does not penetrate the cuticle and none of the host tissue is invaded. Reproduction is by either primary or secondary sporangiospores formed in basipetal succession from the thallus apex. These two spore types serve different functions. Primary sporangiospores are produced by thalli situated at the posterior of the hindgut and pass from the gut to infest other individuals. Secondary sporangiospores are formed by thalli attached to the pyloric sphincter region and on release germinate and attach within the same individual. Sexual reproduction is not known. The relationship between host and fungus, and between the fungus and other gut microbiota is not known. No species has been cultured and the association appears obligate.

Studies on intra- and inter-specific hybridization for strain improvement of a tropical *Agaricus*

V. S. Pahil, J. F. Smith & T. J. Elliott

Horticulture Research International, Littlehampton, West Sussex, BN17 6LP, U.K.

The common button mushroom, *Agaricus bisporus* (Lge.) Imb. (syn. *Agaricus brunnescens* Peck) is principally a species of temperate climates, as it grows at 25°C and fruits at 16–18°C. A second species, *Agaricus bitorquis* (Quél.) Sacc., morphologically similar to *A. bisporus*, grows at 30°C and fruits at 25°C, but it has had limited success commercially because of poor yields and an irregular flushing pattern. We have attempted to identify further strains with potential for growth in the tropics and subtropics. Wild and cultivated strains of *Agaricus* spp., mainly *A. bitorquis*, and single spore isolates of strains designated W19, W20 and W2-F of a tropical *Agaricus* species, were recovered from the HRI culture collection. Intra- and inter-strain mating tests of strains W19, W20 and W2-F confirm that this species has a multiallelic unifactorial heterothallic breeding system. Four different mating-type specifities were recognised. The mycelial growth rate and temperature tolerance of homokaryons, dikaryons and multispore isolates varied considerably. At 28°C, isolates W19 and W20 outyielded all the commercial strains of *A. bitorquis* with which they were compared. Some inter-strain dikaryons produced high yields, better quality fruit-bodies and had a more regular flushing pattern. Such wide variability among these tropical wild *Agaricus* isolates highlights the possibility of further advance by breeding.

Developing a phylogenetic classification for tropical *Colletotrichum* species using ribosomal DNA sequencing

C. Sherriff, J. A. Bailey, M. J. Whelan & G. Arnold

Department of Agricultural Sciences, University of Bristol, AFRC Institute of Arable Crops Research, Long Ashton Research Station, Bristol BS18 9AF, U.K.

Colletotrichum species include important pathogens, which cause economic losses of food and fodder crops. Current taxonomy of *Colletotrichum* relies heavily on spore shape and host. This is inadequate because there are few diagnostically useful morphological characters available. Alternative characters are thus required to distinguish species groups. A molecular approach based on ribosomal DNA sequencing was chosen for establishing a phylogenetic taxonomy. r-DNA was amplified in two polymerase chain reactions. The D1 and D2 regions of the 28S gene, the internally transcribed spacer 2 and the 5.8S gene were sequenced. Results indicate that r-DNA sequencing offers great scope for clarification of the taxonomy of this important genus.

Colletotrichum isolates from legumes and other hosts were sequenced to obtain a measure of the variation within the genus. A similarity matrix was calculated for all sequences and a cluster analysis carried out from which dendrograms were obtained. Results indicate that several existing species groupings are unreliable, and that present taxonomy is sometimes misinterpreted. For example, isolates of *C. lindemuthianum* (Sacc. & Magn.) Br. & Cav., *C. malvarum* (Braun & Casc.) Southworth, *C. orbiculare* (Berk. & Mont.) Arx and *C. trifolii* Bain & Essary should be considered as a single species. In contrast, *C. gloeosporioides* (Penz.) Sacc. comprises several very distinct forms. Furthermore, some isolates designated as *C. truncatum* (Schw.) Andrus & Moore have similar sequences to *C. capsici* (Syd.) Butl. & Bisby.

POSTERS

Occurrence and distribution of decay fungi in different substrata from the Manaus Region, Amazonas, Brazil

Maria Aparecida de Jesus

Instituto Nacional de Pesquisas da Amazônia, CPPF/INPA, Caixa Postal 478, Manaus/AM, 69011, Brazil

A total of 890 samples of decay fungi were studied with regard to their distribution on different substratum types, abundances and economical importance. The samples came from the collection of the Wood Pathology Laboratory of CPPF/INPA, in Manaus, Brazil. Altogether 157 species and 72 genera were identified, including *Trametes* with 8 species, *Auricularia* and *Panus* with 6 species each and *Rigidoporus* and *Hymenochaete* both with 5 species. Some fungi are evenly distributed on all substrates such as *Auricularia delicata* (Hook.) Henn., *Coriolopsis occidentalis* (Klotz.) Cunn., *Rigidoporus lineatus* (Pers.) Ryv., *Panus rudis* Fr., besides others. The most common species occurred on both naturally felled trees and logs, e.g. *Kretzschmaria clavus* (Fr.) Sacc., *Pachykytospora alabamae* (Berk. & Cke) Ryv. and *Polyporus infernalis* Berk., while the species *Grammothele fuligo* (Berk. & Br.) Ryv., *Hyphoderma capitatum* Erikess. & Hjorst and *H.* cfr. *clavigerum* (Bres.). Donk occurred on live trees.

Vegetative incompatibility and sexual systems of *Armillaria* from tropical Africa

J. J. Guillaumin, S. Abomo-Ndongo & C. Mohammed*

*Centre INRA de Clermont-Ferrand, Unité de Mycologie, 12 Avenue du Brézet, 63039, Clermont-Ferrand, France and *Oxford Forestry Institute, South Parks Road, Oxford, OX1 3RB, U.K.*

The results of culture pairing tests with 25 isolates (originating from 15 different African countries) divided them into 4 groups: *Armillaria heimii* Pegler - an aggregate group of 17 isolates from Central, West and East Africa; the African counterpart to the European *A. mellea* (Vahl.: Fr.) Kummer, a small but distinct group of 4 isolates; 3 isolates from high altitude Kenya; and 1 isolate, also from high altitude Kenya.

Isolates separated by a black line were considered as belonging to different groups. Two isolates of the same group intermingled either without any visible separation (i.e. groups 1 and 2) or with a zone of scarce mycelium (i.e. group 3). Isolates of groups 1 and 2 fruited in culture and single spore isolates were obtained from the basidiomes.

Single spore isolates of *A. heimii* from Gabon, Cameroon and Liberia all had a cottony haploid-like morphology (well known in European *Armillaria).* In pairing tests cottony single spore isolates of different geographical origin were compatible giving rise to a crustose isolate typical of a diploid. Pairings between haploid isolates from the same basidiome suggested a heterothallic bipolar sexual system. Single spore isolates from both *A. mellea* and *A. heimii* of Congolese and East African origin were crustose. This and other research (including cytological studies) has led the authors to propose the existence of homothallism in African *Armillaria.*

Fruiting in culture of *Armillaria* isolates from tropical Africa

M. Intini[1], J. J. Guillaumin[2], S. Abomo-Ndongo[2] & C. Mohammed[3]

[1]*Centro di Studio per la Patologia delle Specie Legnose Montane, Piazzallee delle Cascine, 28, 50144, Florence, Italy,* [2]*Centre INRA de Clermont-Ferrand, Unité de Mycologie, 12 Avenue du Brézet, 63039, Clermont-Ferrand, France, and* [3]*Oxford Forestry Institute, South Parks Road, Oxford OX1 3RB, U. K.*

Twenty nine *Armillaria* isolates from 15 different African countries were included in artificial fruiting trials in both Clermont-Ferrand, France and Florence, Italy. These isolates belonged *a priori* to three main groups: (1) the African form of the European species *A. mellea* (Vahl:Fr.) Kummer; (2) *A. heimii* Pegler, a large group with considerable variability; (3) isolates from the highlands of Eastern Africa different from those of the two previous groups. Twenty-three out of the 29 isolates have fruited in culture and all belong to the *A. mellea* or *A. heimii* groups. No isolates of the Eastern highlands group have fruited.

In substratum trials orange segments were found to be a better substratum for obtaining basidiomes than pieces of wood. The different environments tested were equally successful for the induction of basidiomes. African *A. mellea* basidiomes have a morphology closely resembling that of the prominently ringed, honey coloured European *A. mellea*. *A. heimii* basidiomes obtained in culture can be very variable although distinguishable from that of *A. mellea* isolates by the poor development or absence of a ring and a more delicate cap speckled with scales or warts. The most typical morphology of fruiting isolates of this *A. heimii* group conforms with the description of *A. heimii* given by Pegler.

Electron microscope studies on a litter trapping fungus from tropical forest

John Hedger & Ruth Baxter

Department of Biological Sciences, The University of Wales at Aberystwyth, Aberystwyth, Dyfed SY23 3DA, U.K.

In tropical moist forest in Papua New Guinea and Ecuador a white mycelium commonly trapped litter in the understorey by aerial mycelial transfer. SEM studies showed that the fungus colonised the litter surface by a clamp-bearing mycelium which soon differentiated large numbers of dichohyphidia with branched arms up to 10 μm in length and 0·1–0·3 μm in diameter, placing the fungus in Dichostereaceae, possibly the genus *Vararia*. In older parts of the colony the mycelium and bases of the dichohyphidia were immersed in a mucilaginous matrix. Mucilage secretion occurred in two stages: a primary mucilage layer was produced 50–100 μm behind the colony margin and a secondary thicker layer was then produced as coalescing lobes associated with the bases of the dichohyphidia. We suggest that the dichohyphidia and the mucilage layers help to buffer this mycelium against water stress in the understorey canopy.

The origin and development of brown root-rot of Salmwood [*Cordia alliodora* (Ruiz. & Pav.) Cham.] caused by *Phellinus noxius* (Corner)

M. H. Ivory, G. Daruhi* & J. Tungon*

*Oxford Forestry Institute, Department of Plant Sciences, University of Oxford, South Parks Road, Oxford OX1 3RB, U.K. and *Research and Technical Support Section, Department of Forestry, PMB, Port Vila, Vanuatu*

Brown root-rot was noticed on Salmwood in Vanuatu soon after extensive planting began in 1975. Studies have since shown that *P. noxius* occurs as a pathogen or a saprotroph on certain native trees, and that it spreads on to the roots of exotic trees, e.g. Salmwood, which directly contact affected trees or tree residues. Subsequently it spreads between Salmwood trees, once root mingling has occurred (about 2 years after planting at 3 m x 3 m), causing up to 50% mortality by 7 years. Each disease focus contains a distinct fungal clone which originates from an infected stump or log. This suggests that certain indigenous trees become colonized by basidiospores, rather than by contact with fungal mycelia. Foci subsequently enlarge vegetatively on either native or exotic trees via mycelia from the focus origin. Provenance trials and inoculation studies on seedlings and larger trees show no variation in disease susceptibility between Salmwood provenances, or virulence differences between fungus clones.

An integrated approach to the taxonomy of *Trametes* species collected from Zimbabwe

A. Mswaka & N. Magan

Biotechnology Centre, Cranfield Institute of Technology, Cranfield, Bedford MK43 OAL, U.K.

The taxonomic relationship among 8 species of the genus *Trametes* Fr. was investigated using a range of morphological, cultural, biochemical and electrophoretic characteristics. All 8 species had pileate basidiomes, a trimitic hyphal system and thin-walled, hyaline, ellipsoid spores. Major macroscopic differences were in shape and colour of basidiomes. All species produced extracellular phenoloxidases. Mycelial mat textures varied from cottony, crustose to subfelty. Species from the hot and dry regions of Zimbabwe had high temperature growth optima (*T. cervina* (Schw.) Bres. 45-50°C, *T. cingulata* Berk. 40-45°C and *T. socotrana* Cooke 40-50°C), grew slowly at 55°C and survived exposure to 60°C, suggesting that they are thermophilic. Nineteen enzymes were assayed using API-ZYM strips, each species producing a distinctive pattern. Preliminary results of isozyme analysis using starch-gel electrophoresis indicated large interspecific variation compared with total soluble protein analysis by polyacrylamide-gel electrophoresis. The results will be used for numerical taxonomy of the group.

A mycobiotic survey of the polyporaceae associated with the indigenous vegetation of Zimbabwe

A. Mswaka & L. Ryvarden*

*Biotechnology Centre, Cranfield Institute of Technology, Cranfield, Bedford MK43 0AL, U.K. and *University of Oslo, Department of Biology, Division of Botany, P.O. Box 1045 Blindern, N-0316 Oslo 3, Norway*

A total of 1375 specimens were collected from The Moist Evergreen Forests, Miombo Woodlands, Mopane Savanna Woodlands, Mukwa Woodlands, Zambezi Teak Forests and Afromontane Forests, from which 150 species in 58 genera were found. Four species new to science were described, *Datronia concentrica nov. sp. ined., Dichomitus uapakaii nov. sp. ined., D. longispora nov. sp. ined., and Irpex stereoides nov. sp. ined. Phaeotametes decipiens* (Berk.) Wright was collected for the first time outside Australia and South America.

The majority of the fungi collected caused white-rot and only 4% of the total number of species collected were brown-rot fungi. The highest number of species were collected from the Moist Evergreen Forests and the smallest number came from the Afromontane forests. The survey showed that the distribution of polyporaceous fungi is largely dependent upon host distribution. Taxonomic keys were produced for the families, genera and species encountered during the survey.

A comprehensive checklist of the Polyporaceae of Zimbabwe's indigenous vegetation was produced using data from this survey and from previous records.

Basidiomycetes in the canopy and on twigs at ground level: a comparison

Maria Nuñez & L. Ryvarden

Department of Botany, University of Oslo, P.O. Box 1045 Blindern, N-0316, Oslo 3, Norway

During an expedition to Campo (Cameroon) in the dry season, organized by ELF Aquitaine, basidiomycetes in the canopy and on twigs at 1.5 m above the ground were collected and compared to deduce ecological adaptations. The fungi belonged to Tremellaceae s.l., Agaricales and Aphyllophorales. K-strategists (*Vararia, Dichostereum, Hymenochaete*) had perennial basidiomes strongly adapted to drought and were mostly sterile. It is hypothesized that they sporulate when rains come. Fungi with very thin basidiomes exhibited an r-strategy. After short showers, the basidiomes would spread their spores and then disappear. *Trechispora, Leptosporomyces, Tremella* belong here.

K-strategists were more abundant than r-strategists. Remarkably, many polypores were present in the canopy: *Dichomitus africanus* Ryv., *Megasporoporia cavernulosa* (Berk.) Ryv., *Rigidoporus destrinoideus* Johan. & Ryv., *Schizopora flavipora* (Cke.) Ryv., *Phellinus callimorphus* (Lev.) Ryv. and *Phellinus inermis* (Ell. & Everh.) Cunn.

Some resupinate polyporaceous fungi of Papua New Guinea

E. Quanten

Department SBG, Limburgs Universitair Centrum, Universitaire Campus, B-3590 Diepenbeek, Belgium

The resupinate polyporaceous fungi of Papua New Guinea have previously been overlooked and poorly studied. Hymenochaetaceae not considered, although about 30 species belonging to 16 genera can be listed. Several new generic and specific records for Papua New Guinea can be added.

Three new generic records, all represented by one species, are: *Megasporoporia* Ryv. & Wright [*M. setulosa* (Henn.) Wright], *Pachykytospora* Kotl. & Pouz. [*P.* cfr. *alabamae* (Berk. & Cooke) Ryv.] and *Skeletocutis* Kotl. & Pouz. [*S. nivea* (Jungh.) Keller]. *Megasporoporia setulosa* has been collected several times in the Madang Province and the characteristics of this species were illustrated. The characters of two *Pachykytospora*-collections were compared with those of *P. alabamae* (Berk. & Cooke) Ryv. and *P. papyracea* (Schw.) Ryv. and of two *Skeletocutis*-collections were compared with those of *S. nivea* (Jungh.) Keller and *S. novae-zelandiae* (Cunn.) Buch. & Ryv. *Rigidoporus cystidioides* (Lloyd) Corner, already recorded for Papua New Guinea, is characterized by having acicular extrahymenial setae. The apex of these elements can be smooth or encrusted by crystals visible in SEM photographs.

An investigation into witches' broom disease of cocoa using mannan and chitin assays

Danny Penman[1], Keith Hardwick[1], Hamish A. Collin[2], Susan Isaac[2] & George Britton[3]

[1]Department of Environmental & Evolutionary Biology, [2]Department of Genetics & Microbiology & [3]Department of Biochemistry, University of Liverpool, PO Box 147, Liverpool L69 3BX, U.K.

The basidiomycete *Crinipellis perniciosa* (Stahel) Singer causes witches' broom disease in cocoa (*Theobroma cacao* L.) which severely limits cocoa production in large areas of South America and the Caribbean. The disease cycle includes a primary biotrophic and a secondary saprotrophic phase. In order to study the progression of the disease the possibility of quantifying the amounts of both phases in host tissue, over time using changes in mannan:chitin ratio (fungal cell wall polymers), was investigated.

An existing GLC (gas chromatographic) method for mannan estimation was improved and a new highly specific and sensitive GLC method for chitin estimation was developed. The latter was 1500 times more sensitive than previous spectrophotometric methods and can quantify low levels of fungus in as little as 100 μg of host tissue, previously 25 mg was required. The chitin method should be of great value in other studies of fungal invasion of higher plant hosts.

An ultrastructural examination of the genus *Savoryella*

S. J. Read, E. B. G. Jones & S. T. Moss

School of Biological Sciences, University of Portsmouth, King Henry Building, King Henry I Street, Portsmouth, Hampshire, PO1 2DY, U.K.

The genus *Savoryella* Jones & Eaton is characterized by having four-celled ascospores with melanised central cells and smaller hyaline polar cells. Specimens were collected in Malaysia. At the light microscope level the asci of both *S. appendiculata* Hyde & Jones and *S. longispora* Jones & Hyde have an apical apparatus comprising an apical ring with a central pore. However, at the TEM level the inner margin of the apical ring of *S. longispora* is rounded inwards with a clearly defined central pore whereas in *S. appendiculata* the inner margin of the apical ring tapers outwards and the central pore is occluded by a plug which later dissolves away.

In *S. longispora* a fibrillar mucilage is present around the ascospores but little mucilage is present on the central cells of ascospores of *S. appendiculata*. Mucilage, characteristically arranged as clusters of 'starbursts' is present on the hyaline polar cells of both *S. appendiculata* and *S. paucispora* (Cribb & Cribb) Koch. *Savoryella appendiculata* is the only appendaged species and it is shown that the appendages are formed endogenously and are not present prior to release from the ascus. Although many minor differences are observed between the three species, none is sufficiently important to warrant splitting the genus at this stage.

Paraliomyces lentiferus: an ultrastructural study of a little known marine ascomycete

S. J. Read, E. B. G. Jones & S. T. Moss

School of Biological Sciences, University of Portsmouth, King Henry Building, King Henry I Street, Portsmouth, Hampshire, PO1 2DY, UK

The ultrastructure of *Paraliomyces lentiferus* Kohlm. from Taiwan, Republic of China has been examined at the TEM level as part of a review of the taxonomy of the marine Ascomycotina. The ascospore wall comprises a mesosporium, episporium and a well developed extracellular mucilaginous sheath (exosporium?). There is a single, gelatinous, lateral appendage adjacent to the central septum. The appendage comprises electron-opaque fibrils which in immature ascospores are connected to the ascospore wall by fine electron-opaque strands and larger electron-opaque aggregates of material. The origin of the appendage is discussed.

Fungal pathogens as biological control agents of tropical weeds: a practical application of biodiversity

Marion K. Seier & Carol A. Ellison

International Institute of Biological Control, Silwood Park, Ascot, Berks. SL5 7TA, U.K.

Two examples of classical biological control were presented: *Mimosa pigra L.* (Mimosaceae) and *Lantana camara L.* (Verbenaceae). Both these neotropical shrubs are target weeds in Australia, where they pose an agricultural as well as an ecological threat. *Phloeospora* sp. nov. (Coelomycetes), with its teleomorph *Sphaerulina* sp. nov., and the monotypic rust *Diabole cubensis* (Arthur & J. R. Johnson) Arthur, are considered to be potential control agents for *Mimosa pigra.* The rust *Prospodium tuberculatum* (Speg.) Arthur and a web-blight fungus (? *Ceratobasidium* sp.) are being evaluated for control of *Lantana*. An example of the mycoherbicide or inundative approach using a highly specific fungal isolate from the genus *Colletotrichum* to control *Rottboellia cochinchinensis* (Lour.) W.D. Clayton, a pantropical noxious grass weed, was outlined.

Spore ornamentation makes a nice difference: *Daldinia eschscholzii* and *Daldinia concentrica*

Katleen van der Gucht

Laboratory of Plant Morphology, Systematics and Ecology, University of Ghent, K.L. Ledeganckstraat 35, 9000 Ghent, Belgium

Daldinia eschscholzii (Ehrenb.) Rehm., confined to tropics and subtropics, and *D. concentrica* (Bolt.: Fr.) Ces. & de Not., its counterpart in more temperate regions, are closely related species. According to some authors the two taxa are not worth specific rank because of inconstant differences in taxonomic characters.

The ascospores are of similar size (range 11–13 (14·5) x (5) 5·5–7 (7·5)μm for *D. eschscholzii* and 12·5–15 (16) x 6–7·5μm for *D. concentrica)* and appear to be smooth when examined by light microscopy. When examined by SEM, the ascospore surface of *D. eschscholzii* is ornamented with transversely oriented fibrils. The ascospores of *Daldinia concentrica* are almost smooth. Only very faint ridges, laterally around the spore, are visible. Spore ornamentation as revealed by SEM therefore provides a reliable diagnostic character between these two taxa.

Observations on subtropical mangrove fungi

L. L. P. Vrijmoed & E. B. G. Jones*

*Department of Applied Science, City Polytechnic of Hong Kong, 83 Tat Chee Avenue, Kowloon, Hong Kong and *School of Biological Sciences, University of Portsmouth, King Henry I Street, Portsmouth, Hants. PO1 2DY, U.K.*

Our knowledge of marine and mangrove fungi from subtropical locations is fragmentary as compared with the large volume of publications on tropical mangrove fungi which have appeared in the last decade. A recent survey of the mangroves in the Pearl River Estuary along the South China Coast, viz. Hong Kong, Macau and Shenzhen (China) reveals an equally rich fungal diversity resembling that found in the tropics. The common species included *Trichocladium linderii* Shearer, a sphaeropsidaceous fungus, *Phomopsis* sp., *Lignincola laevis* Hohnk, *Marinosphaera mangrovei* Hyde and *Hypoxylon oceanicum* Schatz; all of which were collected more than 20 times in about 300 driftwood samples. Other less common species which occurred between 10 and 20 times included *Aniptodera chesapeakensis* Shearer & Miller, *Halosarpheia abonnis* Kohlm. and *Periconia prolifica* Anastasiou. A number of species new to science have also been found and were illustrated on the poster, e.g. *Massarina armatispora* Hyde, Vrijmoed, Chinnaraj & Jones, *Aniptodera haispora* Vrijmoed, Jones & Hyde, *Pleospora* sp. nov. and *Myelospilea* sp. nov.

Use of a resampling procedure to analyze the contributions of fungal diversity and growth medium diversity in an industrial screening campaign for secondary metabolites

H. G. Wildman

Natural Products Discovery Department, Glaxo Group Research Ltd., Greenford Road, Greenford, Middlesex UB6 0HE

It is often difficult to determine the individual contributions of organism diversity and growth medium diversity to the discovery of novel natural products (secondary metabolites) in industrial screening campaigns. Analyses of the number of samples progressing past primary screening tests or those yielding novel structures require the use of data from many weeks of screening and require careful experimental design to avoid confounding factors during the analysis period (e.g. screen, organism and medium changes). The use of a resampling or bootstrap analysis, however, allows a statistical comparison of the contributions of organism diversity and growth medium diversity using data from a single screening week, and thus avoids the aforementioned confounding factors. A resampling procedure was used to show the significant contributions of fungal diversity and growth medium diversity to the numbers and types of screen positives obtained using a batch of 80 predominantly tropical fungi.

Index

A

B

D

G

H

M

N

O

P

Q

R

T

Printed in the United States
By Bookmasters